STRATEGY OF PROCESS ENGINEERING

STRATEGY OF PROCESS ENGINEERING

Dale F. Rudd
Charles C. Watson

The Chemical Engineering Department
The University of Wisconsin
Madison, Wisconsin

JOHN WILEY & SONS. INC., NEW YORK | LONDON | SYDNEY

10 9 8 7 6 5 4 3

Library of Congress Catalog Card Number: 67-31212
GB 471 74455X
Printed in the United States of America

Dedicated to the Department of Chemical Engineering at the University of Wisconsin and to the continuation of its stimulating environment.

PREFACE

Within the last decade great improvements have been made in the methods of process design and operation in certain segments of industry. We look forward to even greater progress. Problems in process engineering that formerly were considered unsolvable are now routinely dealt with as part of the day-to-day operation of the process industry—greatly improving its efficiency. Almost every activity of the process engineer has been examined critically as a strategy of process engineering continues to evolve.

An example of benefits arising from the successful wedding of theory and practice is the problem of determining the optimal operation of petroleum refineries, a problem that was considered practically unsolvable only a few years ago. The advent of the theory of linear programming provided the means for solving this problem, and now nearly all of the world's petroleum refineries are partially optimized by the solution of linear programs. Our ways of thinking have been altered and clarified by the advent of this useful theory. This is only one example of the vastly improved methods of process engineering.

In this text we attempt to provide a balanced introduction to the strategy of process engineering, consistent both with industrial practice and with the developing means for improving that practice. In this way the book should be of use to both the engineering student as a point of departure for his entry into industry and to the practicing engineer for his continuing education.

The chapters are divided into three parts that encompass the major problems encountered in process engineering. Part I, *The Creation and Assessment of Alternatives*, deals with concepts involved in transforming ill-defined processing problems into processing plans that can be polished into efficient commercial systems. Part II, *Optimization*, explains methods for adjusting the design of a specific system to its most useful mode of operation. Part III, *Engineering in the Presence of Uncertainty*, embodies methods for accommodating to the persistent lack of necessary information to accomplish the optimal design.

During the more than five years spent on this text, which included the preparation of a set of class notes entitled *Strategy of Process Engineering* in August 1966, a close contact has been maintained with engineering teachers and practitioners in an attempt to provide both teachable and useful material.

This text was used in its various stages of development in the university class room by:

Professor D. R. Woods, McMaster University.
Professor P. B. Lederman, Polytechnic Institute of Brooklyn.
Professor M. C. Molstad, University of Pennsylvania.
Professor J. A. Bergantz, New York State University, Buffalo.
Professor G. H. Geiger, Department of Minerals and Metals Engineering, University of Wisconsin.

We have also used this text over the years in a senior design course and in a graduate course in advanced topics in process engineering.

Industrial contact was maintained through short courses for the process-design sections of these corporations:

The Union Carbide Corporation.
The Humble Oil and Refining Company.
Esso Research and Engineering.
Mobil Research and Development Corporation.

Also, industrial contact was maintained through other less formal contacts with industry. We wish to acknowledge the valuable suggestions of A. H. Masso, J. W. Hackney, R. D. Stief, R. C. Wahrmund, A. S. Foss, C. J. King, P. T. Shannon, L. B. Koppel, G. L. Glahn, E. N. Cart, and W. N. Zartman.

Perhaps the largest audience for this text will be found in the traditional senior design course. Our seniors use this text as a base of operations while attacking practical design problems and, in this way, broaden their background in the strategy of process engineering. The course is conducted as a design laboratory with the students being assigned responsibility for the complete design of small processing systems. Frequent lectures are given from the text as they discover the need for the material. However, this book can be used as an introductory course to precede a case study or projects course. The short problems herein are provided for this purpose.

For the practicing engineer, who needs no formal introduction to actual industrial problems, and the graduate student, who may desire to study the strategy of process engineering as a discipline by itself, the usual lecture and problem-working mode of presentation is suggested.

Although there is little doubt that further improvements could be made, we believe that the text, in its present form, will enable the student to make a good beginning in process engineering.

DALE F. RUDD
CHARLES C. WATSON

January, 1968

CONTENTS

1
INTRODUCTION

This first chapter is an introduction to the field of process engineering and serves as a preview of this text. The process engineer has the responsibility for creating processing systems which will economically transform raw material, energy, and know-how into useful products. We shall outline the problems encountered during the design of commercial processing systems and discuss the ways of thinking required to cope best with those problems.

Typical Problem

There is a need for some hundred thousand pounds per month of phenobarbital for use as a sedative. How best can a design group develop the specifications for a process to manufacture this drug profitably? This is the kind of problem which attracts the attention of the process engineer.

1.1 THE SCOPE OF THIS TEXT

We propose to examine the kinds of thinking which lead to the creation of commercial processing systems. This activity is one of the most responsible and challenging in industry, and the economic health of a significant portion of our society depends upon the skill of the process designer.

A processing system is a collection of equipment which effects the transformation of materials through chemical reaction, phase transition, heating and cooling, agglomeration, size reduction, separation, extraction, combustion, and so forth. For example, by means of properly designed processing systems:

Air, water, and electric power can be transformed into fertilizer.

Sulfuric acid can be produced from sulfur, air, and water to meet the needs of industry.

Crude oil can be refined into the many petroleum products our modern society demands.

Magnesium can be removed from the sea profitably.

Polio vaccine can be produced in mass quantities.
Stainless steel can be manufactured from scrap steel and other materials.

Processing systems form the backbone of our modern industrial society. Over one half of the nation's 500 largest companies deal with some kind of processing. The du Pont Company, one of the largest chemical processors, supplies less than 8 per cent of the chemical market, yet has annual sales expressed in the billions of dollars. That modern society is geared to the productivity of processing systems is evidenced by the fact that the average American annually consumes, in the form of food, drugs, fiber, fertilizer and other processed commodities, about 60 pounds of ammonia, 60 pounds of caustic soda, 80 pounds of chlorine, and 200 pounds of sulfuric acid. It has been reported that 20 per cent of the gross national product can be traced to that portion of process engineering which involves the application of catalysis.

The creator and keeper of the processing systems which support this large sector of our economy most commonly is the chemical or metallurgical engineer. The talents of the engineer are called upon as soon as a processing need is suspected, and it is his responsibility to attempt to prepare the complete specifications for a manufacturing system which is capable of satisfying that need most economically. It may take a skilled design engineer a week to design a small processing system, or a team of engineers may spend a year designing a large processing system, say, for the manufacture of a new plastic. Regardless of the size of the design task the same mental processes must be in action.

1.2 AN ONRUSHING TECHNOLOGY

The process engineer finds himself at the interface between an onrushing technology and economic obsolescence. A processing concept is a perishable commodity, and timing plays an important role in determining its success. On the one hand, a process too long in the planning stage will be obsolete before it reaches the stage of production, and the need for which it was designed may be satisfied by some other means or by a competitor. On the other hand, a hastily engineered process may not be able to satisfy the need or may not be efficient enough to survive economic competition. This peculiar position in which the process designer finds himself does much to determine the kinds of skills which are deemed valuable.

The constant pressure to realize the benefits of advancing science in the form of new technology and new products forces the engineer to work beyond the supporting sciences. For example, processes for recovering magnesium from the sea were designed and operated profitably long before all of the scientific details of the phenomena occurring were available. This environment forces the engineer to take calculated risks on the basis of inadequate information.

The desire to compress the time scale between research discovery and commercialization is quite strong since the saving of only one year might reap profits measured in the millions of dollars. Figure 1.2-1 shows the time scale for

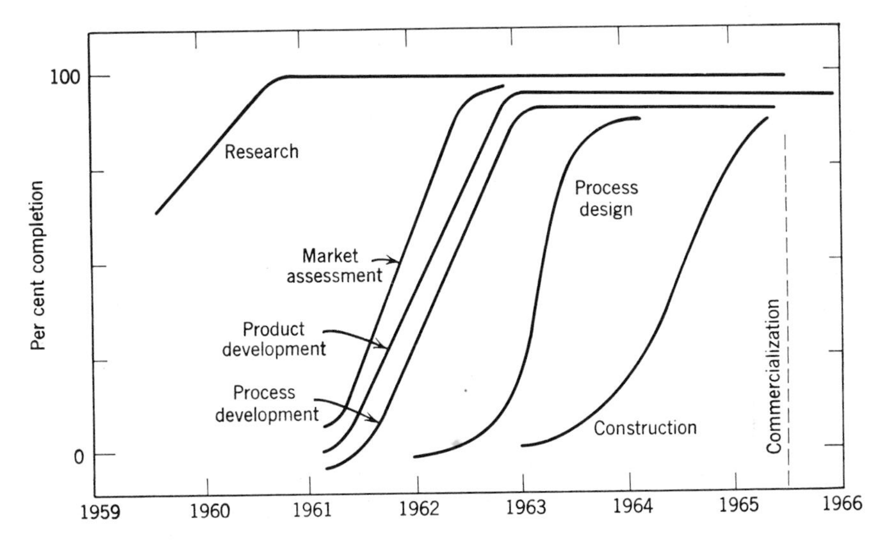

Fig. 1.2-1 The compression of the time scale for the development of a process for the manufacture of "KAPTON" polyimide film (from "This Is Du Pont," E. I. Du Pont De Nemours and Co., 1966).

the commercialization of a process for the manufacture of "Kapton," polyimide film used for electrical insulation, fireproof clothing, printed circuits, magnetic and pressure-sensitive tapes, and other special products. Less than five years after Kapton was developed in the laboratory, the first commercial process was in operation. To achieve this rapid schedule it was necessary to initiate the process development and design well before the market was assessed and even before the product was completely developed. The process itself had to be partially designed even before the processing concept was developed in detail. Construction began when the process was only partially designed. Had each of these steps been initiated only after all the necessary information was available, the time scale would have been expanded from less than five years to more than nine years. This would have cost the manufacturer millions of dollars in lost profits and the public years of inconvenience for want of the product. It is the hope of winning such a payoff in an engineering gamble that motivates the study of the strategy of process engineering.

It appears that this component of uncertainty, coupled with the social responsibility for all the consequences of the proposed system, is a normal or steady-state condition of engineering. And furthermore, it appears that this condition will persist indefinitely, being motivated by man's impatience and willingness to gamble on the early application of his discoveries.

The strategy of process engineering aids in making better use of the limited time available during a process design and aids the engineer in making critical decisions based on limited information.

Process engineering continues after the design and construction of a new plant have been completed. Process engineers on the staff of the operating department, or loaned as consultants by a technical service group, carry out a program of process adjustment to secure the best operating conditions. Such adjustments commonly involve the redesign of parts of the process to improve the quality and quantity of production. Thus many of the strategies of process engineering are applicable in production as well as in design.

1.3 FROM THE PRIMITIVE TO THE SPECIFIC

Having outlined the environment in which the engineer must function, we now sketch through the mental processes which lead to the creation of a processing system. In Section 1.4 we examine the obstacles which hinder the creation of the optimal system, and the remainder of the text deals with the ways and means for overcoming these obstacles.

Every processing system begins as a *primitive problem*. A primitive problem is an ill-defined statement of a need such as:

There is a local need for oxygen.

A shortage of nitrogenous fertilizer exists in the upper Mississippi valley.

The cost of hydrogen is too high in our plant.

The fumes issuing from the stack at the heating plant are polluting the air.

What can we do with all this sulfur?

Our abundant but inferior Wisconsin iron ore requires beneficiation before it can find use in steelmaking.

The Athabaska tar sands offer an enormous source of hydrocarbons if only they could be removed economically.[1]

These primitive problems may be thought of as expressions of a gap which exists between the available raw materials, energy, and know-how and the local needs of society.

In an attempt to bridge this gap, the engineer seeks *specific problems* the solution of which might solve the original primitive problem. For example, the primitive problem

[1] See, "Tar Sands Yield to Technology," *Chem. Eng. Progr.*, No. 11, **63**, 42–43 (1967).

a shortage of nitrogenous fertilizer exists in the upper Mississippi river valley

might be solved by the solution of any one of the following specific problems:

Design a processing system to synthesize ammonium nitrate and urea in pellet form in Northern Illinois.

Ship liquid ammonia up the Mississippi river from a chemical manufacturing center on the Gulf Coast, and store the ammonia in a specially designed tank farm for later distribution throughout the agricultural community.

Convert the ammonia shipped from the Gulf Coast into urea or ammonium nitrate to put the fertilizer in a solid form rather than in the liquid form of pure ammonia.

Rather than store the liquid ammonia, which boils at −28°F at atmospheric pressure, design a system which will convert the ammonia into an aqueous solution to be applied directly as liquid fertilizer.

The solution of any one of these and many more specific problems might satisfy the local need for nitrogenous fertilizer. However, the engineer must know which of these specific problems, when solved, will *best* solve the primitive problem.

Therein lies the basic dilemma of process engineering. How can we know that the best specific problem has been created? How can we be sure that the proper specific problem is being considered during process design? Knowing how to create the proper problem to solve is equally as important as being able to solve the problem and, in many situations, far more important.

This synthesis of plausible alternative problems is the first and most critical step the engineer takes towards the solution of a primitive problem. The alternatives then are screened carefully and there remain the more promising specific problems to be solved.

In well-established areas of technology the solutions to specific processing problems may be available in the marketplace. The designs for standard processes for the manufacture of sulfuric acid, ammonia, inert gases, steam, and deionized water, to name just a few, are available from vendors through license arrangements. The vendor of a process may arrange for the construction, start-up, and service during operation and usually will provide a guarantee of performance.

Such processes which are available in the marketplace may be thought of as bits of available technology, which along with pumps, heat exchangers, compressors, distillation towers, and so forth, then become the components for the process engineer to use in assembling a solution to a specific processing problem. A *system* may be formed from these *components*, tailored by the engineer to best solve the specific problem.

Once a system has been synthesized in concept there remains the challenging task of establishing the design detail which will result in safe and economical manufacture. The tools brought to bear are those of engineering science, economics, and mathematics. Material and energy balances establish the demands by the process for raw materials, energy, cooling water, and so forth. Thermodynamics is brought to the solution of vapor-liquid and chemical equilibrium problems and to the solution of compression, liquefaction, and other thermodynamic problems. The principles of unit operations and transport phenomena are employed to firm-up segments of the processing concept. The methods of process control and dynamics are used to insure the operability of the concept. The engineer must draw on a wide variety of fields of knowledge, either through his own competence or through consultation with experts, to shape the specific problem into a specific solution.

There result the detailed specifications of a processing system, which we hope, will provide a satisfactory solution to the primitive problem. Engineers who were engaged in the design of a process are often enlisted to supervise construction, start-up, and operation, taking advantage of the intimate knowledge of the process which only the designer possesses.

1.4 THE OBSTACLES ALONG THE WAY

We have sketched briefly the development of a solution to a primitive processing problem, and were there no severe obstacles along the way to the optimal system there would be no need to entertain the serious study of the strategy of process engineering. However, sizable obstacles do appear, and the direct and unguided application of the principles of science, economics and mathematics to processing problems will lead to unnecessary difficulties. An examination of the obstacles with which the engineer must contend exposes the particular skills which take on great value; the skills are those of synthesis, analysis, optimization, and decision making in the face of uncertainty. These must be part of the skill of the successful engineer.

The Creation and Assessment of Alternatives, (Part I of this text). To overcome the severe obstacle of not having specific problems to solve requires a high degree of inventiveness and creative skill. The need to synthesize new situations and quickly assess their potential arises throughout all phases of process engineering. Skill in this area can be developed through experience in problem solving, much in the same way a physical skill such as tennis is developed through coaching, conditioning, and practice.

Optimization, (Part II of this text). A simple calculation identifies the severe problems which arise when the engineer attempts to adjust the design of a process to its most economical form. Suppose that the engineer is free to adjust ten variables in a design to achieve economy of operations; for example, the variables might be the number of trays in a distillation tower, the area of a heat-transfer surface, or the holding time in a reactor. Further, suppose that the engineer wishes to select one of ten levels for each variable.

There exists a total of $(10)^{10}$ unique combinations of these 10 design variables; the problem of finding the optimal design is overwhelming. Even a high-speed computer is of little use if haphazardly employed, for if the computer could be programmed to evaluate the alternative designs at a rate of one design per second, the evaluation of every possible design would take $(10)^{10}$ sec or on the order of 300 yr! And a commercial process such as an oil refinery can be orders of magnitude more complex than this example.

The methods of optimization afford efficient and orderly means for overcoming and circumventing the enormous obstacle of an overwhelming number of alternatives.

Engineering in the Presence of Uncertainty, (Part III of this text). Even if the optimal processing concept can be synthesized and a means for optimizing the design established, it is generally impossible to proceed directly to the optimal design. Wide gaps appear in the information required to firm-up the optimal design. Will future changes in the economy and in the demand for the services of the process appear, greatly reducing the usefulness of the proposed design? How can the design of a distillation tower component be computed when the available data on tray efficiencies may be in error by 20 per cent? What happens to the process if by chance it should be improperly operated or if a pressure vessel should burst? What influence will the erratic arrival of raw materials have on the process? The engineer must shape the system to possess some degree of immunity to factors which are unpredictable or about which insufficient information is available.

1.5 CONCLUDING REMARKS

In the remaining chapters we shall examine the kinds of thinking required to handle the three types of obstacles discussed in Section 1.4. Certain of the methods to be discussed are highly organized and well developed such as those which find their origins in the mathematical theory of optimization. Other areas are less clearly defined and remain primarily within the domain of human rather than scientific skills; the area of problem synthesis is a good example. Still other areas are in a state of transition between the empirical and the scientific, as are the areas of process safety and designing with incomplete data.

That this heterogeneous variety of knowledge must be drawn upon to solve the problems in process engineering attests to the dynamic and vital nature of this field of engineering practice.

In closing this introductory chapter, we must remark that our attention will be focused on ways of thinking which may lead to plans of action, rather than on the means for implementing plans of action. For example, a problem in implementing a plan of action might be that of designing a condenser to liquefy 20 per cent of a hydrocarbon vapor stream. Our attention will be focused on the problem of determining if the liquefaction of 20 per cent of the vapor stream is the optimal plan of action to implement. The problem of implementing a well-defined plan of action falls under the domain of *process equipment design*, a topic about which much has been written and about which we shall not attempt to say more.

PART I

The Creation and Assessment of Alternatives

Part I is a series of five chapters dealing with the mental activities which lead to the creation of specific processing problems.

Chapter 2, *The Synthesis of Plausible Alternatives*, is concerned with the creative skills upon which all of engineering must rely. The engineering sciences generally find use only on well-defined problems, and a first step in engineering is that of creating the proper problems.

Chapter 3, *The Structure of Systems*, deals with the identification of the degrees of freedom which exist in a proposed process. A careful examination of the structure of the processing problem can ease the assessment of the process enormously.

Chapter 4, *Economic Design Criteria*, contains the economic principles which enable the engineer to use intelligently the degrees of freedom in a proposed design in order to achieve the most useful process. We are here concerned with what is meant by the term *the most useful process.*

Chapter 5, *Cost Estimation*, bridges the gap between the technical details of the proposed system and the economic design criteria which define the desired goals to be reached.

2
THE SYNTHESIS OF PLAUSIBLE ALTERNATIVES

The single most important step in process engineering, that of creating plausible processing alternatives, is the step for which only tenuous guidelines can be offered to the initiate. The engineer must rely heavily on his own creative abilities and practical experience. Given a *primitive* problem, the engineer must create a variety of *specific* problems, the solution of which might solve the primitive problem. This initial step is of critical importance since the methods of engineering analysis and optimization can only polish specific processing concepts. We must be certain that the specific processing alternatives created have the quality to polish into a practical solution to the primitive problem.

Typical Problem

There is a local need for 10,000 tons of ammonium nitrate in solution form for agricultural purposes. How might we go about creating a number of plausible methods of satisfying that primitive problem, as a first step towards the engineering of a commercial system?

2.1 THE CREATION OF ALTERNATIVES

A recipe for roast raccoon from a New England cookbook begins with characteristic Yankee common sense, "First catch a raccoon." The process engineer faces a similar situation when venturing into fields for which there is no established technology or in which he suspects that the established technology is not the best alternative for the solution of his processing problem. In this situation the engineer must first "catch" a processing system.

The synthesis of new and novel processing alternatives requires creative and inventive skills that are not nurtured solely by the study of science and mathematics. A knowledge of these cognate fields is necessary, but not sufficient. It appears to many that creative skills in engineering can be acquired or at least developed through practice. For this reason it is important that the initiate in process engineering try his hand at a variety of practical problems, such as those included at the end of this chapter.

However, it should be realized that useful ideas can come only to the well-prepared mind and that even the experienced engineer must do his homework before attempting to create plausible alternatives for the solution of a new problem. The public literature contains a wealth of information which, when combined with the private information available within an engineering and scientific group and with the professional training of the engineer, provides the base for his creative excursions.

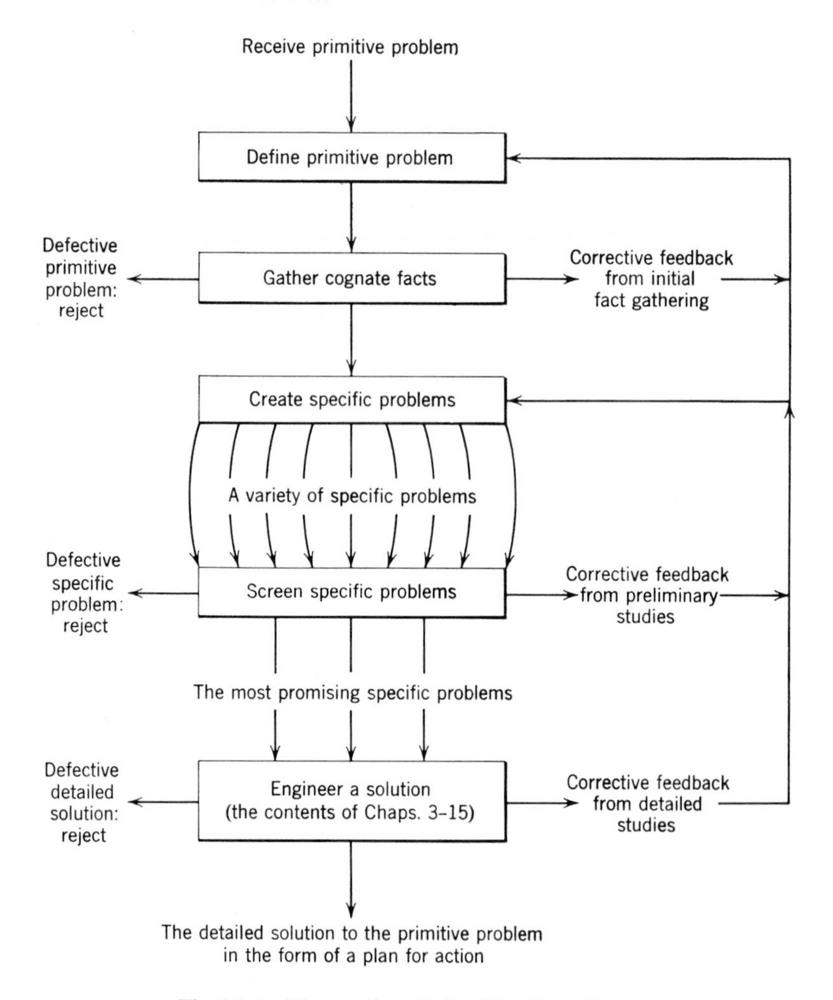

Fig. 2.1-1 The creation of plausible alternatives.

In this chapter we attempt to give the initiate in process engineering some of the concepts upon which he may begin to build his experience in the creation of plausible alternatives.

The steps to be followed, as outlined in Fig. 2.1-1, are:

Define primitive problem. State what the need is that demands the engineer's attention.

Gather cognate facts. Review the situation with those directly concerned, and obtain pertinent business, technical, and scientific information.

Create specific problems. Devise several specific problems the solutions of which could solve the primitive problem.

Screen specific problems. Test the proposed specific problems for obvious defects in logic, cost, and safety.

Engineer a solution. Perform the detailed selection of the equipment and determine the optimum operating conditions required for the best solution of the most promising specific problems.

2.2 DEFINE THE PRIMITIVE PROBLEM

The assignment of a process engineer rarely takes the form of a specific processing system to be designed. Rather, the engineer commonly receives an ill-defined statement of a need. This we call the primitive problem. In this section we examine the nature of a typical primitive problem, showing how it must be altered and redefined as the engineer attempts, with the cooperation of the business community, to determine the proper specific engineering problems upon which to focus attention.

Consider the following primitive problem. *A local expansion of agricultural activity is anticipated, suggesting a potential market for agricultural chemicals.* Obviously, this is a primitive and ill-defined engineering problem, and cognate facts must be gathered to define the problem more clearly. After a detailed study of the circumstances surrounding this problem perhaps the following modified primitive problem might be formed. *A local need for some 10,000 tons per year of nitrogen fertilizer is anticipated; the potassium and phosphorus components of a well-balanced fertilizer are presently available in an aqueous solution form.*

Further studies of the primitive problem may result in the following modification. *An investigation is to be undertaken of methods for the production of 10,000 tons per year of nitrogen in the form of an aqueous solution of ammonium nitrate to be blended with existing fertilizer solutions.*

Such modifications of the primitive problem normally occur as the engineering group attempts to match its particular skills to the needs of society. If such a match seems possible and practical, specific engineering problems are to be created.

The thought processes which result in the redefinition of the original primitive problem correspond to the first feedback loop in Figure 2.1-1. This kind of mental activity may occur during conferences with the technical management of the engineering group, with business and technical leaders of the organization which may be the recipient of the engineering services, and with others who may possess special knowledge of importance. Attempts are made to uncover the following information.

1. *The peculiar circumstances surrounding the primitive problem.* There is no general process engineering problem to which a standard solution can be applied. Each problem has peculiar circumstances surrounding it which must be contended with and taken advantage of, and these circumstances are not known to the casual observer. The exact reasons for the origin of the primitive problem must be known, as well as the leeway allowed in the definition of a satisfactory solution. The availability of water, power, land, raw materials, and human talent must be determined. The capability of the local transportation system of roads, railroads, barges, and so forth, must be assessed, as well as local pollution and safety requirements. Possible markets for by-products must be explored since frequently a process for satisfying a particular need is partially financed by the sale of by-products or energy. Also, the general economic structure into which any process must integrate must be known, and the trends must be predicted as well as possible.

2. *Technical information.* The chemical and physical properties of all materials which might be involved in the solution to the primitive problem must be available. Particular attention is focused on the chemical and physical reactions among the materials, for these reactions form the basis of processing concepts and give rise to possible hazards by the generation of toxic or explosive materials. We must be familiar with the standard equipment and processes which are available in the marketplace to achieve these transformations. Especially when new technology is to be developed, the desired chemical and physical data will frequently be missing. Many times such data may be estimated with sufficient accuracy by generalized correlation and prediction methods (to be discussed in Section 2.5).

2.3 THE SPECIFIC PROBLEMS

With this background information, the engineer begins to lay out a plan of action by creating a variety of specific engineering problems, the solutions of which plausibly could solve the primitive problem. The success of an engineering campaign depends critically on the quality and variety of the specific problems created.

A selection of plausible specific problems for the primitive fertilizer problem includes the following alternatives.

Alternative 1. We might purchase the ammonium nitrate solution at some distant chemical manufacturing center to be shipped to a site near the agricultural center to accumulate for local distribution. This might only require the design of storage facilities for the solution. If the transportation and storage problems could be solved economically, the primitive problem would be solved.

Alternative 2. Rather than pay shipping costs for the water component in the solution, we might purchase the solid ammonium nitrate and then design a storage and dissolving facility to prepare the solution, using local water and power sources.

Alternative 3. It was observed that there is a local supply of nitric acid. Perhaps ammonia could be purchased in tank-car quantity to be shipped to a process in which the ammonia and nitric acid are reacted, and the concentration of the resulting solution adjusted to the desired quality.

Alternative 4. Expanding on alternative 3, we might effect a considerable saving in capital investment if the reaction between the ammonia and nitric acid could be performed in the tank cars which are used to deliver the raw materials. For example, we might slowly pump the ammonia from its tank car into a tank car with nitric acid, after designing some kind of a heat-exchange device to remove the heats of solution and reaction. From a tank car of ammonia and one of nitric acid, two tank cars of fertilizer might be manufactured ready for shipment. If this alternative is technically plausible, it may offer economic advantages.

Alternative 5. There might be quite a demand for ammonia and perhaps it might pay to manufacture our own ammonia for the reaction with the available nitric acid and sell the surplus ammonia as a fertilizer itself. Processes are available for the manufacture of ammonia once nitrogen and hydrogen are available. The design of a process to separate nitrogen from air and a process to tear hydrogen from either the water molecule or some hydrocarbon such as methane might then be part of this alternative.

Alternative 6. If a process is to be designed to manufacture ammonia, perhaps it would pay not to use the available nitric acid but to manufacture the acid from surplus ammonia. In this way a larger ammonia synthesis process might be designed to achieve a saving in the cost of ammonia per ton. This saving may be sufficient to warrant not using the local nitric acid, which must be purchased and which might not be available in future years. There is an advantage to having complete control of raw material sources.

Clearly, these six alternatives are but a sample of the large number of alternatives that must be proposed before we can hope to have a proper base upon which to begin the engineering of a process. It is also clear that these initial concepts can only be plausible, and that it is only after a considerable investment in engineering effort that the true nature of the alternatives becomes apparent. The defective alternatives must be weeded out early so that this investment will not be wasted on schemes that cannot possibly be shaped into a commercial system. Herein lies a most critical difficulty in process engineering. On the one hand, we cannot afford the investment in time and talent to go into the engineering detail to properly assess the plausibility of all the alternative methods of solving a given problem. On the other hand, we cannot properly limit attention to one or a few of the alternatives without having first made a detailed assessment of all the alternatives. There is the great risk of eliminating the optimal concept along with the majority of concepts which are not even

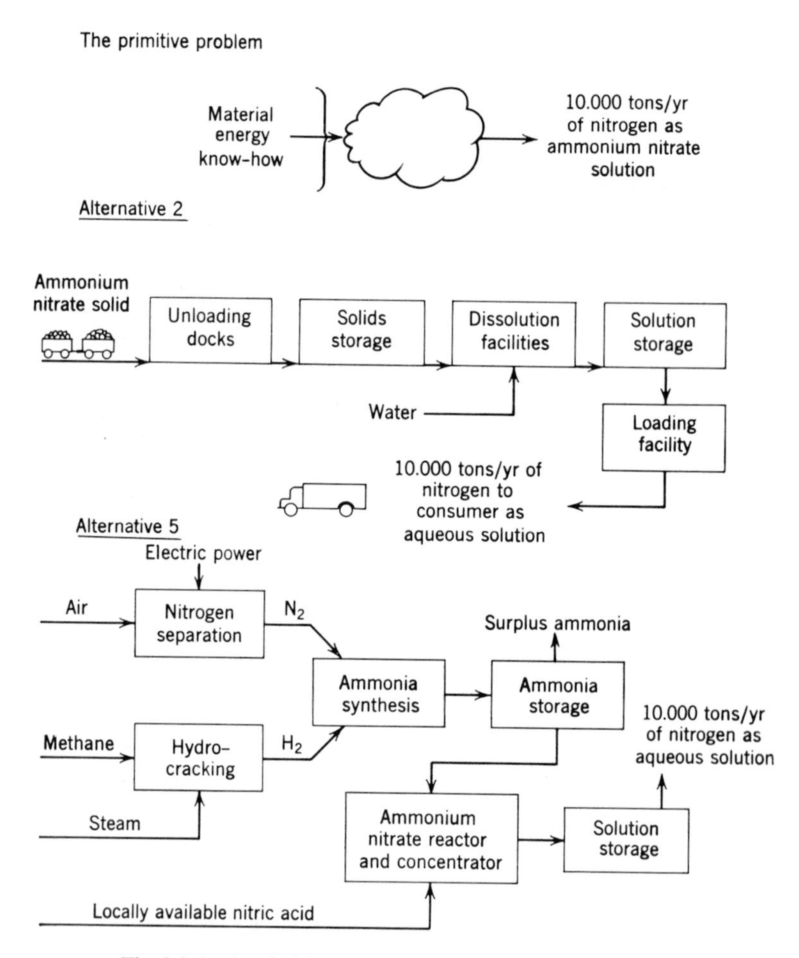

Fig. 2.3-1 A primitive problem and two specific alternatives.

plausible. At this stage in process engineering, the specific problems take the form of a rough flow sheet, as shown in Fig. 2.3-1 for alternatives 2 and 5 above.

2.4 THE PRELIMINARY SCREENING OF ALTERNATIVES

Clearly the creation of processing alternatives should lead to a large variety of possible engineering problems differing in concept and detail. However, it is impossible to carry each alternative through to a complete design to make a detailed comparison among the alternatives. The design of a single system may be an enormous task, and this effort cannot be wasted on alternatives which are defective and cannot be shaped into commercial systems.

In this section we demonstrate how the variety of alternatives must be screened to eliminate those alternatives which are inferior. Preliminary screening is of critical importance since only those alternatives which survive are candidates for the final processing system. On the one hand, there is the danger that an inferior system will survive only to be detected after the waste of valuable engineering effort. On the other hand, there is the danger that the optimal system inadvertently may be screened out.

During preliminary screening, the engineer asks the following questions of each alternative.

1. *Is the concept illogical?* A specific processing problem may violate common sense or the principles of technical logic. For example, an involved processing concept may be rejected if it can be shown to violate a law of thermodynamics.

2. *Can the concept be shown to be inferior to one of the other alternatives suggested?* A paired comparison of all alternatives often can be used to screen alternatives. For example, suppose that two alternative processing schemes differ only in the method of obtaining hydrogen, one using the electrolysis of water and the other the thermal cracking of methane. In the economic structure present in the United States it is known that thermal cracking of methane is the better concept. Thus, the alternative involving the electrolysis of water can be eliminated since it is necessarily inferior to the other alternative. However, this does not imply that the other alternative will not be eliminated for some different defect.

3. *Can the concept be shown to be equivalent or inferior to a known inferior processing concept?* Certain methods of processing are known to be inferior, and often the engineer can eliminate a number of alternatives by showing that they are equivalent to or less valuable than these known inferior methods of processing. Often this comparison is accomplished by tentatively setting the established technology as the superior concept.

4. *Can the concept be shown to require too much technical or economic extrapolation from existing technology, thus involving too high a risk?* A processing concept involving the use of a heat exchanger at 4,000°F might be rejected as being too much of an extrapolation from existing technology, since it has not been established that such a heat-exchange device could operate in a commercial environment at such extremes of temperature. If it can be shown that the economic advantages derived from the use of this concept are sufficiently great, a small scale experimental study would be required before a definite decision could be made on the plausibility of this concept.

5. *Is the concept unsafe?* A process for the manaufacture of nitroglycerine in the form of an emulsion was rejected in spite of possible economic advantages, since the process was thought to be unsafe. For the same reason, in the early 1950's a fertilizer plant in Iceland was designed using a crystallization process for the manufacture of ammonium nitrate solids rather than a then new high-temperature prilling process. Now that this prilling process has been shown to be relatively safe, there is a good chance that if preliminary screening were to be carried out today the crystallization process would be eliminated as inferior to the prilling concept.

6. *Does the concept suggest a better alternative?* A given alternative may be rejected when an examination of the concept suggests a modification which will lead to a better concept.

7. *Does the concept involve special technical competence which our group does not possess or which we cannot acquire through normal business channels?* For example, an engineering firm may reject a processing concept which involves the use of a plasma jet reactor for fixing nitrogen on the grounds that this particular technology is beyond the competence of our engineers and may not be purchasable by consultation with experts in this particular situation.

Clearly these are just a few of the methods used to screen the inferior alternatives.[1] In fact, the creation and preliminary screening of alternatives often occur simultaneously. We illustrate this by sketching out the design of a storage facility for ammonia.

Example: Storage of 60,000 Tons of Ammonia

The engineering division of a midwest engineering firm has contracted to investigate means for storing 60,000 tons of ammonia, which will accumulate each year, for sale in smaller amounts to the local agricultural community. We now outline a portion of a conference in which a solution to this problem is being shaped.

[1] It is expected that the modern theories of model discrimination may provide the engineer with useful quantitative tools for efficiently guiding the screening and modification of alternatives. See for example, G.E.P. Box and W. J. Hill, "Discrimination Among Mechanistic Models," *Technometrics*, No. 1, 7 (1965).

ENGINEER 1: Ammonia is a vapor under normal conditions, we might store it in large gas holders as city gas was stored a few years back.

ENGINEER 2: How much space is required to store the ammonia this way?

E1: The volume of the ammonia at standard temperature and pressure is approximately:

$$V = (60{,}000 \text{ tons})(2{,}000 \text{ lb/ton})(1/17 \text{ lb-moles/lb})$$
$$(370 \text{ SCF/lb-mole}) \cong 2.9 \times 10^9 \text{ ft}^3$$

E2: Even a storage vessel 100 ft high and a mile in diameter doesn't have sufficient capacity. That alternative is out!

E1: Possible alternatives are to consider liquid or solid phase storage. The ammonia might be stored as:

A liquid at ambient temperature under its vapor pressure.

A liquid at atmospheric pressure refrigerated to its boiling point.

A solution in some solvent, say water, which lowers the vapor pressure and allows less elaborate containment.

A solid compound, from which the ammonia might be released as demanded, for example, by heating.

Solid ammonia at atmospheric pressure under refrigeration.

Another alternative is to change the stated problem and deliver something other than ammonia to the consumer. For example, perhaps the consumer could use an ammonia solution for fertilizer, or perhaps the nitrogen could be tied up in some other compound, such as solid urea, rather than in ammonia. In this way we might be able to bypass this troublesome storage problem.

E2: Limiting attention to the storage of pure ammonia, which of the two phases, solid or liquid, seems the more easily stored?

E1: The vapor pressure of liquid ammonia is not excessive at low temperatures.

Liquid Ammonia

Vapor Pressure	*Temperature*
1 atm	−28°F
5 atm	41°F
10 atm	78°F

whereas the difficulty of handling the solid might be the source of problems. The storage of pure solid ammonia might thus be rejected.

E2: The temperatures at atmospheric pressure to which you refer are quite low. Would not the heat-transfer rates into any vessel be enormous?

E1: A quick order of magnitude calculation will answer that question. Should the 60,000 tons of ammonia be stored as a liquid at its normal boiling point, a temperature difference on the order of 100°F might exist on a summer day. Now, the primary resistance to heat transfer would be offered by the tank-air interface with heat-transfer coefficient on the order of 10 Btu/hr-ft^2-°F and by any insulation we might elect to install with a conductivity of 0.15 Btu/hr-ft-°F-in. for urethane foam insulation. The surface area exposed to the atmosphere on a tank of equal height and diameter would be on the order of

$$A = \frac{5\pi}{4}\left(\frac{4V}{\pi}\right)^{2/3}$$

where

$$V = (60{,}000 \text{ tons})(2{,}000 \text{ lb/ton})\left(\frac{1}{50} \text{ ft}^3/\text{lb}\right)$$
$$= 2.4 \times 10^6 \text{ ft}^3$$
$$A \cong 8.5 \times 10^4 \text{ ft}^2$$

The heat-transfer rate can then be estimated as

$$Q = UA\,\Delta T = \left(\frac{1}{10} + \frac{x}{0.15}\right)^{-1} (8.5 \times 10^6) \text{ Btu/hr}$$

Heat Transfer to Tank

x, Insulation Thickness, In.	*Q Btu/hr*
0.0	8.5×10^7
6.0	2.1×10^5
12.0	1.1×10^5

E2: Even with a foot of good insulation you would expect about 10^5 Btu/hr to be transferred into the tank, and that would result in the boiling of nearly 200 lb of ammonia per hour.

$$(10^5 \text{ Btu/hr})(1/590 \text{ lb vaporized/Btu}) = 1.7 \times 10^2 \text{ lb/hr}$$

We cannot allow such a pollution of the atmosphere even if we could afford the loss.

E1: We have a choice of investing in more insulation, disposing of the vapors in some way, or perhaps we could reliquefy the vapors and return them to the tank. I recall that the older refrigerators used ammonia as a working fluid, perhaps we could install a compressor and a heat exchanger and use the tank as a flashing vessel in a refrigeration loop to recover the liquid ammonia. The selection among these alternatives would be based on cost considerations.

E2: Let me pose a few more questions.

Will water vapor condense in the insulation to form ice?

Ammonia concentrations of 16 to 27 per cent in air are flammable. Would it be hazardous to fill the tank?

Is there a construction material that can last for years at −28°F?

Can anyone build a tank that big?

Doubtless, the ground will freeze beneath the tank. Will this cause trouble?

Why not store the ammonia in a large underground cavern as Dow Chemical Company stores ethylene?

Could we freeze a section of earth and then excavate a large storage area over which a top could be placed, eliminating the need for a large tank? This has been done in the storage of liquid methane.

Liquid petroleum products are often stored under pressure. What led to the use of such storage?

Is the cost of storage more than the value of the ammonia? How are you going to transfer the ammonia from storage to the consumers' vehicles?

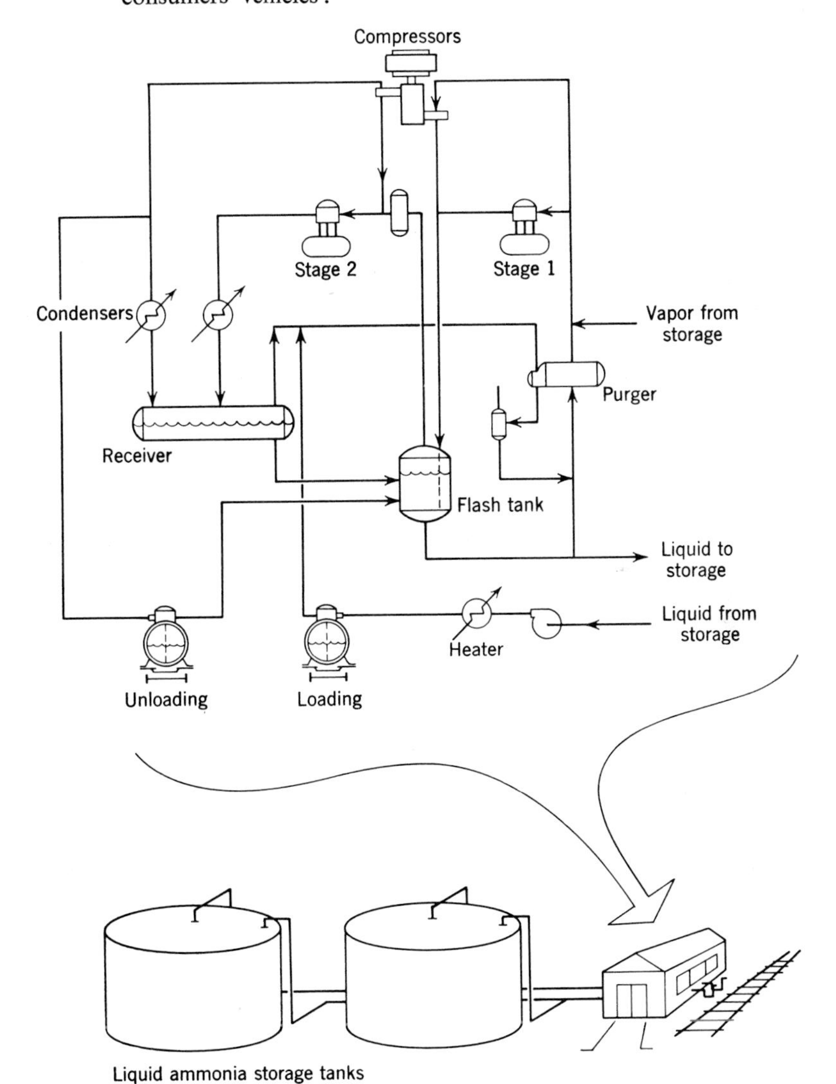

Fig. 2.4-1 An atmospheric pressure ammonia storage system.

Could the presence of this amount of material present a hazard to the surrounding community in case of tornado, hurricane, or earthquake? We cannot infringe upon the safety of the community.

AUTHOR: From this creation and screening of alternatives a reasonable solution to the problem evolves. We leave the engineering group at this point, to return later as their report is being presented to management.

E1: Our engineering group suggests, after careful consideration, that the liquid ammonia be stored in two 30,000 ton tanks of ASTM A-201 steel plate, insulated with 6 in. of cellular glass and incorporating the simple refrigeration loop (shown in Fig. 2.4-1) to recover the vapor losses. We have devised a start-up purge plan which will eliminate the hazard of explosion during filling, have made adequate provisions for loading and unloading, and provided sufficient safeguards against the sources of disaster which we could foresee. The tank and associated equipment will cost between 1.5 and 2.0 million dollars or about $30.00 per ton of capacity. We expect an operating cost of about $4.00 per year per ton of ammonia stored.

AUTHOR'S COMMENT. The above conversation, while imaginary in detail, is historical in general content. The actual embodiment of such deliberations may now be seen at the new ammonia fertilizer plants throughout the country, in the form of large insulated storage tanks for liquid ammonia.[2]

Example: The Development of a Separation Process

We further illustrate the creation and screening of plausible alternatives by presenting the development of a process to separate hexane from peanut oil. This section is based on the class notes of Professor D. R. Woods of McMaster University, Hamilton, Ontario.

Problem. Ten tons per hour of miscella (15 wt per cent peanut oil in hexane) leaves a peanut processing plant at 35°C. It is necessary to propose a processing scheme which can separate the hexane from the oil so that the final oil contains less than 0.01 per cent hexane and such that the temperature never exceeds 80°C.

The mental processes in action. We now sketch through the development of a process which might perform the task. Engineer E is in charge of the project and DA is an engineer serving as the devil's advocate.

DA: How can you separate the hexane from the heavy oil?

E: I might begin with what seems to be the most economical process: distillation.

DA: What makes you so confident that distillation is plausible?

E: The vapor pressure of peanut oil is very small because of its high molecular weight, whereas hexane is light and volatile.

DA: What about the constraint on concentration and temperature? The vapor pressure of hexane in a solution containing 0.01 per cent hexane in peanut oil

[2] See F. L. Applegate, *Chem. Eng. Progr.* No. 1, **61**, (1965)

at 80°C is only 0.6 mm Hg. Normally, even vacuum distillation towers do not operate at pressures less than 100 mm Hg.

E: I can handle that problem by making up for the difference between the tower pressure and the vapor pressure by use of some inert gas such as steam. This is the concept of steam stripping, and the amount of steam required to achieve the separation might be estimated by

$$\text{moles steam/moles hexane} = (p - 0.6 \text{ mm Hg})/0.6 \text{ mm Hg}$$

where p is the tower pressure.

DA: How are you going to choose the tower pressure?

E: If I elect to condense the hexane under vacuum, the vapor pressure of hexane in the condenser must be such that hexane will condense on the heat-exchange surface. Cooling water is available at 25°C and the dew point for hexane is above 25°C for pressures above 150 mm Hg. Thus, I might be able to condense the hexane using available cooling water if the condenser operates at 150 mm Hg, and since I envision the condenser as connected directly to the tower this might be an estimate of the tower operating pressure.

DA: The pressure you cite is a lower bound, since theoretically you could achieve the separation at any pressure merely by supplying sufficient inert. How would you estimate the highest pressure you might want using steam as the inert?

E: If the steam is to act as an inert gas, it must not condense in the distillation tower. The temperature during separation is not to exceed 80°C and at 400 mm Hg steam condenses at 85°C. This suggests that the distillation tower should operate somewhere in the range of 150 to 400 mm Hg.

DA: For the sake of argument, fix the still pressure at 200 mm Hg and assume that the still can be designed. What other problems arise?

E: A condenser will be needed to condense the hexane under vacuum, a steam jet ejector might be used to provide the vacuum, and a decanter could be used to separate the hexane-water condensate since they are immiscible. The plausible processing system is shown in Fig. 2.4-2.

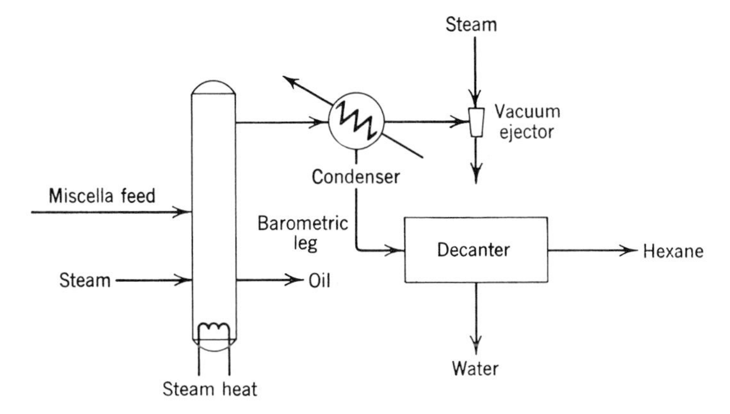

Fig. 2.4-2

DA: You sketched out the design of the distillation tower by focusing attention on the reduction of the hexane concentration to 0.01 per cent, but yet the feed is 85 per cent hexane. This is a big gap, and I doubt if a system designed with only one extreme in mind can be efficient. Look at the other extreme.

E: There is no problem at all in removing part of the hexane from the peanut oil: the boiling point of pure hexane is below 80°C at 1 atm pressure. This suggests that the load on the vacuum still could be reduced by some initial flashing of the hexane either at atmospheric pressures or at some slightly reduced pressure.

DA: Indeed, this is a plausible alternative. Vapor liquid equilibrium calculations reveal that a flash of the hexane-peanut oil solution at 75°C and 1 atm will remove 85 per cent of the hexane, and at 75°C and 300 mm Hg 98 per cent of the hexane will be removed.

E: The alternative processes shown in Figs. 2.4-3 and 2.4-4 should be far superior to our initial alternative in Fig. 2.4-2.

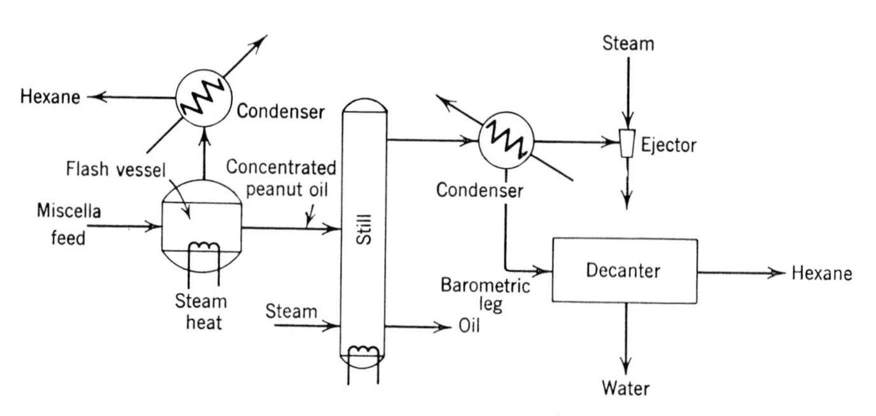

Fig. 2.4-3

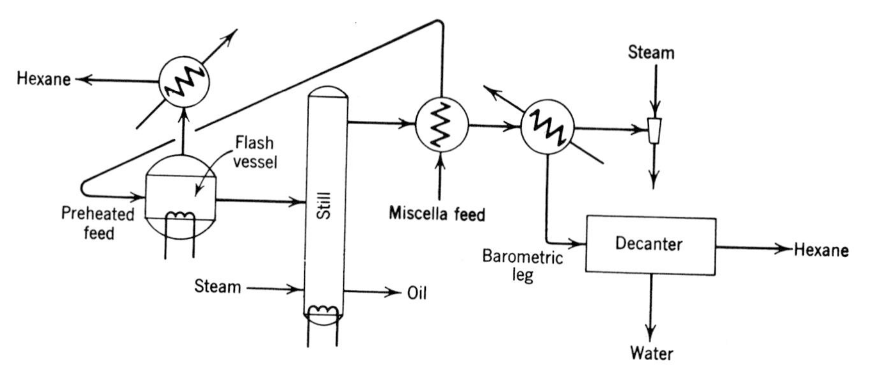

Fig. 2.4-4

DA: Let us go back to the very beginning of this process development program to the point where you first established the pressure of the vacuum still by stating: "If I elect to condense the hexane-steam mixture under vacuum." This statement focused all our attention on the concept of running the condensers under vacuum. Could we put the steam ejector before the condenser and then condense the hexane at atmospheric pressures? I see all sorts of questions that have to be answered before I'll be satisfied.

E: Obviously we could spin our wheels with hundreds of minor modifications and the real question to be answered is: which system does the job and does it most economically?

AUTHOR'S COMMENT. Here we have witnessed the development of the first ingredient of a commercial process, the plausible processing alternatives. However, this is but the first ingredient, it is impossible to go further until the more quantitative techniques of process engineering are available for use.

2.5 COMMON SOURCES OF BACKGROUND INFORMATION

In this section we give a brief outline of the sources of background information upon which the process engineer commonly draws. More detailed source material will be referenced in certain following chapters. Part of the information comes from public sources, while other sources are private and only available within a particular firm. The private sources of information constitute part of the know-how of an engineering firm, which places the firm in a better competitive position. The outsider is unable to tap these private sources.

1. PRIVATE SOURCES OF INFORMATION
   (a) *The file on the project.* The project file should contain all correspondence leading up to the formation of the primitive problem. Frequently this gives clues concerning the peculiar circumstances surrounding the problem. Also, vital sources of further information are often referred to in such correspondence.
   (b) *The company report file.* A company usually develops competence in several specific areas of engineering, and special information is stored in the form of company reports which are the private property of the company. These will include research reports, design reports, operating reports, and compilations of special data.
   (c) *The design questionnaire.* Special forms are usually filled out by both the engineer and the customer concerning the nature of the environment into which the process must integrate. Figure 2.5-1 shows part of the data thus sought by the Universal Oil Products Company, an engineering firm specializing in the design of processing systems.

UOP PROCESS DIVISION
30 ALGONQUIN ROAD • DES PLAINES, ILLINOIS 60016

PROJECT SPECIFICATION
BASIC ENGINEERING DESIGN DATA

NO.	REV.
- 102 -	
SHEET	OF
BY	APP'D
DATE	

EQUIPMENT DESIGN INFORMATION

INSTRUMENTS

1. Shall the control board be graphic? ______

2. Shall the control board be semi-graphic? ______

3. Shall the control board be non-graphic? ______

4. Shall instruments be miniature? ______

5. Shall the multi-point temperature indicator be mounted on a console desk? ______

6. Shall instruments be electronic? ______

7. Shall control valves be operated pneumatically? ______

8. Shall extent of instrumentation be minimum required for operation? ______

9. If extent of instrumentation is not to be minimum: ______

   9.1 Shall all process charge and product streams be measured with flow recorders? ______

   9.2 Shall charge and product stream flow recorders be integrating meters? ______

   9.3 To what extent shall process streams be measured with flow recorders? ______

   9.4 To what extent shall heat exchangers be equipped with temperature points to measure their performance? ______

   9.5 If heat exchangers are to be equipped with performance measuring temperature points, describe the temperature measuring device. ______

   9.6 Shall all utility flow rates be metered and recorded as process unit totals? ______

UOP PROCESS DIVISION
30 ALGONQUIN ROAD • DES PLAINES, ILLINOIS 60016

PROJECT SPECIFICATION
BASIC ENGINEERING DESIGN DATA

NO. - 102 - REV.
SHEET OF
BY APP'D
DATE

NOTE – THESE DATA ARE CONFIDENTIAL AND THE PROPERTY OF UNIVERSAL OIL PRODUCTS COMPANY AND SHALL NOT BE DISCLOSED TO OTHERS OR REPRODUCED IN ANY MANNER OR USED FOR ANY PURPOSE WHATSOEVER EXCEPT BY WRITTEN PERMISSION OR AS PROVIDED IN A SIGNED AGREEMENT WITH UNIVERSAL OIL PRODUCTS COMPANY RELATING TO SUCH DATA.

9.7 Shall utility stream flow recorders be integrating meters?

9.8 Shall provisions be made for any type of individual utility flowrate metering, such as fuel to heaters or water to exchangers?

9.9 Shall all process levels be shown on the control board?

9.10 Shall high and low level alarms be used?

9.11 Shall continuous stream analysers be used?

10. Other remarks.

9A

Fig. 2.5-1 One of a library of forms useful in the gathering of design data.

(d) *Site visitation.* Frequently the engineer will spend several days on the site of the primitive problem in an attempt to detect features which may have been missed in the more formal methods of data gathering.

(e) *Personal contact.* It is most important for the engineer to examine personally all the evidence leading up to the problem, so that he may decide for himself what the situation really is. Thus he may find it worthwhile to interview public officials, suppliers, sales engineers, customers, other process engineers, plant operators, research chemists, and so forth.

2. PUBLIC SOURCES OF INFORMATION. The mounting literature available to the public is potentially a source of vital information in process design; however, it is becoming increasingly difficult to search effectively for specific bits of information, and it is difficult to assess the accuracy of the data found. For example, significant errors in the vapor pressure of light hydrocarbons continued unchallenged in some standard physical property tabulations until recently. The original errors were made fifty years earlier. In the future, the science of information retrieval will play a more significant role.

(a) *Existing processing systems.* Tabulations are occasionally made of processing systems which are available by purchase or license. See, for example, the annual petrochemical process review published by *Hydrocarbon Processing and Refiner*, which gives a flow diagram and brief description of scores of processes. See also the process flow sheet included as a special feature in each issue of *Chemical Engineering.* Advertisements in any number of trade journals and those compiled in the *Chemical Engineering Catalog* may initiate valuable contacts with process vendors. *Thomas' Register of American Manufacturers* provides an exhaustive listing of the makers of manufactured items of all kinds.

(b) *Physical and chemical property tabulations and correlations.* A few of the public sources of data on the properties of materials are the following. *The International Critical Tables; The Chemical Engineer's Handbook; The Handbook of Chemistry and Physics; The Chemical Rubber Handbook;* and *The Merck Index.*

Many studies have been made, relating physical and chemical properties of substances to convenient parameters on an empirical or theoretical basis. Some of these correlations appear in textbooks, such as Reid and Sherwood, *The Properties of Gases and Liquids*, McGraw-Hill, New York, 1967: as computer programs, such as the AIChE's "The Computer Estimation of Physical Properties," or only as private company documents.

(c) *Manufacturer's data.* The manufacturers of certain chemicals prepare brochures on physical and chemical data, toxicity, fire and explosion hazards, and methods of safe handling and storage. Some brochures even suggest applications and special uses for the chemicals and tend to be authoritative.

(d) *Data on safe practices.* A number of organizations offer information on safe practices. For example, the AIChE has a continuing symposium series on Safety in Air and Ammonia Plants, and The National Fire Protection Association has a number of publications of interest. Codes and laws are often imposed, such as the ASTM specifications for materials of construction, the AISC and AWS codes for safe structural design, the ASME code for unfired pressure vessels, and the ASA code for pressure piping (as will be detailed in Chapter 13).

(e) *Economic and business data.* Such data are available from a number of sources, including *The Chemical Business Handbook*, and *The Chemical Statistics Handbook*. The *Chemical and Engineering News* quarterly lists the prices of common chemicals. Cost data are published in various trade publications, including *Oil, Paint and Drug Reporter*, and *Oil and Gas Journal.*

(f) *Governmental codes and regulations.* Federal, state, and local codes and regulations may prescribe allowable limits for the design of transportation systems and manufacturing plants, where the public safety is affected. Often the codes of national technical organizations are adopted wholly or in part into these legally enforced regulations.

In summary, there is a wealth of data with which the engineer must be familiar and which is required to define more clearly the primitive problem assigned the engineer.

2.6 CONCLUDING REMARKS

We have been describing the creative skills upon which our modern industrial economy was founded. In the early days of chemical and metallurgical manufacture it was sufficient merely to create a single plausible solution to a processing problem, for profit margins were high and almost any process which was capable of running might well end up as a profitable and useful process. This is not the situation now, except in a few areas such as the production of certain pharmaceuticals, where the costs of research, development, and marketing outweigh the costs of manufacture. The present high level of competition and of technical competence requires that a solution to a processing problem be close to the best possible solution. The engineering of such required excellent solutions to difficult problems within a limited time has forced the industry to call upon new theoretical and practical research developments as soon as they come into being. And, in fact, the processing industry has been the motivating influence for the development of many of the areas of mathematics and science.

We have indicated the important role played by the engineer in acting as an "operator" upon the primitive problem to produce engineering tasks in the form of specific problems. It is only natural to wonder how this creative operation of the process engineer can be carried out most efficiently. A good deal has been written on this subject. Techniques have been described with more or less involved rituals comprising group interaction, sometimes called "brain storming," which purport to increase the creative output of the engineering team. Such organized routines certainly have useful applications in generating at least the raw material for creative thought, and in reducing the chance that some potentially attractive approach to the problem will be ignored. However, informal group thinking sessions have been held down through history. And the finally effective group for advancing to a sound idea for the solution of the problem is still the group of one: the individual engineer wrestling with the remaining gap between the desired result and the available techniques. That the result of this struggle may be a new and valuable concept, not predictable or to be expected of everyone, is attested by the existence of the patent system, which recognizes the property right which the individual may have in his problem-solving concept.

It should be noted that the primary objective of the process engineer is to get the most satisfactory and economic solution to the primitive problem, whether or not this includes patentable ideas of his own. The optimum solution may indeed be to utilize or adapt the ideas of others which are available in the forms of standard equipment, licensed designs and engineering services of specialists. The creative act then becomes the synthesizing of a new and useful system from elements that are in existence.

On balance, we must conclude that there appears to be no automatic procedure for group effort in turning out process concepts. We recommend that the initiate begin his creative excursions with the primitive problems provided at the end of this chapter and plan to make the sharpening of the required skills a lifelong project.

References

The classic books of Polya should be studied to sharpen skills in attacking ill-defined problems.

G. Polya, *How to Solve It*, Anchor, Doubleday, New York, 1957.

———, *Mathematical Discovery*, Vol. 1, 1962; Vol. 2, 1965; Wiley, New York.

For more details on the practice of process design see:

T. K. Sherwood, *A Course in Process Design*, MIT Press, 1963.

M. S. Peters and K. D. Timmerhaus, *Plant Design and Economics for Chemical Engineers*, McGraw-Hill, New York, 1968.

F. C. Vilbrandt and C. E. Dryden, *Chemical Engineering Plant Design*, McGraw-Hill, New York, 1959.
J. Happel, *Chemical Process Economics*, Wiley, New York, 1958.
H. F. Rase and M. H. Barrow, *Project Engineering of Process Plants*, Wiley, New York, 1957.

Methods are abuilding for the systematic synthesis of system designs:

D. F. Rudd, "The Synthesis of Process Designs, I. Elementary Decomposition Principle," *AIChE J*, **14** (1968).
A. H. Masso and D. F. Rudd, "The Synthesis of Process Designs, II. Heuristic Structuring," *AIChE J*, **15** (1968).

An excellent discussion of the interplay of ideas leading to a process is presented in:

H. K. Eckert and G. A. Cain, "Profile of a Profitable Project," *Chem. Eng. Progr.*, No. 3, **60** (1964).

PROBLEMS

In the following problems you are to prepare plausible solutions to primitive processing problems. The solutions should take the form of a rough process flow diagram, such as in Fig. 2.4-2, in which the major items of process equipment and required raw materials are identified.

2.A. Prepare a rough flow sheet of a process for the desalination of seawater by selective freezing. The seawater is to be partially frozen, the ice washed free of salt water and melted, yielding potable water. Identify the major operations and equipment needs for your system. What source of power is to be used? What technical difficulties may be expected?

2.B. Could the system proposed in Problem 2.A be modified to achieve economy, for instance, by using the warm incoming seawater to melt the ice? How might the latent heat of freezing be used? Make three major modifications in the original system which might improve its efficiency.

2.C. Butane boils just below the freezing point of water. Sketch out a process for selective freezing of seawater based on this fact.

2.D. Rather than use freezing to achieve the salt removal, consider the opposite extreme, boiling. Prepare a preliminary flow sheet for a process system which might desalinate seawater by boiling the water and condensing the pure vapors. Might the formation of scale on the heat-transfer surfaces be a source of trouble? Could you eliminate the conventional heat-transfer surfaces altogether, for example, by forcing live steam into the brine? Would this work or would it just waste steam? Could you get the necessary heat from the seawater itself and eliminate the need for steam?

2.E. Make three major modifications in the desalination process above to achieve economy of operation. Include in one modification, solar energy collection pools to preheat the seawater. If the solar constant is 400 Btu/hr-ft^2, what is the order of magnitude of the surface area for the collection pools to preheat the seawater by 20°F for a 100,000 gal/day desalination plant? Is this reasonable?

2.F. Devise a process for continuously separating sugar from milled sugar beets by solvent extraction. The sugar is to be prepared in crystal form. What solvent do you recommend? What do you intend to do with the waste pulp and solvent? What energy source are you going to use? What big pieces of process equipment will be needed? Base your design on a plant to process 55 tons/day of sugar.

2.G. When butane is thermally cracked a product of the following composition results:

Component	*Weight Per Cent*
Hydrogen	0.2
Methane	12.0
Ethylene	30.0
Ethane	10.0
Propylene	20.0
Propane	5.8
Butene	12.0
Butane	10.0

Sketch a system to separate the hydrogen and methane for use as a fuel in the butane cracking reactor, the ethylene for use in the manufacture of polyethylene, and the bulk of the higher molecular weight components to be recracked. We suggest some sort of compression and flash vaporization process. Would this be adequate or might a more elaborate separation technique be needed? What units require high-pressure operation, heating, cooling?

2.H. Prepare a flow sheet for a process to recover nitrogen from air for use as raw material for an ammonia converter. Try the following concepts.

(a) Combustion of hydrogen in the air stream followed by cooling and condensation of the water vapor.

(b) Combustion of ammonia in the air to form nitrogen oxides to be removed.

(c) Liquefaction and distillation of air.

2.I. Devise three diverse schemes for separating a small amount of sand from solid paraffin at a rate of 100 tons/month. Sketch the flow sheet and equipment requirements.

2.J. It is proposed to make liquid sulfur dioxide by reacting sulfur with oxygen, which can be obtained on a tonnage basis for as little as 0.4 ¢/lb. What economies in processing would this permit, compared to the conventional processes in which sulfur is burned in air? What special problems would arise? How might these be solved? Prepare a flow diagram.

2.K. Speculate on a process for the manufacture of carbon dioxide, taking advantage of low-cost pure oxygen. Devise processes utilizing three different sources of the carbon atom.

2.L. Given the following size distribution for coke available within an integrated steel company, devise two flow sheets for achieving the indicated desired distribution for blast furnace feed. For information on crushing and screening operations see *Chemical Engineers Handbook*, Perry, ed., McGraw-Hill, New York, 1963.

Initial Distribution		*Desired*	
>6 in.	10 per cent	>6 in.	0 per cent
4 × 6 in.	15	4 × 6 in.	10 per cent or less
3 × 4 in.	30		
2 × 3 in.	25		
1 × 2 in.	12		
<1 in.	8	<1 in.	10 per cent or less

What are the major operations involved? What effect does each flow sheet have on the coke plant productivity? What technical difficulties might be encountered? Are there any aspects of the processes which will require attention to social constraints?

2.M. In Problem 2.L, it was assumed that the initial coke was produced by the usual water quenching process. Often, however, the fluctuating residual moisture is detrimental to blast furnace operation and the question may be raised: would it be practical to consider quenching the coke with N_2 gas, which is a by-product from the O_2 plants? Make some simple rough calculations to test this possibility. Are there any safety hazards involved?

2.N. In a process now being used to separate niobium and tantalum from a highly insoluble ore, the digestion of ammonium bihydrofluoride leads to the release of 12,000 lb/hr of a saturated 5 per cent by weight ammonia-steam system. An earlier stage in the process requires a 20 per cent by weight ammonia solution. An examination of processes for the concentration of the ammonia-water system is to be made. Special consideration in process selection is to be given to the problems of disposal and pollution associated with the respective processes.

3

THE STRUCTURE OF SYSTEMS

In Chapter 2 the principles were exposed which lead to the creation of arrangements of equipment to perform a given processing task. We now begin to build a framework by means of which the technical details of a design can be adjusted to produce the most useful process. In every design, there appear certain free *design variables* which are at the disposal of the engineer, and these are to be adjusted to tailor the process to the processing need. However, to make use of this freedom, it is necessary to delve into the *information flow structure* of the design equations which describe a process; and if this can be done properly significant simplifications result.

Typical Problem

A design group is beginning the detailed optimization of the design of a refinery, to determine the best flow rates, temperatures, pressures, and product distributions. The system consists of twenty component processes, with scores of design relations, numerous constraints on temperatures, pressures and flow rates, and nearly one hundred variables. The information about the system is embodied in the design equations, performance tables, a process flow diagram, computer subroutines, and economic reports. How can we organize this problem and reduce it to something definite and manageable?

3.1 A SYSTEM AND ITS SUBSYSTEMS

The modular structure of a process system is its most obvious feature. Large systems consist of a number of easily identifiable components or subsystems, such as stills, heaters, acid plants, dryers, boilers, vessels, heat exchangers, extracting towers and washers, which interact with each other to perform some larger function. Such systems are not just *agglomerates* of component parts which operate independently of each other in no unusual or unexpected way. Rather, critical interactions occur in which the performance of each component is strongly dependent on the other components, and the performance of the system as a whole is more than just the sum of the performance of its component parts.

Useful performance characteristics of large systems arise from such interactions. For example, allowing the product stream from a process to interact with the incoming feed via heat exchange can greatly increase the efficiency of the process. The interactions are a double-edged sword, however, since they confound the engineer during the design phase of process engineering and introduce considerable difficulties in analysis as well as in the control of the operating process. A haphazard approach to the optimization of an interacting system can completely overwhelm the engineer and his associated computing facilities.

A clue to the efficient approach to large engineering problems appears to be the *information flow structure*. Information is passed from component to component within a processing system by variables common to several components, the output from one component being an input to others. This transfer of information traces out an information flow structure of the system, which provides a skeleton upon which we can organize an orderly computation strategy.

In this chapter we show how we identify the core or dominant problem which must be solved to specify the optimal system design. Once the core structure has been identified, a proper optimization strategy can be tailored for the problem at hand.

3.2 SYSTEM INTERACTION

To illustrate the effects of interactions in a system, consider two processing units, a heat exchanger and a catalytic reactor. The reactor and the exchanger perform quite normally when separated (as shown in Fig. 3.2-1). Now suppose the heat exchanger is connected to the reactor to recover the heat lost in the product stream by heating up the feed stream (shown in Fig. 3.2-2).

The temperature of the stream leaving the reactor must equal the temperature of the stream entering the exchanger since these variables are now one and the same. This must also be true of the stream leaving the exchanger and entering the reactor. Plotting the performance curves for the two components on the same graph reveals that there are *three* points which satisfy this condition. The upper level of operation (system ignited) and lower level (system extinguished) may be *stable*, in the sense that, after any small disturbance, the system will return to one of these states. The middle state, however, is *unstable* and the system will never settle down to that mode of operation.[1]

[1] See, for example, H. Kramers and K. R. Westerterp, *Elements of Chemical Reactor Design and Operation*, Chap. IV, Academic Press, New York, 1963.

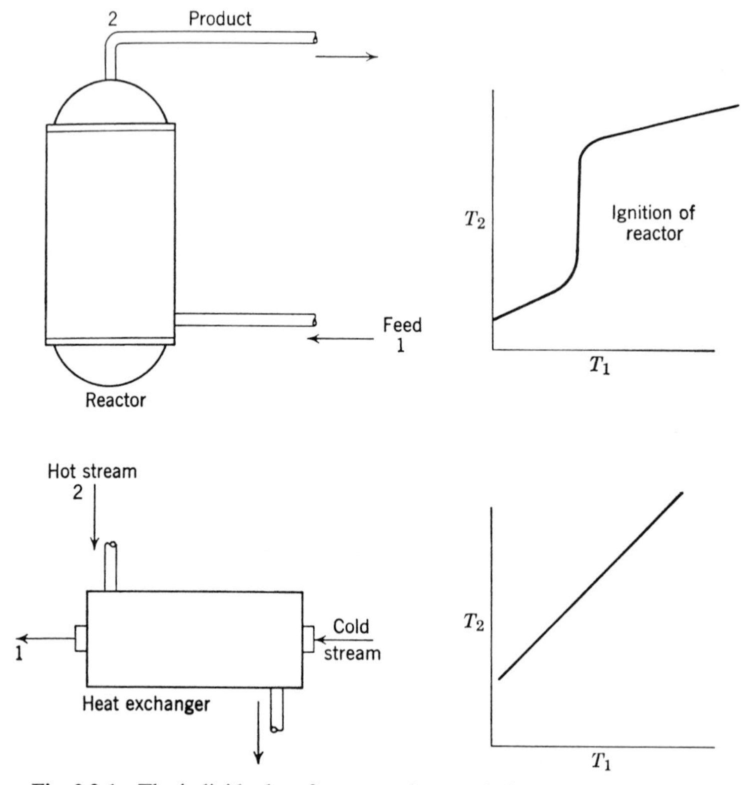

Fig. 3.2-1 The individual performance characteristics of two subsystems.

This illustrates the concept that a system can exhibit its own individuality, and characteristics may appear in a system which are not obvious in its component parts. The reactor alone cannot operate ignited with cold feed, but with the heat exchanger the higher level of performance is accessible. The adjustment of such a system to best accomplish a processing task must be approached carefully, using the principles to be exposed in the remainder of this chapter.

3.3 DEGREES OF FREEDOM IN A SYSTEM

We now begin the systematic dissection of processing systems to reveal an underlying skeleton upon which the strategy of process optimization might be fastened.

In nearly every design the values of certain of the variables are not specified. These variables are free to be adjusted to achieve a more profitable process. The first problem in design is to identify these free *design variables*, the *number* of which is called the economic *degrees of freedom*.

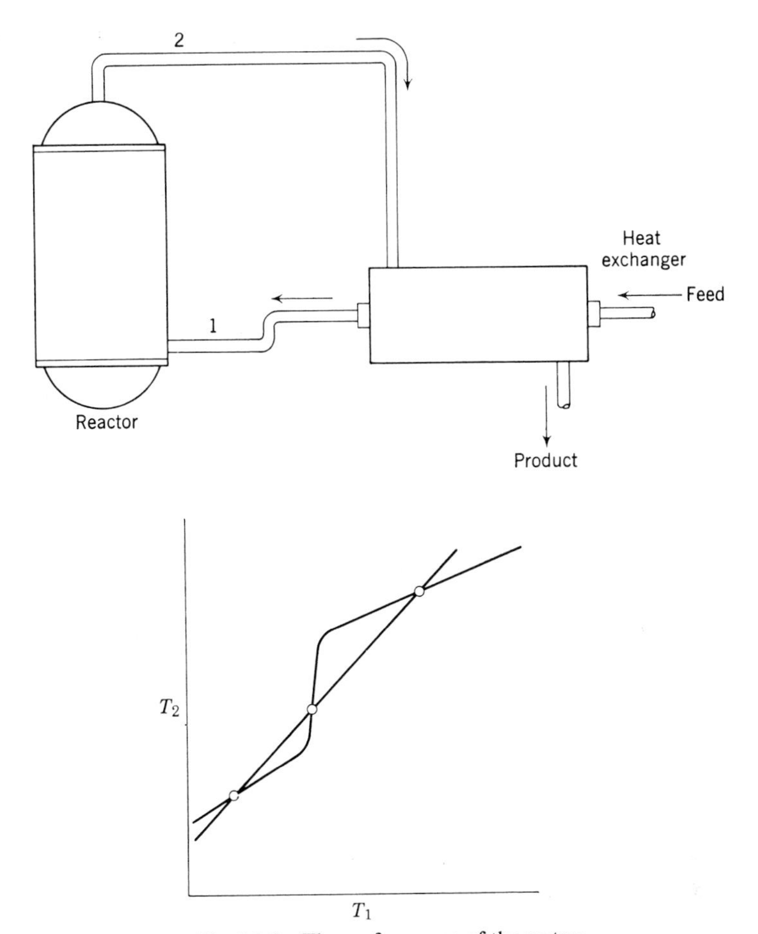

Fig. 3.2-2 The performance of the system.

We begin by listing the design variables for each unit in the process and the relations which connect them. The behavior of each component can be described approximately by equations, computer programs, manufacturer's recommendations, tables and graphs, pilot plant scale-up data, the judgment of an experienced engineer, or any combination and number of sources of information. All of these bits of information about the nature of the process components and their connections are called the *system design relations*.

The design relations consist of N sources of information about the system, including reference to M variables x_j (the equipment size, operating conditions, connecting stream condition, and so forth). The design relations must constitute *independent* sources of information, and any relation which can be derived from the others should be eliminated in the tally of equations.

Case I: Contradiction $N > M$

When there are more independent design relations than variables the design problem is not well formulated and it is generally not possible to find values for all of the variables which satisfy the design relations. The mathematical formulation, the physical nature of the problem or both are suspect.

In a trivial example consider the problem of blending two streams in a mixer to form a third stream as illustrated in Fig. 3.3-1.

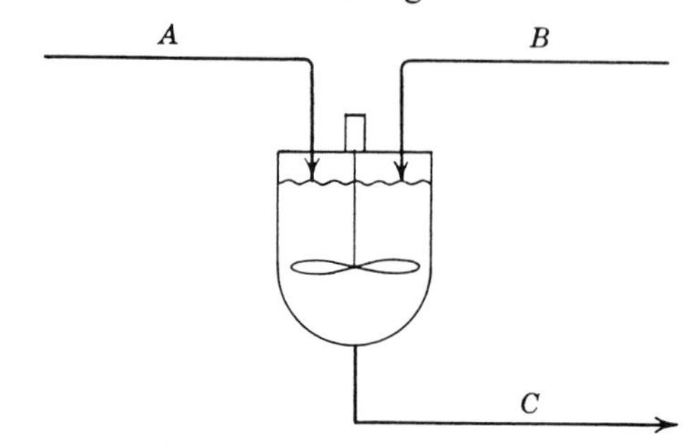

Fig. 3.3-1 A mixer portion of some larger processing system.

A material balance over the mixer gives

1. $A + B = C$

The quality of the blend, K, is defined as the ratio of B to A, by

2. $K = B/A$

Now, if A, C and K are to be of specified value so that the mixer can fit into a larger system

3. $A = 1{,}000$ lb/hr
4. $C = 2{,}000$ lb/hr
5. $K = 4.0$

The number of design relationships $N = 5$, and the number of variables $M = 4$, (A, B, C, and K); $N > M$. It is not possible to find a value for B, the unspecified variable, which satisfies all the conditions of the problem. The problem is not well formulated.

Case II: No Freedom $N = M$

When there are as many design relations as variables in a well-formulated problem, there exists no freedom or latitude in the selection of the values of the variables. Only certain definite values of the variables satisfy the design relations.

For example, if the connecting relation on C is relaxed in the mixing problem above, the number of design relations, $N = 4$

1. $A + B = C$
2. $K = B/A$

3. $A = 1{,}000$
4. $K = 4.0$

equals the number of variables, (A, B, C, K), $M = 4$. The values of the remaining variables are fixed by the solution of the design relation at $B = 4{,}000$, and $C = 5{,}000$, and there is no freedom for B and C to take on other values.

In Cases I and II no optimization problems occur, since there exist no alternative conditions in the process to choose among.

Case III: Degrees of Freedom, $N < M$

When a well-formulated design problem contains more variables than design relations, there exist variables in the design, the values of which are not specified. These variables may assume a number of values, thereby offering alternative operating conditions. The existence of such alternatives is the essential feature of an optimization problem.

For example, if the connecting relation defining K, the blend quality, is dropped in the mixing example, then $N = 3$ and $M = 4$. There are a number of values of B, K, and C which satisfy the design relations, so freedom exists. This freedom could be used to reach an optimization goal such as " blend the streams to maximize the value of the blend minus the cost of the raw materials A and B."

In general, the number of degrees of freedom equals the difference between the number of pertinent variables and the number of independent design relations, $F = M - N$. Then F of the system variables (x) may be *elected* as *design variables* (d) the values of which are *free* to be adjusted, and the remaining variables are called *state variables* and denoted by (s). Once the engineer has assigned specific values to the design variables, the values of the state variables are obtained by the solution of the design relations.

The design relations take the symbolic form

$$f_i(d_j, s_k) = 0 \tag{3.3-1}$$

$i = 1, 2, \ldots N$	number of design relations
$j = 1, 2, \ldots F = M - N$	degrees of freedom; number of independent *design variables*
$k = 1, 2, \ldots N$	number of dependent *state variables*

The expression for the degrees of freedom F in a process design problem involving M pertinent variables and N independent design relations

$$F = M - N \tag{3.3-2}$$

is identical in concept to the phase rule of Willard Gibbs used in thermodynamics. Gibbs' phase rule states that the number of intensive degrees of freedom F of a system of C components in P phases in equilibrium is

$$F = C + 2 - P. \tag{3.3-3}$$

Gibb's rule is restricted to thermodynamic equilibrium, and in process design we must puzzle out the degrees of freedom for each specific case, since each design is commonly unique.

In a practical problem, the degrees of freedom are consumed in two ways.

1. Certain variables are assigned definite values to provide a connection between the process and its environment. For example, the coolant temperature required by the process may be fixed at the temperature of the available supply of cooling water.
2. The remaining degrees of freedom are consumed in the selection of variables to be adjusted in maximizing the system profitability. For example, the designer might be free to adjust only the volume of a reactor and the operating temperature to achieve the most profitable operation of a chemical reactor system. We might call these the *economic degrees of freedom.*

3.4 THE DEGREES OF FREEDOM IN A SINGLE HEAT EXCHANGER

A schematic diagram for a heat exchanger is shown in Fig. 3.4-1. The major variables number thirteen.

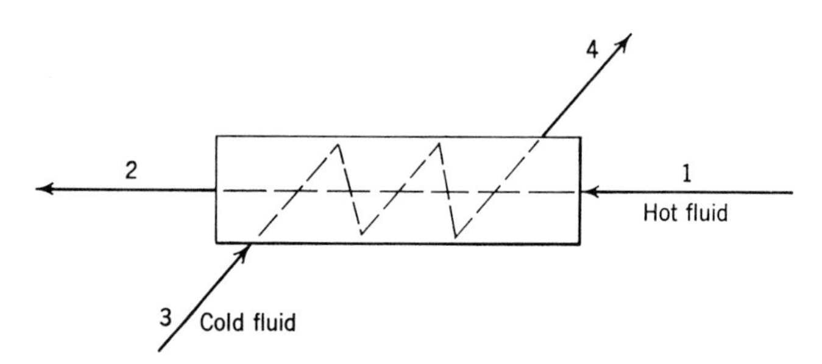

Fig. 3.4-1 A countercurrent flow heat exchanger.

1. K, the kind of heat exchanger (finned tube, double pipe, countercurrent, cocurrent, etc.).
2. Q heat transferred
3. A exchange area
4. U overall heat-transfer coefficient
5. W_1
6. W_2
7. W_3
8. W_4

(5–8: fluid mass flow rates)

9. t_1
10. t_2 } temperatures
11. t_3
12. t_4
13. $(\Delta t)_{\text{lm}}$ log mean temperature difference

The design relations number seven.

1. $Q = UA(\Delta t)_{\text{lm}}$ definition of heat-transfer coefficient
2. $(\Delta t)_{\text{lm}} = \dfrac{(t_1 - t_4) - (t_2 - t_3)}{\ln \dfrac{(t_1 - t_4)}{(t_2 - t_3)}}$ definition of log mean temperature difference
3. $W_1 = W_2$ } conservation of mass
4. $W_3 = W_4$ }
5. $Q = W_1 c_p(t_1 - t_2)$ } energy transferred between the two streams
6. $Q = W_3 c_p(t_4 - t_3)$ }

   U is a correlated function of the flow rates and fluid properties (which are temperature dependent) and the kind of the exchanger. Thus:
7. $U = U(W_1 \ldots W_4, t_1 \ldots t_4, K)$

   Total degrees of freedom $F = M - N = 13 - 7 = 6$

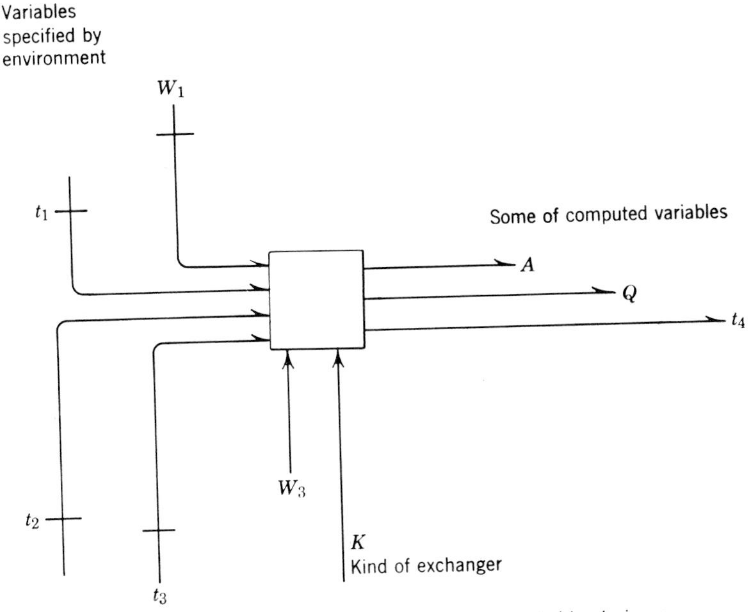

Fig. 3.4-2 An information flow diagram for the heat exchanger.

Now, for example, if the heat exchanger is to perform a special service to a larger system by cooling $W_1 = 1{,}000$ lb/hr of a hot fluid from $t_1 = 200°$F to $t_2 = 100°$F, using cooling water available at a temperature $t_3 = 60°$F, four of the six degrees of freedom would be consumed integrating the exchanger into this environment. The remaining two free variables, for instance, the kind of exchanger and the coolant rate W_3, could then be adjusted by the designer to achieve economy, consuming the remaining degrees of freedom and allowing the solution of the design equations for the other variables of interest, among them the area of the heat exchanger.

This direction of calculation might be represented by the *information flow* diagram shown in Fig. 3.4-2. The externally specified variables, W_1, t_1, t_2, and t_3 are denoted by crossed arrows, the design variables, W_3 and K, by full-headed arrows, and the dependent computed variables by half arrows. This diagram summarizes the flow of information through the equations which describe the performance of the heat exchanger.

3.5 REVERSAL OF INFORMATION FLOW

In this section we demonstrate that a reselection of design variables results in a reversal of information flow through the component of interest, and that such reversals influence the ease of computation by changing the number of equations that must be solved simultaneously. We also demonstrate that certain combinations of variables in a given problem cannot assume the role of design variables.

For the moment, consider the equilibrium stage extractor shown in Fig. 3.5-1. We wish to determine the wash-solvent rate and kind of solvent which maximize the following economic objective for the extractor.

$$\underset{\substack{\text{(kind of solvent)}\\ \text{(solvent rate)}}}{\text{Maximize}} \quad [(\text{value of the solute extracted}) - (\text{cost of the wash solvent})]$$

$$\max_{(s=A \text{ or } B)(w)} \quad [C_e Q(X_f - X_0) - C_w W] \tag{3.5-1}$$

where C_e is unit value of solute in wash phase, and C_w is the unit cost of the wash solvent.

The design equations for this unit are, assuming the usual simplified physical picture

$$QX_f = QX_0 + WY_0 \qquad \text{Material balance over solute phase} \tag{3.5-2}$$

$$\phi(X_0, Y_0) = 0 \qquad \text{Equilibrium relation between effluent phases: depends on solvent selected} \tag{3.5-3}$$

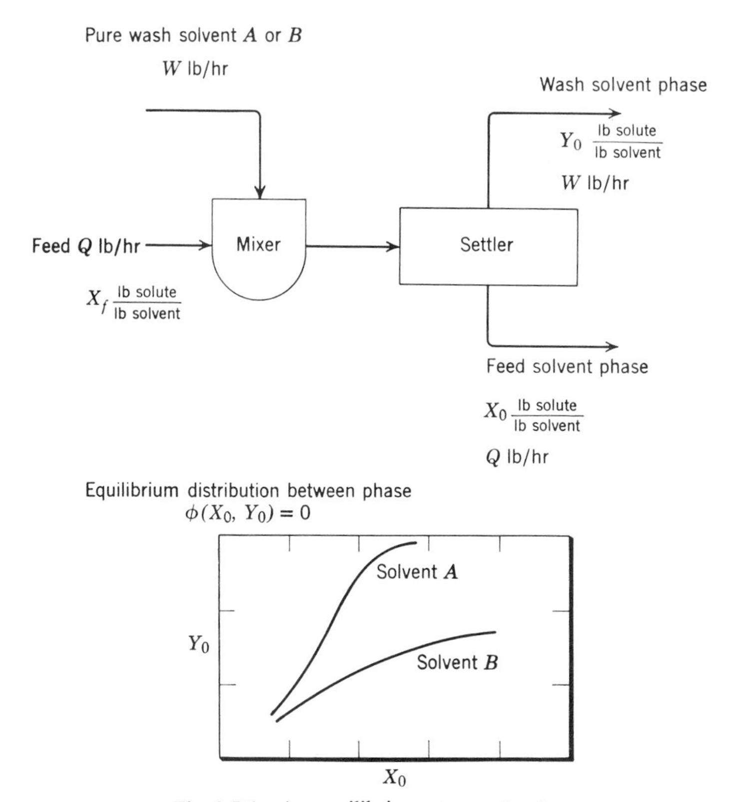

Fig. 3.5-1 An equilibrium stage extractor.

where Q and X_f, the feed conditions, are specified in the definition of the problem.

Thus, if we should specify the kind of solvent ($S = A$ or B), and a value of the wash solvent rate W, all other variables can be computed; in this case, by the *simultaneous* solution of the material balance and equilibrium equation. This is illustrated pictorially in Fig. 3.5-2(a), an information flow block for this extraction unit. This extractor optimization problem possesses two economic degrees of freedom.

However, observe: were we to elect to specify as a design variable the desired effluent concentration X_0 instead of the wash solvent rate, the computations would be much easier. We could first determine the equilibrium concentration in the feed solvent phase Y_0 by reference to the equilibrium diagram, and then solve directly for the wash solvent rate W through the material balance. This reversal of information flow simplifies the process computations, allowing us to gather the same kind of information on the performance of the extractor in one pass through the design equations, avoiding the complicated

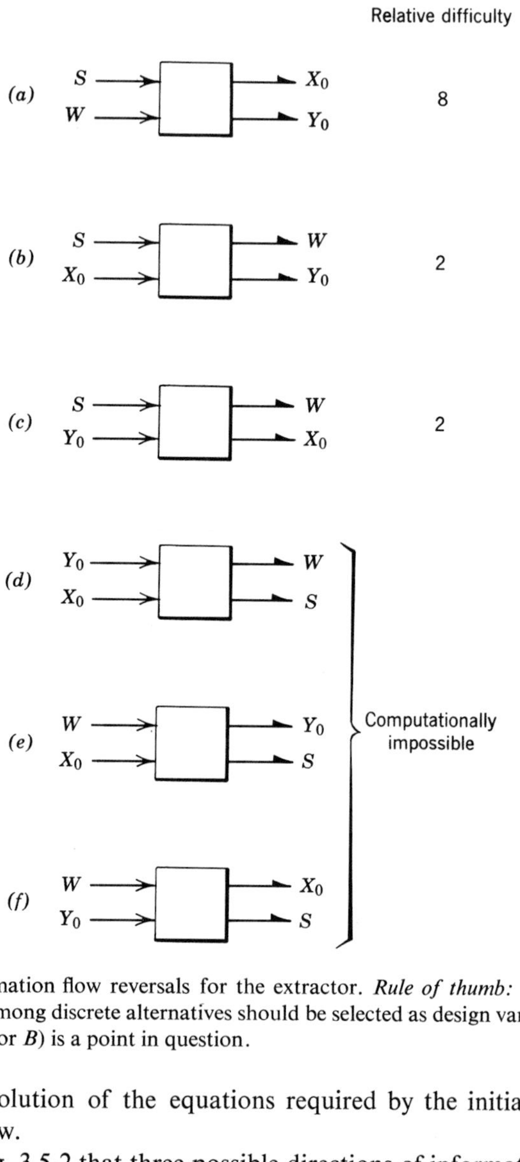

	Inputs	Outputs	Relative difficulty
(a)	S, W	X_0, Y_0	8
(b)	S, X_0	W, Y_0	2
(c)	S, Y_0	W, X_0	2
(d)	Y_0, X_0	W, S	Computationally impossible
(e)	W, X_0	Y_0, S	Computationally impossible
(f)	W, Y_0	X_0, S	Computationally impossible

Fig. 3.5-2 Information flow reversals for the extractor. *Rule of thumb:* Variables which involve a choice among discrete alternatives should be selected as design variables. The kind of solvent $S = (A \text{ or } B)$ is a point in question.

simultaneous solution of the equations required by the initial direction of information flow.

We see in Fig. 3.5-2 that three possible directions of information flow exist in this unit calculation, two of which are of equivalent difficulty and one of which is unnecessarily difficult. We would expect that the optimization problem in Eq. 3.5-4 would be much simpler than that initially posed in Eq. 3.5-1, particularly since the range over which the design variable is allowed to vary is limited in the new formulation. This is suggestive of the effects to be achieved on a grander scale in more complicated processes by reordering computations.

$$\max_{(S=A \text{ or } B)(X_0)} [C_e Q(X_f - X_0) - C_w W] \tag{3.5-4}$$

where

$$0 < X_0 < X_f$$

Observe also that certain reversals of information flow through the system are not possible. For example, we could not expect to select arbitrary values of the effluent concentration X_0 and the wash rate W and compute the kind of solvent to be used; the technology is here limited to one of two solvents A or B. Such a reversal is technically and computationally impossible. This suggests a rule-of-thumb: *when a variable involves a choice among a number of discrete alternatives it is usually wise to elevate that variable to the role of a design variable.*

Notice further that a relative difficulty of computation has been ascribed to each of the three possible directions of information flow. This is based on the observation that frequently the difficulty of solving a set of equations increases as the *cube* of the number of equations that must be solved simultaneously. Thus, the computations required to implement the information flow in Fig. 3.5-2(a), involving the solution of two simultaneous equations, might be four times as laborious as the computations to implement the information flow in the two other cases, which involve the solution of equations one at a time [$(2)^3$ as compared to $1 + 1$].

3.6 A DESIGN VARIABLE SELECTION ALGORITHM

In the preceding examples, the number of degrees of freedom in a design, the best design variables to consume those degrees of freedom, and the resulting information flow structure were obtained by inspection. As the processes become more complicated the analysis by inspection becomes impossible and a more systematic approach must be taken.

In the extractor example, considered in the previous section, there were two equations and two variables to be solved for. The number of ways in which the extractor design equations can be ordered for solution is $(2!)^2 = 4$; this can be handled by inspection.

But, suppose a process is described by one hundred design equations which must be solved for one hundred state variables. The problem merely of arranging the equations and variables in the proper order for solution is overwhelming, for there are $(100!)^2 = 10^{300}$ ways in which one hundred equations and one hundred variables can be ordered.

Fortunately some very simple design variable selection algorithms have been developed to overcome such severe combinatorial problems, and these algorithms are now part of industrial practice. It is common practice, especially in the large oil and chemical companies, to submit new designs to the computer for dissection by the methods to be discussed next. These methods are also very useful as an aid to hand computation.

It is convenient to prepare a table which describes the structure of the design equations which are to be dissected. This is called the *structural array*, the columns of which correspond to all the variables which enter into the design, and the rows of which correspond to all the design equations. An × is placed whereever a variable appears in an equation. The structural array for the heat exchanger of Section 3.4 is shown in Fig. 3.6-1.

Equations \ Variables	[K]	Q	A	U	(W_1)	W_2	W_3	W_4	(t_1)	(t_2)	(t_3)	t_4	$(\Delta t)_{lm}$
1		X	X	X									X
2									X	X	X	X	X
3					X	X							
4							X	X					
5		X			X				X	X			
6		X					X				X	X	
7	X			X	X	X	X	X	X	X	X	X	

○ Fixed by environment
▭ Preferred as design variable

Fig. 3.6-1 The structural array for the heat exchanger.

In Fig. 3.6-1 four variables are fixed:

$$W_1 = 1{,}000 \text{ lb/hr}$$
$$t_1 = 200°\text{F}$$
$$t_2 = 100°\text{F}$$
$$t_3 = 60°\text{F}$$

to fit the heat exchanger to the heat-exchange task it must perform, and the variable K, the kind of heat exchanger to be used, has been selected as a design variable free to be assigned by the engineer. This last selection is based on the rule of thumb which evolved in Section 3.5 concerning variables of discrete choice. We now concern ourselves with the systematic selection of the design variable to consume the remaining single degree of freedom. The following algorithm is proposed for this task.[2]

[2] W. Lee, J. H. Christensen and D. F. Rudd, *AIChEJ.*, No. 6, **12** (1966).

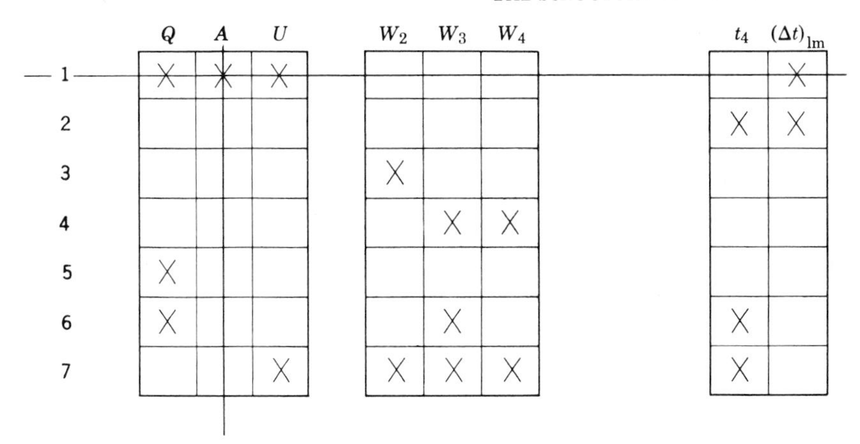

Fig. 3.6-2 First pass through the design variables selection algorithm.

1. Locate a column which contains only one × and delete the column and corresponding equation.
2. Repeat step 1 until all equations have been eliminated.

The remaining variables are the design variables which result in the precedence ordering of the equations, and the *precedence order* is the reverse of the order in which the equations were deleted. The *precedence order* is the order in which the equations are to be solved.

Should this algorithm not eliminate all equations, a recycle loop has been detected which cannot be eliminated by the proper selection of design variables, and certain of the equations must necessarily be solved simultaneously instead

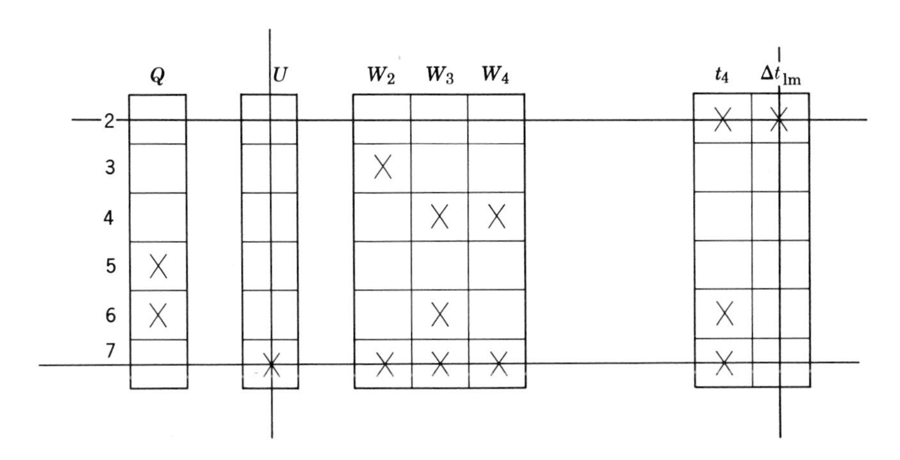

Fig. 3.6-3 Second pass.

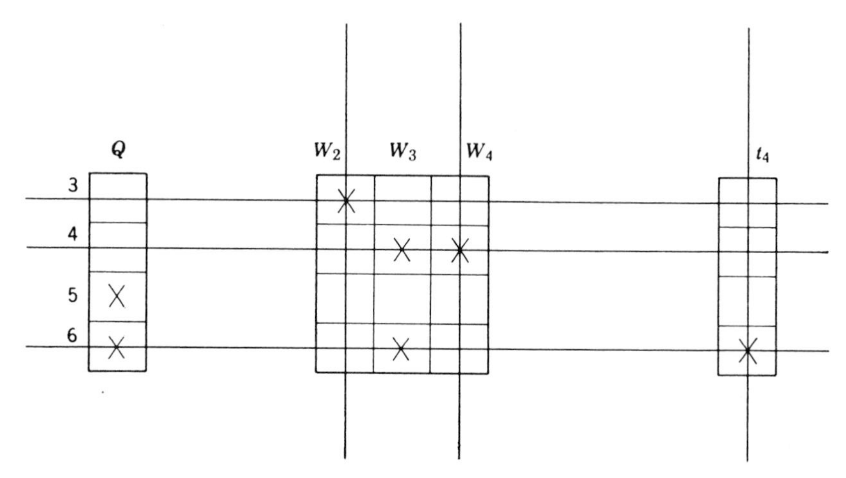

Fig. 3.6-4 Third pass.

of sequentially in spite of our attempts to eliminate such difficulties. This simple algorithm is applied to Fig. 3.6-1 in Figs. 3.6-2 through 3.6-5.

In Fig. 3.6-6 the precedence order of the calculations is shown. If W_3 is selected as the design variable, as the algorithm prescribes, the seven design equations can be solved one at a time in the precedence order: eventually computing the area of the exchanger. It is interesting to notice that the area of the heat

Fig. 3.6-5 Fourth pass.

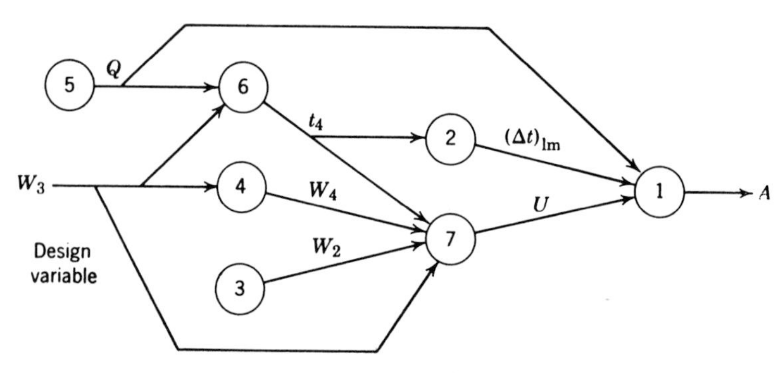

Fig. 3.6-6 The precedence order of calculation.

exchanger is not the recommended design variable, and that the use of the area of the exchanger as the design variable in this problem merely forces the engineer into an unnecessarily difficult design computation in which design relationships must be examined simultaneously.

3.7 A SINGLE EQUILIBRIUM STILL

We further illustrate the computation of degrees of freedom by considering the equilibrium still shown in Fig. 3.7-1. The liquid feed stream enters at an

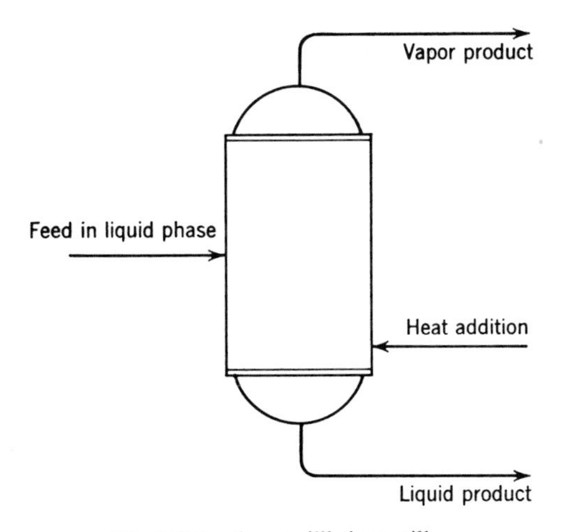

Fig. 3.7-1 An equilibrium still.

unspecified temperature and pressure, containing C components. Heat is added to (or removed from) the still to achieve partial vaporization (or condensation) and the liquid and vapor removed. The still is to be operated at a yet unspecified pressure and temperature, P_S and T_S, which are the same for the two effluent streams.

To determine the number of degrees of freedom we first list the variables.

		Number of Variables
Feed:	Flow rate of each component, temperature, pressure, enthalpy.	$C + 3$
Vapor:	Flow rate of each component, enthalpy.	$C + 1$
Liquid:	Flow rate of each component, enthalpy.	$C + 1$
Still:	Temperature and pressure.	2
Still:	Heat addition rate.	1
	Total	$3C + 8$

Next the relations among the variables are listed.

	Number of relations
[Given the temperature, pressure and composition[3] of any stream, we can compute the enthalpy by thermodynamic methods. This amounts to one equivalent equation for each of the three streams.]	3
[For a given temperature and pressure in the still,[3] we can compute the vapor and liquid compositions; an equilibrium relation exists between the vapor and liquid phase of each component.]	C
Mass balance over the still for each component.	C
Energy balance over the still.	1
Total	$2C + 4$

Total degrees of freedom $F = (3C + 8) - (2C + 4) = C + 4$

For example, if the flow rate for each component entering the system is given, and the pressure and temperature of the feed are given, a total of $(C + 4) - (C + 2) = 2$ variables are free to be manipulated. The designer might attempt to regulate the pressure in the still and the heat addition rate to achieve the economically optimal separation of the components.

3.8 INFORMATION FLOW THROUGH THE SUBSYSTEMS

With this background in estimating the degrees of freedom and information flow through the design equations of some elementary process components we address the problem of system information flow. To illustrate the effects which might arise when components are connected to form a system, we consider the process shown in Fig. 3.8-1, consisting of a heat exchanger, an extractor, and a still, all previously analyzed as separate units.

A feed solvent containing a solute enters the mixing portion of the extraction unit along with a separate wash solvent stream. The feed rate and solute concentration are specified by the statement of the problem. As this mixture attains equilibrium, it is fed to the settler with the waste feed solvent going to the sewer, and the solute rich wash solvent going to the still for further concentrating. This extractor operates at ambient conditions and requires cool wash

[3] An interesting exercise is to demonstrate that these statements are consistent with Gibbs' phase rule for systems in thermodynamic equilibrium.

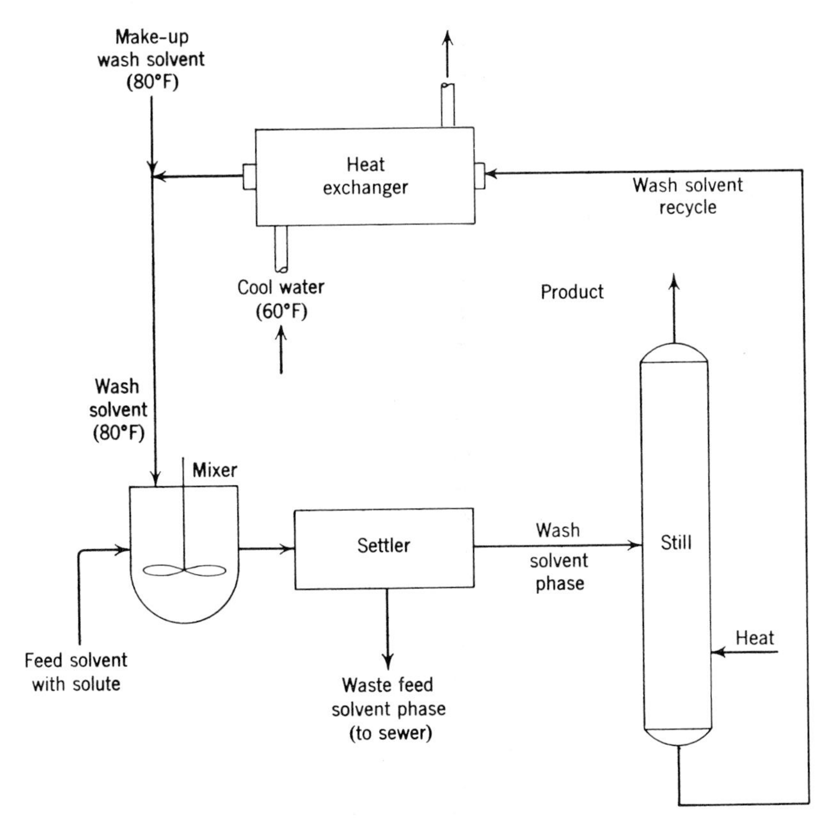

Fig. 3.8-1 A process material flow diagram.

solvent. The solute rich overhead vapor phase from the still is the product from the system, and the wash solvent rich phase is recycled to the mixer from the bottom of the still. The recycle wash phase must be cooled to 80°F before entering the mixer. Additional pure solvent is added to make up for any losses of solvent in the product stream. Suppose we limit our attention to one particular wash solvent.

We now demonstrate how the system information flow structure can be constructed from the structure of the several blocks or subsystems which compose the system and the special conditions imposed on the system operation.

The Extractor. The information flow block for the extractor is shown in Fig. 3.8-2(a), with specified variables denoted by crossed arrows.

The still. Two components enter the still, the wash solvent and valuable solute. The temperature and pressure at this feed are known, since the feed comes from the extractor as a liquid at ambient conditions. Thus, two variables

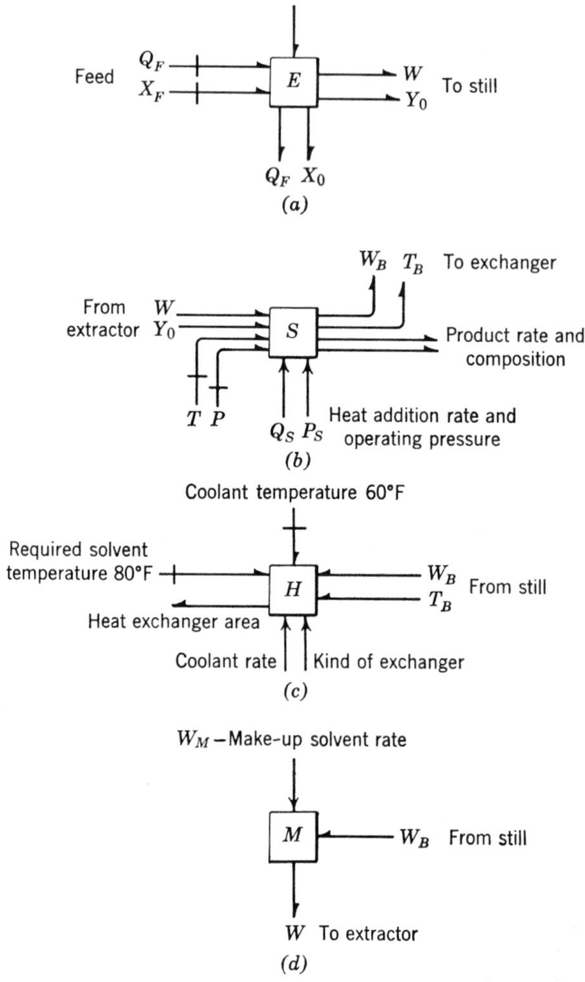

Fig. 3.8-2 Information flow through the components. (*a*) Extractor—1 local degree of freedom. (*b*) Still—4 local degrees of freedom. (*c*) Exchanger—4 local degrees of freedom. (*d*) Make-up feed—2 local degrees of freedom.

are free to be manipulated; for example, the operating pressure of the still and heat addition rate, in Fig. 3.8-2(*b*). The bottom stream is essentially pure wash solvent.

The heat exchanger. Two outside conditions are imposed on the exchanger, the coolant temperature is fixed at 60°F and it is required by the extractor that the effluent from the exchanger be at 80°F. This is illustrated in Fig. 3.8-2(*c*).

The make-up solvent. An additional block is needed to account for the make-up solvent. Specifying the amount of make-up solvent and the amount of solvent coming from the exchanger, enables the computation of the wash solvent rate going to the extractor.

The information flow structure which obtains is shown in Fig. 3.8-3. Notice

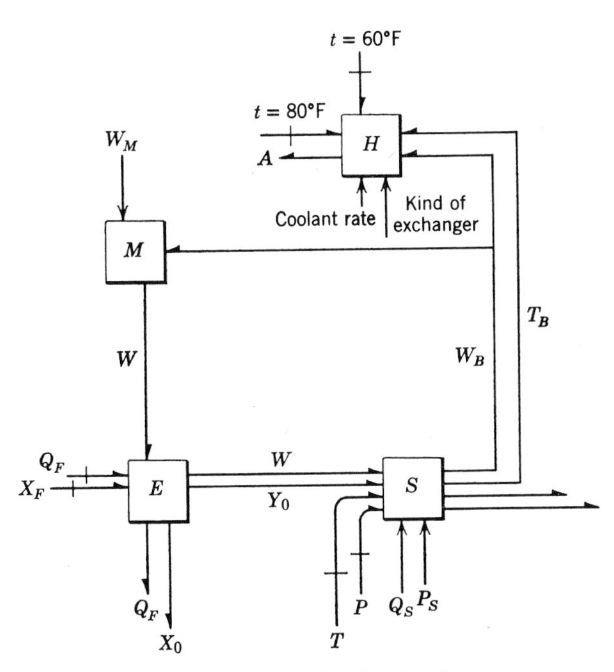

Fig. 3.8-3 System information flow.

how the heat exchanger is left out of the recycle loop. This results from the fact that the effluent temperature requirement of 80°F must be met by the exchanger, and how this is met is irrelevant to the rest of the system. Information on the design of the exchanger is not fed on to the extractor. Connecting the components has consumed degrees of freedom; the sum of the local degrees of freedom for the components being $1 + 4 + 4 + 2 = 11$, and the economic degrees of freedom for the system being 5. In general, the number of degrees of freedom for a system equals the sum of the degrees of freedom for its components F_i minus the number of connecting relationships needed to tack the system together.

$$F_S = \sum F_i - \text{number of connecting relations} \tag{3.8-1}$$

In Fig. 3.8-3, there are 6 connecting relations, matching inputs from one component to outputs from other components and $F_S = 11 - 6 = 5$.

3.9 SYSTEM INFORMATION FLOW REVERSAL

We saw in Section 3.5 how the reversal of information flow in the extractor simplified the analysis of that unit. Now we consider information reversal on the system level. Notice in Fig. 3.8-3 how information on a change in the amount of make-up solvent added to the solvent recycle stream is passed on to the extractor, to the still, to the heat exchanger, and even back to the site of the change. A complex recycling of information requires that all units be considered simultaneously when design modifications at a given unit are being entertained, a troublesome situation. Need this be?

Suppose that we selected the two parameters, W, and W_B as independent design variables. This would require information reversal at the extractor. Such a reversal would pass on to the make-up feed block: The reversal of W_B would require a reversal at the still: perhaps, we might compute the heat addition rate Q_S to achieve the desired wash rate W_B. A new information flow diagram is shown in Fig. 3.9-1.

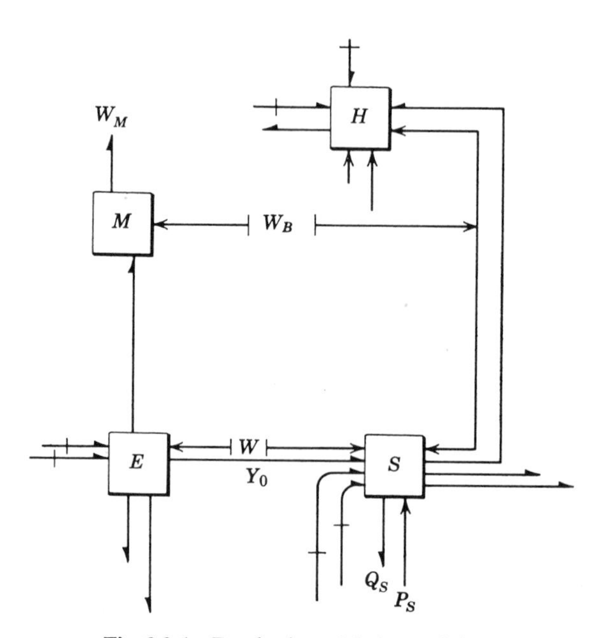

Fig. 3.9-1 Reselection of design variables.

Notice how in *theory* the units can be computed without recycle, in the *precedence order*.

1. Extractor
2. Make-up
3. Still
4. Exchanger

A reselection of independent design variables has changed the direction of information flow and reduced the number of units which must be considered simultaneously during the evaluation of a particular design, a happy situation.

This reversal, one of several, may not be possible to implement should it involve *impossible* information flow through any component. For example, is it possible to compute the required heat addition rate to generate a given overhead vapor rate from a given still feed? Should this be impossible to compute, the reversal shown in Fig. 3.9-1 cannot be implemented.

We next present a simple method of design variable selection which incorporates partial consideration of allowed information reversals at the unit level.

3.10 THE STRUCTURAL EFFECTS OF DESIGN VARIABLE SELECTION

Certain apparently minor features in the information flow structure of a process can be a source of trouble to the design engineer. For example, a recycle loop connecting the end of a process to its beginning may force the engineer to undertake iterative recycle calculations to close the material and energy balances. Without that loop the material and energy balances might be calculated component by component in one sweep through the system.

Fortunately, the process engineer is not at the complete mercy of any particular information flow structure, since by a reselection of design variables the structure can be changed. We illustrate this kind of thinking by considering the process whose information flow structure is shown in Fig. 3.10-1 and by extending the concepts of Section 3.6.

The design variables in this system are denoted by the full arrows; they may be considered as sources of information to the system from the process engineer. The half arrows denote state variables. For example, component A has two input variables, one a design variable the value of which is assigned by the process engineer and the other a state variable coming from component G. Once the values of these input variables are known, the two output variables can be solved for, passing information on to components B and D. The number of input arrows to a component equals the local degrees of freedom for the component, a number that is preserved during the manipulations which follow.

Suppose now that the values of the design variables have been specified by the engineer, and the state variables are to be solved for. The structure shown in

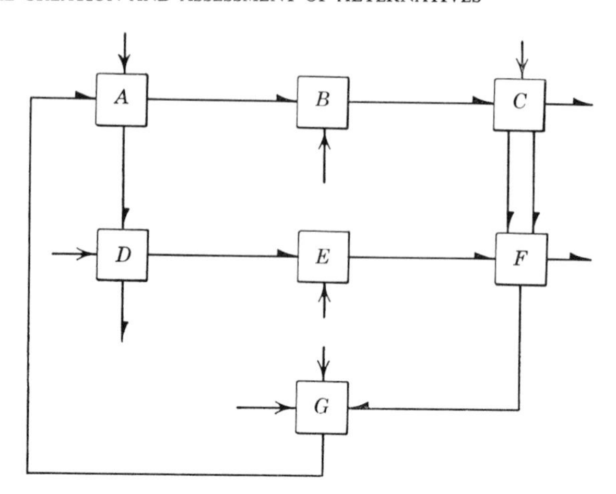

Fig. 3.10-1 An initial information flow structure.

Fig. 3.10-1 is troublesome indeed, since the design equations for the entire system must be solved simultaneously when the value of any design variable is changed. The recycle loops within the system are the source of this trouble.

Fortunately, the engineer is not forced to contend with the problem in this form. Rather, a reselection of design variables can be entertained with the hope of avoiding a structure so intimately tied together by recycle streams. To illustrate this kind of thinking, we apply a second design variable selection algorithm to the problem.[4]

Step 1. Record the local degree of freedom for each component, and remove all heads from the arrows in the initial information flow structure.

Step 2. Note for special consideration any variables which do not connect components in the system. Variables so distinguished are assigned an *outward* direction. This assignment is continued up to the point where the number of unassigned variables equals the local degrees of freedom for the component, at which time the component is deleted from the diagram.

Step 3. Repeat Step 2 on the reduced diagrams until no further reductions occur. The unassigned variables are the design variables.

Figure 3.10-2 results from the application of Step 1 to our example process. Notice that certain design variables have previously been assigned to stages A and E; these may be heavily constrained variables over which close control is required or variables which involve a choice among alternatives. The algorithm allows any preselection of preferred design variables. Figure 3.10-3 shows the results of two passes through Steps 2 and 3 resulting in the assignment of

[4] W. Lee, J. H. Christensen and D. F. Rudd, *op. cit.*

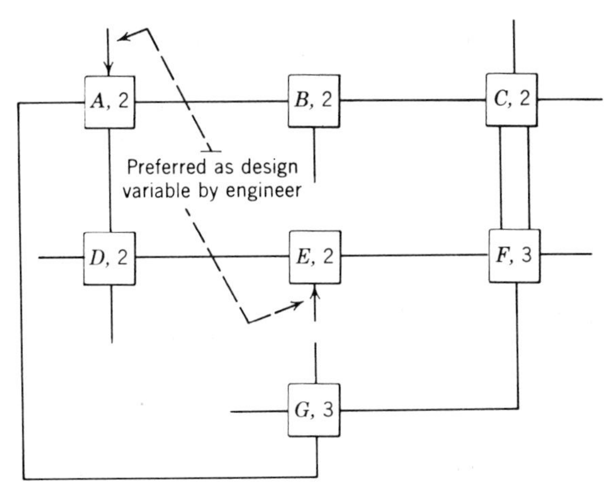

Fig. 3.10-2 The undirected graph with local degrees of freedom affixed.

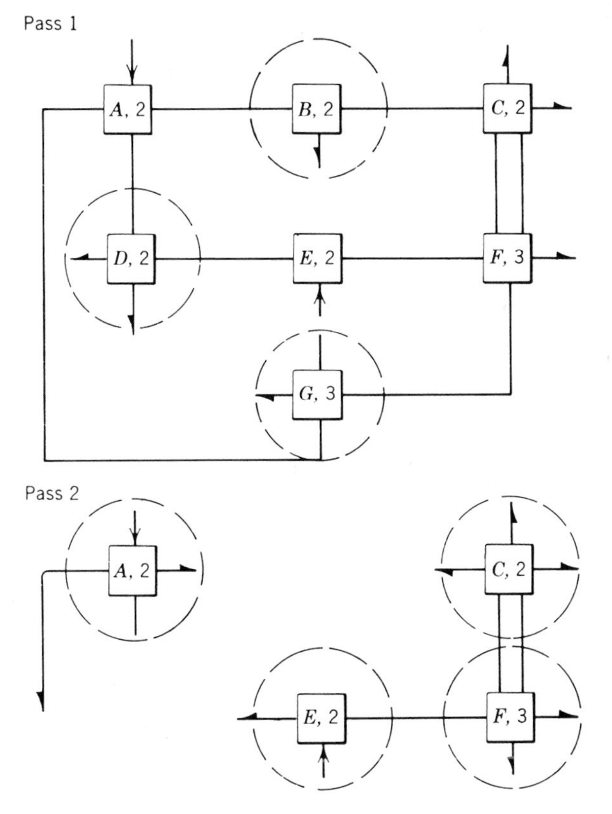

Fig. 3.10-3 The application of Steps 2 and 3.

output variables for each component. Notice that in a number of cases one choice of direction of information flow was made from several possible alternatives. This enables the engineer to assign preferred direction to the information flow, placing his insight into the structure, favoring easy directions of information flow through given blocks. There are several sets of design variables which eliminate recycle loops.

Figure 3.10-4 shows the information flow for the new assignment of design variables. Merely by elevating certain state variables to the role of design variable, the recycle loops have been removed. For example, calculations might begin by specifying the values of the two variables (now design variables) which connect components C and F and solving for the variable connecting unit C to unit B. Once the other input to B is solved for by the design of component A, the output from B can be computed, and so forth through the system. The precedence order of these calculations is also shown in Fig. 3.10-4, indicating that at no time need the engineer consider more than one component.

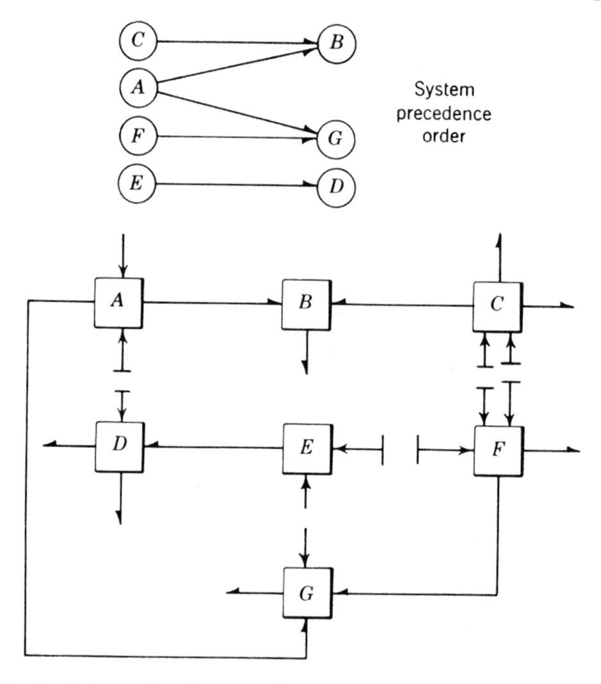

Fig. 3.10-4 An information flow resulting from the reassignment of design variables.

The reversal of state and design variable roles frequently results in a design problem which is orders of magnitude less troublesome than its parent problem. The manipulation of information flow structure is the basis of some of the more powerful methods of optimization, presented in Chapters 9 and 10.

3.11 PERSISTENT RECYCLE

In many chemical processes it is impossible to eliminate all recycle merely by the careful selection of independent design variables. Either the design variables are not sufficiently numerous, the required information flow reversals are difficult to implement, or the desired connections amongst the processing components are too numerous. In such situations we must solve for all the variables within such recycle loops simultaneously: it is impossible to completely precedence order the attack on the system.

The persistent recycle is detected by the design variable selection algorithms, because the selection process will terminate before all equations or units are deleted. For example, in Fig. 3.11-1 the design variable selection (using the algorithm of Section 3.6) terminates before any of the equations are deleted, since there exist no variables which appear in only one equation. This indicates that the design equations represented by Fig. 3.11-1 are not predisposed to precedence ordering and that recycle persists.

Equations	x_1	x_2	x_3	x_4	x_5	x_6
1	X	X				
2	X		X	X		
3		X	X	X	X	X
4				X	X	X

Fig. 3.11-1 A structural array.

Efficient methods for the detection of the design variables in the presence of persistent recycle are still in the developmental stage, however let us describe one primitive method.

Step 1. Apply the selection algorithm of Section 3.6. If it fails to terminate with the deletion of all equations, go to Step 2.

Step 2. Define $k = \min \rho(x_i) - 1$: where $\rho(x_i)$ is the number of equations in which variable x_i appears.

Step 3. Identify sets of k equations which have the property that when the set is deleted there remains an array containing at least one variable which appears in only one equation.

Step 4. Delete one such equation set.

Step 5. Apply the design variable selection algorithm of Section 3.6 to the array which remains.

Step 6. If no precedence order is obtained in Step 5, try another set of k equations.

Step 7. If the deletion of no set of k equations results in an array that can be precedence ordered, increase k by one and return to Step 3.

For example in Fig. 3.11-1, the min $\rho(x) = 2$ and therefore $k = 2 - 1 = 1$, so one equation should be deleted at a time in Steps 3 and following to detect a substructure which can be precedence ordered. The deletion of Eq. 1 yields the substructure in Fig. 3.11-2 which is precedence ordered by the design variable selection algorithm.

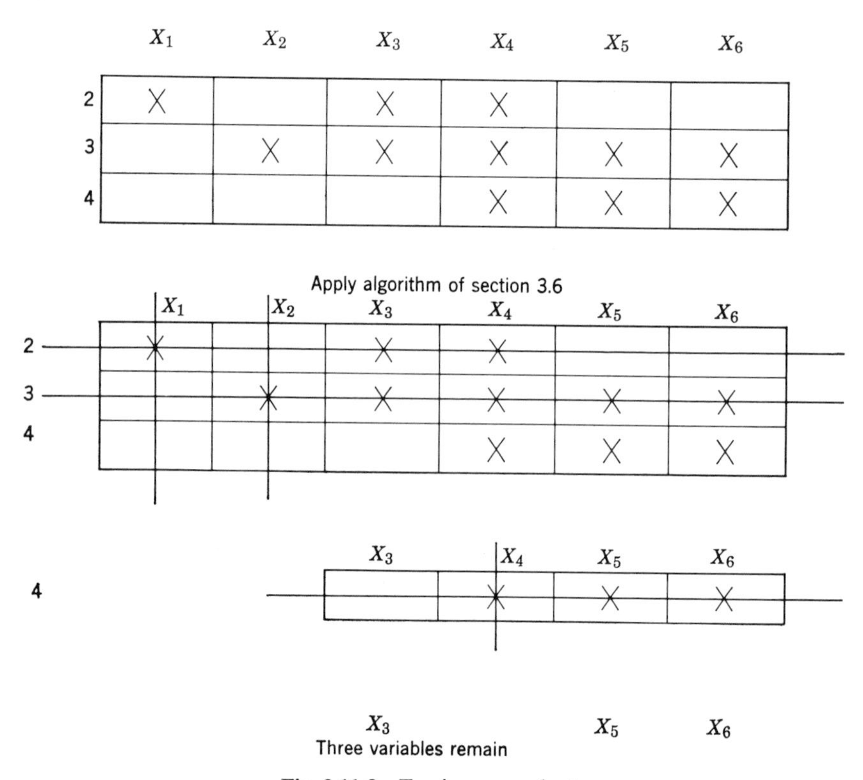

Fig. 3.11-2 Tearing a recycle stream.

Three variables remain undeleted; and since $k = 1$, one of the three variables will be a recycle variable and the remaining two will be design variables. Figure 3.11-3 shows the information flow structure which exhibits this fact.

A common approach to the analysis of recycle problems involves assuming values for the troublesome recycle streams and computing unit-by-unit through the then acyclic process (as shown in Fig. 3.11-3). This yields computed values for the recycle streams to be compared with the assumed values. If the computed

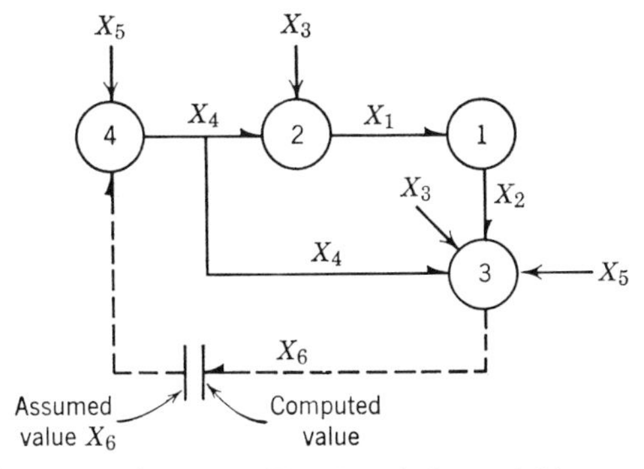

Fig. 3.11-3 Iteration on recycle stream. X_3, X_5 as design variables; X_6 as recycle variable.

and assumed values are sufficiently different, convergence has not been achieved, and new values for the recycle streams are tried. This iterative recycle calculation is continued until the desired match is achieved between the assumed and computed values of the recycle stream. A careful approach to such calculations is of the utmost importance in complex industrial problems.

The first step in a complex recycle problem involves the detection of the simplest set of recycle streams. Observe in Fig. 3.11-4 that even in rather simple processes there are a number of different sets of recycle streams. The simplest set is defined as that set which involves the *minimum number of streams:* Case II in Fig. 3.11-4. This set requires that the values of a minimum number of streams be assumed during recycle calculations and, hence, may speed iterative computation.

It is often extremely difficult to detect the simplest recycle set in a system by inspection, but simple algorithms have recently been developed to aid the engineer.[5] We now show how the simplest recycle set can be easily detected by using these algorithms.

We begin by tracing all the cycles in the system and constructing a *cycle table*. Steward[6] and Norman[7] suggested systematic ways for detecting cycles in a complex process, when the mere tracing of cycles becomes cumbersome. The rows in the cycle table correspond to the cycles, and the columns correspond to the streams. An entry appears wherever a stream appears in a cycle. The cycle table for a process is shown in Fig. 3.11-5. Notice that several streams appear in more than one cycle: selecting such streams as recycle streams breaks several cycles.

The degree of a cycle is defined as the number of streams contained therein. Obviously, if a cycle is of degree one, the single stream which appears in that

[5] W. Lee and D. F. Rudd, *AIChE J.*, November 1966.
[6] D. V. Steward, *J. SIAM Numer. Anal.*, Ser. B., No. 2, **2** (1965).
[7] R. Norman, *AIChE J.*, May 1965.

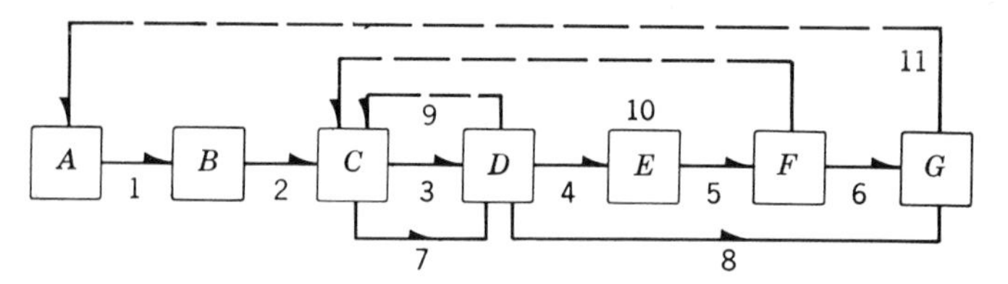

Case I The original structure with three persistant recycle streams denoted by a broken line.

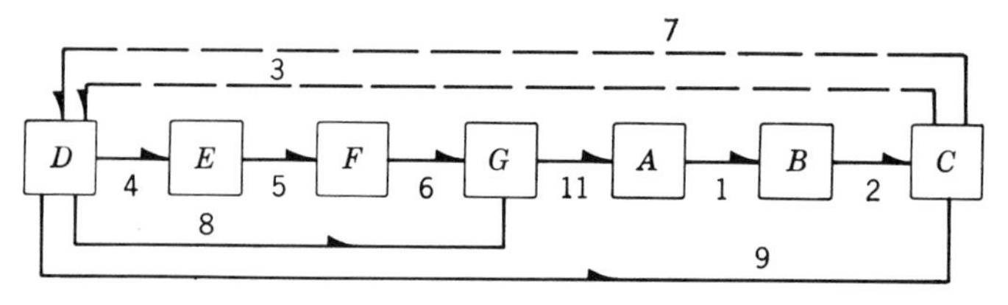

Case II

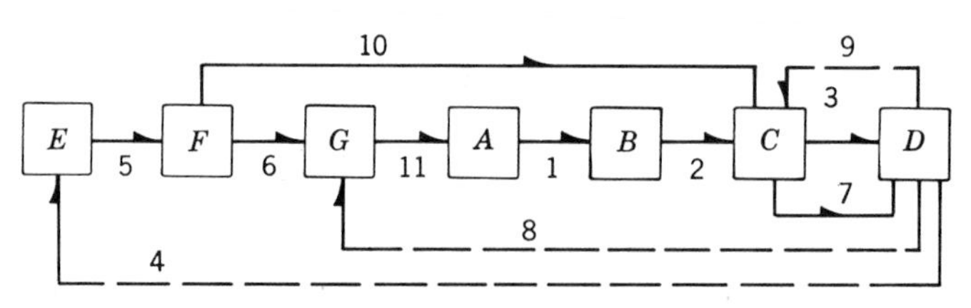

Case III

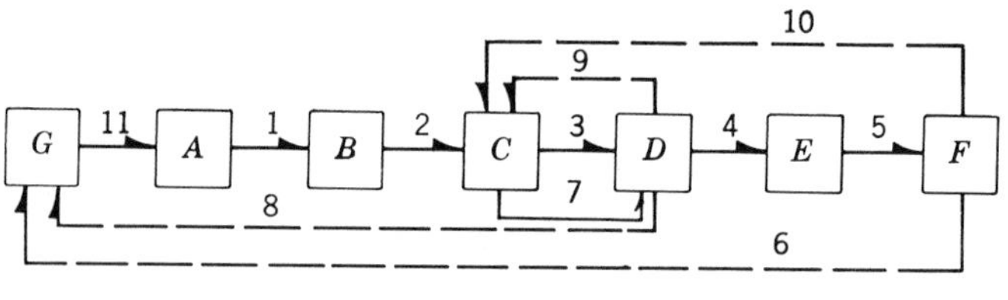

Case IV

Fig. 3.11-4 The ordering of recycle calculations by *selecting sets of recycle parameters.* The direction of information flow is preserved.

cycle must be selected as a recycle stream; that is the only way the cycle can be torn. Should the degree of a cycle be greater than one, we have a choice of several streams and it is not immediately obvious which is the best choice.

A column in the cycle table is said to be *contained* in a second column if selecting the variable corresponding to the second column as a recycle stream

Sample system

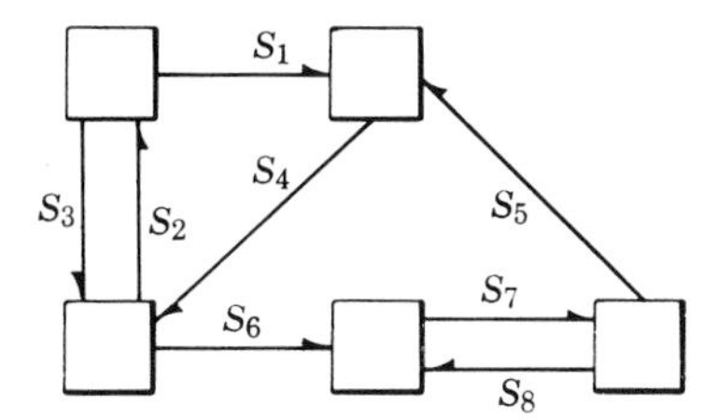

Cycles of the system

Cycles
$A = S_2, S_3$
$B = S_7, S_8$
$C = S_1, S_2, S_4$
$D = S_4, S_6, S_7, S_5$

Augmented cycle table of the system

	S_1	S_2	S_3	S_4	S_5	S_6	S_7	S_8	Degree of cycle
A		1	1						2
B							1	1	2
C	1	1		1					3
D				1	1	1	1		4

Fig. 3.11-5 A process and cycle table.

removes all the cycles in which the first stream appears. For example, in Fig. 3.11-5 columns S_1 and S_3 are contained in column S_2; columns S_5 and S_6 are contained in column S_4; and column S_8 is contained in column S_7. It is obvious that we should not bother with S_1 as a recycle stream to break cycle C, since selecting S_2 breaks cycle A as well as cycle C. This lack of efficiency obtains for any contained columns; these can be disregarded as candidates for recycle streams.

When all contained columns have been eliminated from the cycle table, the remaining columns are said to be *independent*. In Fig. 3.11-6 we see that elimination of contained columns has reduced the number of candidates for the recycle streams. In this case an inspection of the degree of the cycles detects the recycle streams as S_2 and S_7. Note that the selection of these two breaks all cycles in the system. We can see how simply the recycle streams can be found in rather complex situations if an orderly search is undertaken.

	S_2	S_4	S_7	
A	1			1
B			1	1
C	1	1		2
D		1	1	2

Fig. 3.11-6 The cycle table with independent columns. S_2 and S_7 are the simplest recycle streams.

Algorithms are available along the lines illustrated here to find the simplest recycle set in systems of arbitrary complexity.[8] This includes methods for finding the minimum parameter recycle streams, when streams carry information by means of several parameters such as composition, temperature, pressure, flow rate, and so forth. The study of these algorithms is recommended.

3.12 RECYCLE COMPUTATIONS

In the previous sections we have developed methods for organizing the analysis of complex processing systems so that the engineer need not engage in unnecessary calculations merely to determine how changes in the design of parts of a system might propagate through the system. In many processes it will not be possible to precedence order the design equations, and recycle loops will persist. Thus, some sort of iterative calculations will be required to solve the design equations within the recycle loop. This is a troublesome task at best, and we propose to alert the engineer to convergence problems that may be encountered.

Consider the simple problem of numerically closing the recycle loop on the reactor-separator system shown in Fig. 3.12-1. The feed material enters the process and is mixed with the recycle of unreacted feed. This stream enters the reactor where it is partially converted to product (here 50 per cent conversion) and then goes to the separator from which the recycle stream of unreacted feed and the product stream flow.

The equations describing the process are:

1. $A_3 = 0.5A_2$ (50 per cent conversion in reactor)
2. $A_3 = A_5$ (all of unreacted feed is recycled)
3. $A_2 = A_1 + A_5$ (mixing of recycle and feed streams)
4. $A_1 = 1{,}000$ (amount fed to the system)

where A (lb/hr) denotes the flow rate of feed material.

[8] J. H. Christensen and D. F. Rudd, *AIChEJ.*, **15** (1968), R. W. H. Sargent and A. W. Westerberg, *Trans. Inst. Chem. Engrs. (London)*, **42** (1964).

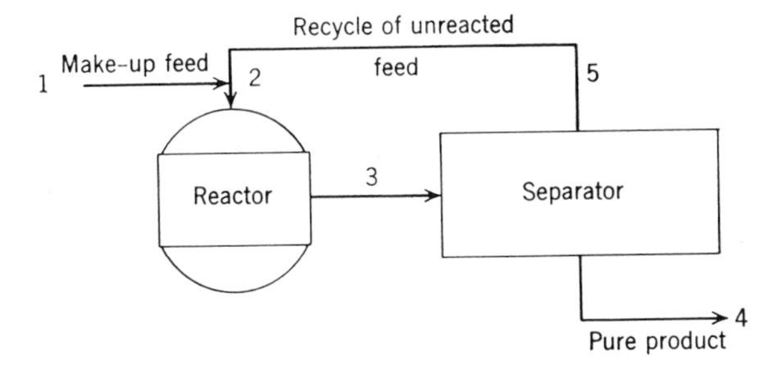

Fig. 3.12-1 A reactor-separator system.

This problem can be solved analytically, to give $A_2 = 2{,}000$, $A_3 = A_5 = 1{,}000$, but in a more complicated case an iterative or successive approximation approach must be taken. However, let us use this simple example to illustrate the importance of a proper strategy of calculation in such cases. Consider two reasonable strategies shown in Fig. 3.12-2 which arise from a study of the structure of the equations, both of which involve *tearing* the recycle stream (i.e., guessing the value of A_2 to get the calculations started), followed by the successive substitution of the updated value of the torn variable into the calculation sequence. Figure 3.12-3 shows the results of these two methods of calculation, one converges and the other diverges. The only difference between the two methods is the direction of calculation.

Problems with the speed of convergence of recycle calculations, and with their divergence, occur time and time again in industry. A large company will generally have programmed on the computer several recycle convergence

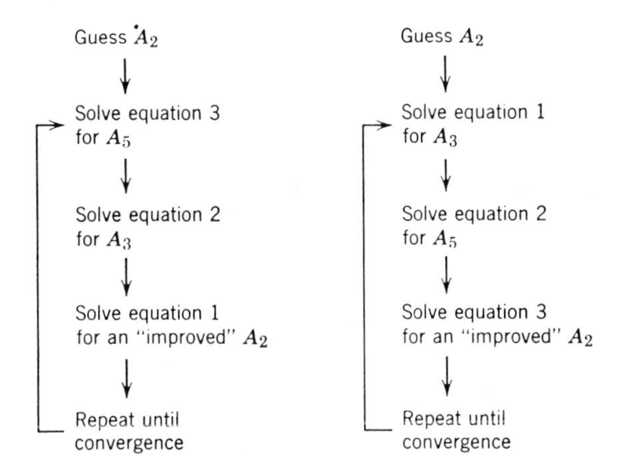

Fig. 3.12-2 Two recycle calculation strategies. The recycle stream is torn by assuming a value for A_2.

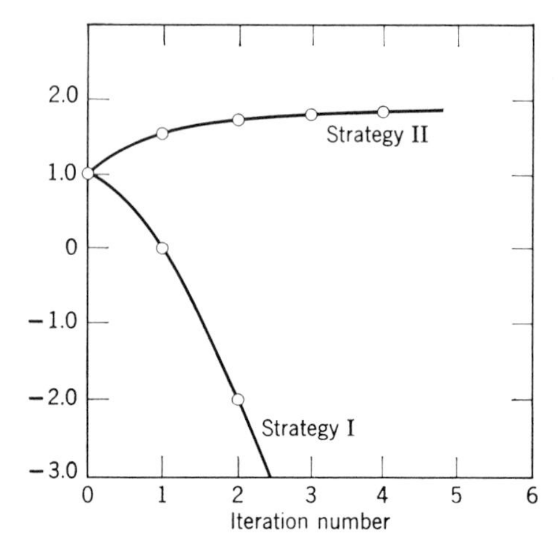

Fig. 3.12-3 The convergence properties of strategies I and II.

accelerator schemes, and it will be necessary to empirically determine the best approach to a given class of problems, for example, the class of design problems involving systems of flash vaporization units. We shall not delve into these methods in this text.

3.13 CONCLUDING REMARKS

There are several thoughts to be carried away from this chapter, the most important of which is that there exist a certain number of degrees of freedom in the design of a proposed process. The designer uses this freedom to adjust the details of a process to accommodate best to the economic design criterion. It has been shown that the whole complexion of a design problem can be altered by a careful selection of design variables, and that the problem of assessing the technical details of a design can be made much easier by the application of a little foresight.

We must remark that the tallying of variables, design equations and information flow structure is not a simple task in industrial design problems, and considerable ground work is required before the anatomy of a design problem becomes distinguishable. The engineer is not told what the design equations are, what the important variables are, or what the desired system structure is. Rather, only a word picture of the problem is available. This forces the engineer to consider the following practical factors.

Hidden and dominant variables. For example, it is often necessary to consider only the key components in distillation calculations and not to consider the trace impurities as variables. Without such a practical compromise the distillation calculations could get out of hand and consume more time than justified. If this trace material, however, has an adverse effect on the bacterial purification of the waste from the process as required for pollution abatement, its presence, which was hidden in the distillation, becomes dominant in the bacterial purification.

Not only may variables inadvertently be hidden but they may well be completely forgotten. Many a process has been designed, constructed, and put into operation, only to have a forgotten factor play an important role. A designer may "forget" or perhaps have no way of knowing that 0.2 per cent of NO_2 in the synthesis gas to an ammonia converter will poison the catalyst. This forgotten variable could prove to be the cause of the failure of some large processing plan.

On the other hand, the design relations may be unduly sensitive to changes in a variable. For example, if a reactor is designed to operate near the ignition point, a change in the feed rate to the reactor may "blow" the reaction out of the reactor, causing process shutdown. This undue sensitivity is not desired, since it may indicate uncertainty in the process or an improper formulation of the design relations, either of which must be changed.

Such value judgments on the relative importance of variables (judgments which may well be the key factor in achieving a solution to a problem) often do not show up immediately on a tally of equations and variables and only appear later when experience is gained about the particular problem at hand. We point to this difficulty so that the engineer may more fully appreciate the complications which may arise.

The occurrence of constraints. The design equations may be thought of as definite requirements for the values that the variables may assume. Only certain very definite values of the variables will achieve the goal of the process. The design equations must be satisfied completely.

Often constraints appear in the form of inequalities which limit the range over which variables may vary. For example, mass fractions may range only between zero and one, pressures may not exceed the rupture pressure of a vessel, temperatures may be constrained to avoid freezing in a pipeline. Limits on acceptable product quality are of this form. Such constraints may be expressed as

$$g_* \leq g(s_j;\, d_k) \leq g^*$$

where g_* and g^* are the bounds on the range of the constrained relationship g.

For the purposes of determining the degrees of freedom in the design problem, these constraints can be ignored, since they do not impose strict requirements on whether or not a variable can be manipulated. But rather, they only limit the range over which the variable can vary, thus limiting the *numerical* values a variable can assume.

The constraints do give a clue concerning the strategy of selecting the design variables, however. Recall that the design variables are those variables that are to be adjusted, and the state variables are determined by the solution of the design equations. Since the constraints must be satisfied, it is politic to use, when possible, the more narrowly constrained variables as design variables. In such a situation, the constrained variable can be adjusted to range only within the constraints during the optimization. For example, if the reaction temperature in a process is limited to a 20°F range, it might be wise to use this variable as a design variable and try different temperatures until satisfactory reactor performance is reached. On the other hand, had this temperature been assigned the status of a state variable, we would be able to check to see if the constraint were satisfied only after the solution of the design equations.

The general rule-of-thumb to follow is: *the design variables should, when possible, be those variables which are most constrained.*

Factors of preference. The constraints introduced in the previous section allow latitude and some freedom of movement of the variables but cannot be violated. There exist weaker constraints which can be violated if need be. These are the factors of preference which cannot easily be placed in quantitative form but enter the design problem in a subjective way.

The factors of preference tend to force the design towards that system that has the right "feel," the feeling of handiness, compactness, ruggedness, accessibility for service and repair, and the like. A design which has a delicate piece of equipment exposed, involves overly elaborate piping, or requires an unusual feature may be rejected just because it has the wrong "feel." These constraints of preference can be violated if the violation can be shown to be worthwhile.

References

A pioneering work in the analysis of the degrees of freedom is:

E. R. Gilliand and C. E. Reed, "Degrees of Freedom in Multicomponent Absorption and Rectification Columns," *Ind. Eng. Chem.*, **34**, 551 (1942).

A thorough analysis of a variety of processing components is available in:

M. Kwauk, "A System for Counting Variables in Separation Processes," *AIChEJ.*, No. 2, **2** (1956).

Other work in the analysis of degrees of freedom includes:

P. L. Morse, *Ind. Eng. Chem.*, **43**, 1863 (1951).
M. O. Larian, *Petrol. Refiner*, **32**, 219 (1953).
L. Bertrand and J. B. Jones, *Chem. Eng.*, **68**, 139 (1961).
B. D. Smith, *The Design of Equilibrium Stage Processes*, McGraw-Hill, New York, 1963.

P. Foldes and I. Nagy, *Periodica Polytech., Chem. Eng.*, No. 2, **10** (1966) Budapest.
G. M. Howard, *Ind. Eng. Chem., Fundamentals*, No. 1, **6** (1967).

An excellent text in mathematical modeling and the structure of design equations is:

D. M. Himmelblau and K. B. Bischoff, *Process Analysis and Simulation*, Wiley, New York, 1968.

See also:

R. G. E. Franks, *Mathematical Modeling in Chemical Engineering*, Wiley, New York, 1967.

The basic references in the theory of information flow structure are:

D. V. Steward, "On an Approach to Techniques for the Analysis of the Structure of Large Systems of Equations," *SIAM Review*, No. 4, **4** (1962).
D. V. Steward, "Partitioning and Tearing Systems of Equations," *J. SIAM Numer. Anal.*, Ser. B, No. 2, **2**, (1956).

The analysis of information flow structure forms the basis for the development of computer programs for the automatic analysis of complex processes as reported by:

R. W. H. Sargent and A. W. Westerberg, *Trans. Inst. Chem. Engrs. (London)*, **42** (1964).

Other work in this area includes:

W. Lee, J. H. Christensen, and D. F. Rudd, *AIChEJ.*, No. 6, **12** (1966).
W. Lee and D. F. Rudd, *AIChEJ.*, No. 6, **12** (1966).
J. H. Christensen and D. F. Rudd, *AIChEJ.*, **15** (1968).

For information on industrial process calculations see, for example:

A. E. Ravicz and R. L. Norman, *Chem. Eng. Progr.*, No. 5, **60** (1964).
E. M. Rosen, *Chem. Eng. Progr.*, No. 10, **58** (1962).

Progress in the use of the computer for the analysis of systems is reported by:

P. T. Shannon, A. I. Johnson, C. M. Crowe, T. W. Hoffman, A. E. Hamielec, and D. R.Woods, *Chem. Eng. Progr.*, No. 6, **62** (1966).
C. M. Crowe, *et al.*, *Chemical Plant Simulation*, Wiley, New York (to appear).

PROBLEMS

3.A. A boiling liquid refrigerant is to be used to cool a process stream in the following heat exchanger (see Fig. 3.A-1).

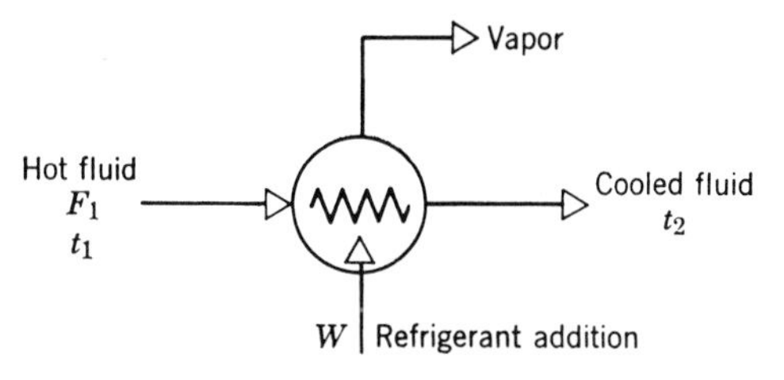

Fig. 3.A-1

where the variables are:

Hot fluid rate, F_1 lb/hr
*Fluid heat capacity, C_p Btu/lb-°R
Inlet temperature, t_1 °R
Outlet temperature, t_2 °R
Refrigerant rate, W lb/hr
*Heat of vaporization, λ Btu/lb
*Boiling temperature, t_ω °R
Heat-transfer area, A sq ft
*Heat-transfer coefficient, U Btu/hr-°R-sq ft

* These values are known in this problem.

(a) Derive the design equations

$$F_1(t_1 - t_2)C_p = \lambda w$$

$$\lambda w = UA(\Delta T)_{\ln}$$

$$(\Delta T)_{\ln} = \frac{t_1 - t_2}{\ln[(t_1 - t_\omega)/(t_2 - t_\omega)]}$$

(b) How many degrees of freedom exist in this design? What design variables do you recommend?
(c) If it is required that $t_2 = 600$°R with the feed conditions $F_1 = 1{,}000$ lb/hr and $t_1 = 700$°R, how many degrees of freedom exist? What design variables do you suggest?

3.B. An adiabatic continuous flow well-mixed reactor is to be used to conduct the first-order chemical reaction, $A \rightarrow B$. The physical chemistry of the reaction has been studied and data are available on the following parameters.

Heat of reaction, ΔH Btu/lb
Heat capacity, C_p Btu/lb-°R
Reaction velocity, constant, k 1/sec
Activation energy, E/R °R

(a) Derive the following design equations for the reactor (see Fig. 3.B-1).

$$F_0 A_0 = F_0 A_1 + VA_1 k\, e^{-E/Rt_1}$$

$$A_0 - A_1 = B_1 - B_0$$

$$F_0 C_p t_0 = VA_1 k\, e^{-E/Rt_1}\, \Delta H + F_0 C_p t_1$$

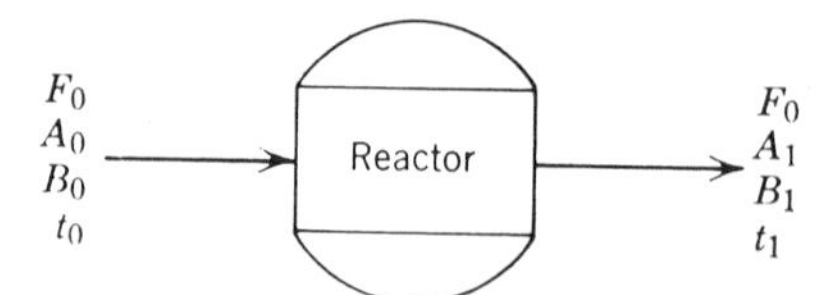

Fig. 3.B-1

The variables are defined as:

Feed flow rate, F_0 lb/hr
Compositions, A_0, A_1 lb A/lb
B_0, B_1 lb B/lb
Temperatures, t_0, t_1 °R
Reactor capacity, V lb of material

(b) If the reactor is to receive pure feed at $t_0 = 500$°R, $A_0 = 0.20$, and $B_0 = 0.00$, how many economic degrees of freedom exist? What design variables do you suggest? What recycle parameters exist?

(c) How does the picture change if a constraint is imposed that the product composition $B_1 = 0.10$?

3.C. It is required that the product from the reactor in 3.B be of a specified composition $B_1 = 0.10$ lb B/lb. This is to be achieved by controlling the reaction temperature with the help of a pump-around cooler. The process shown in Fig. 3.C-1 is assembled from the components of problems 3.A and 3.B.

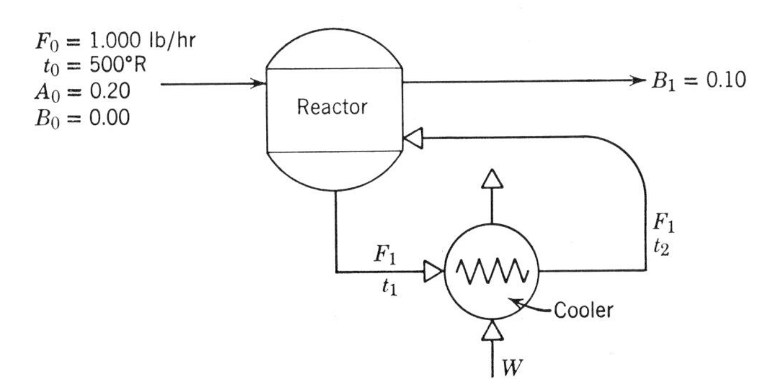

Fig. 3.C-1

(a) Modify the energy balance for the reactor in problem 3.B to include contributions from the stream going through the cooler.

(b) How many degrees of freedom exist? What design variables do you suggest?

(c) What happens if the pump for the cooler is of fixed capacity $F_1 = 1{,}000$ lb/hr and the heat-exchange area is fixed at $A = 500$ sq ft?

3.D. Rather than achieve the required product composition by temperature control solely, the effluent from the reactor is to be blended with pure B as shown in Fig. 3.D-1.

(a) How many degrees of freedom exist in this design? What design variables do you suggest?

(b) Suppose that an available heat exchanger and reactor are to be used, with $V = 600$ lb capacity and $A = 500$ sq ft. What design variables are recommended?

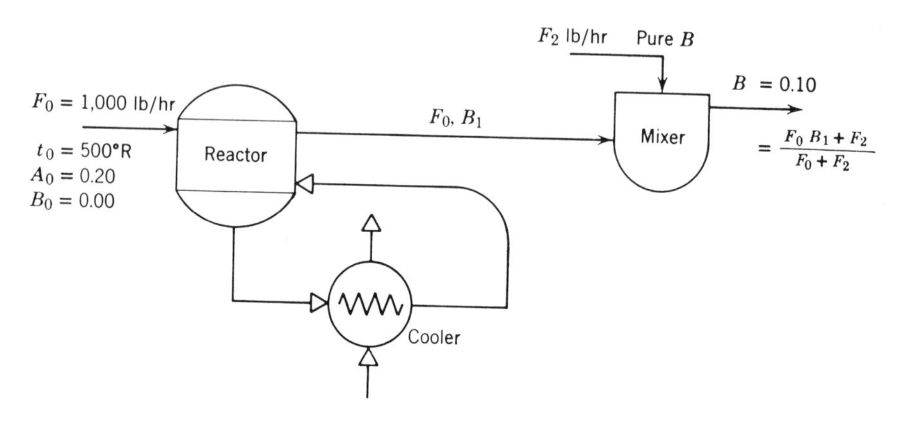

Fig. 3.D-1

3.E. Complex systems of heat exchangers are used commonly to increase the economy of processes.

(a) How many degrees of freedom exist in the design equations for the single heat exchanger shown in Fig. 3.E-1.

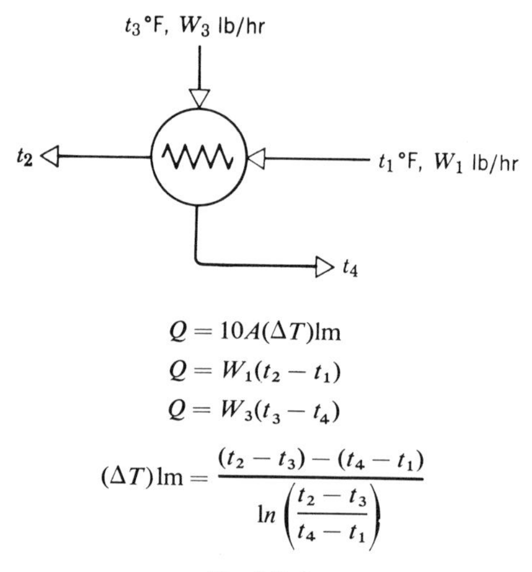

$$Q = 10A(\Delta T)\text{lm}$$
$$Q = W_1(t_2 - t_1)$$
$$Q = W_3(t_3 - t_4)$$
$$(\Delta T)\text{lm} = \frac{(t_2 - t_3) - (t_4 - t_1)}{ln\left(\dfrac{t_2 - t_3}{t_4 - t_1}\right)}$$

Fig. 3.E-1

(b) Using the method summarized in Eq. 3.8-1 determine the degrees of freedom of the heat-exchange system for the reaction process shown in Fig. 3.E-2.

(c) Select the design variables for the heat-exchange system.

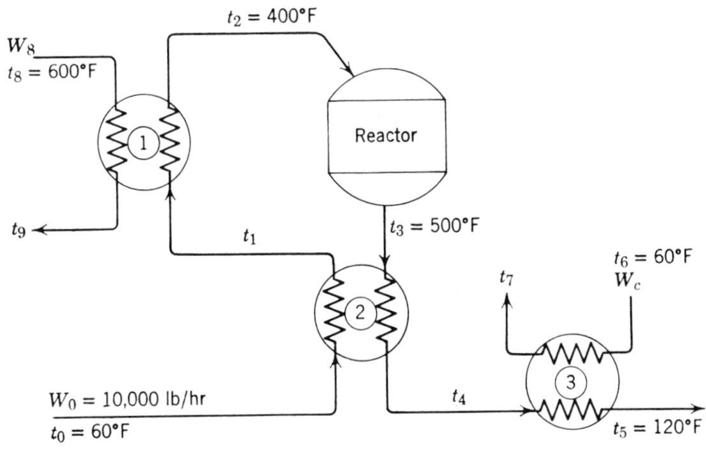

Fig. 3.E-2

3.F. A pure saturated vapor is to be liquefied by compression and cooling. Vapor-liquid thermodynamic data are supplied in the temperature-entropy diagram (see Fig. 3.F-1).

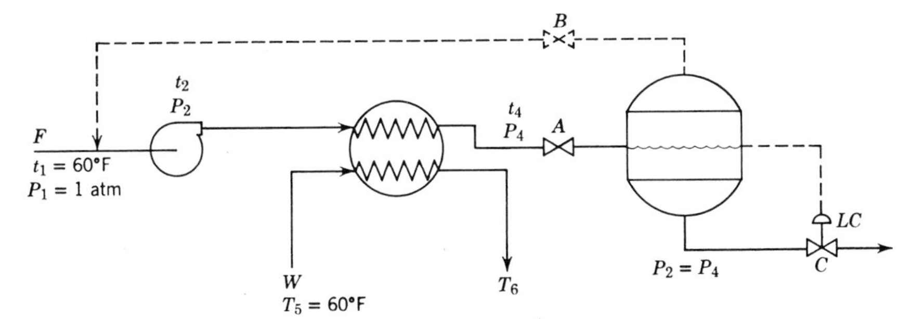

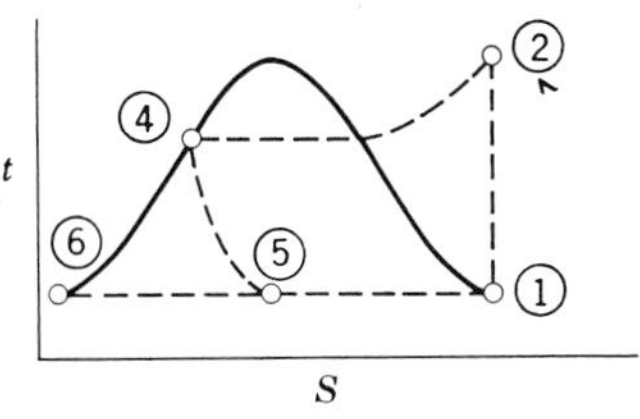

Fig. 3.F-1

Assuming ideal isentropic compression,

(a) Calculate the total degrees of freedom of the system with valve A open and valve B closed.

(b) Suggest suitable design variables.

3.G. In Fig. 3.F-1, if the expansion is to produce a mixture of vapor and liquid at 1 atm (1 and 6 on the *t-S* diagram), find the degrees of freedom of the system. Now valves *A* and *B* may be partially closed.

3.H. The vapor compression cycle is often used in water purification processes. Cold water is boiled using compressed vapor as the heating medium as shown in Fig. 3.H-1.

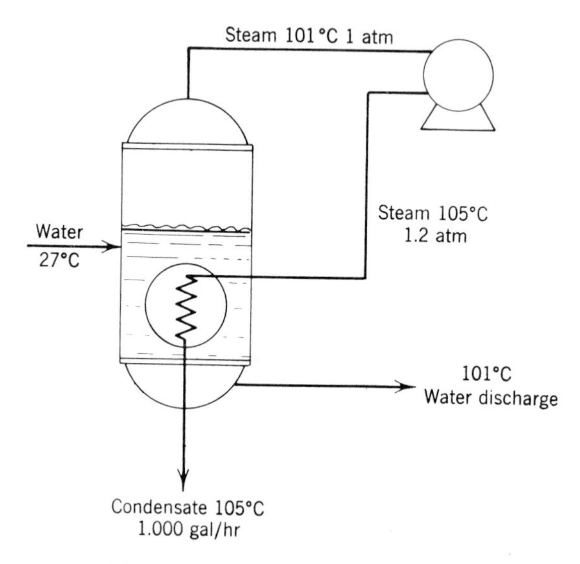

Fig. 3.H-1 A vapor compression still.

From the data given and available in steam tables, compute the amount of energy that must be supplied by the compressor and the amount of water that must be taken in. If the overall heat-transfer coefficient for the condenser is 100 Btu/hr-ft^2-°F, compute the required surface area of the condenser.

3.I. To achieve economy in vapor compression, it is desirable to recover heat from the condensed vapor by preheating the cool water. How many more degrees of freedom appear in the design shown in Fig. 3.I-1 than in the design shown in Fig. 3.H-1.

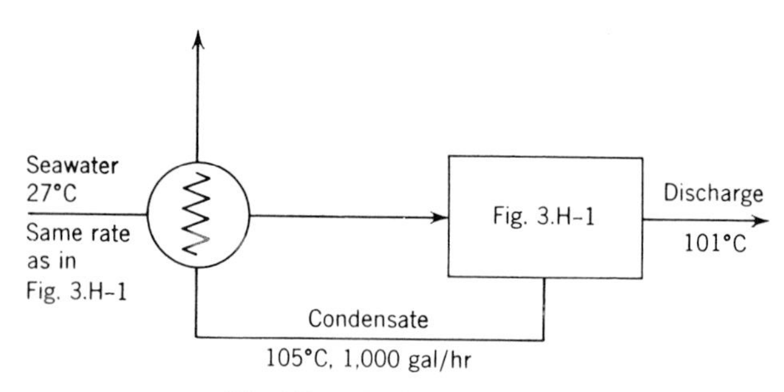

Fig. 3.I-1 An improved still.

3.J. An alternate design for the preheating of the cool water is shown in Fig. 3.J-1 in which a system of two heat exchangers is used. How many more variables can the designer manipulate in Fig. 3.J-1 than in Fig. 3.H-1?

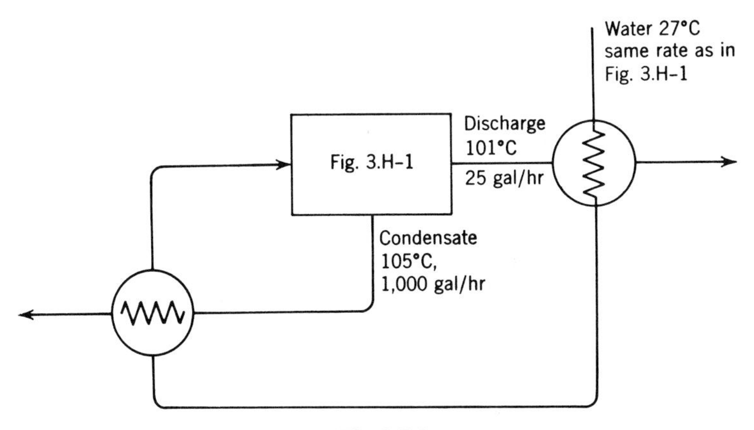

Fig. 3.J-1

3.K. Select suitable design variables for the process shown in Fig. 3.8-3 using the methods of Section 3.10. How many different sets of design variables can be found to precedence order the process components?

3.L. Suppose that the system in Fig. 3.8-1 is to be designed to use an available heat exchanger. In this case, the kind of exchanger, the surface area, and the coolant flow rate are fixed. How does this alter the information flow diagram in Fig. 3.8-3? Apply the design variable selection algorithm to this new situation.

3.M. In this problem the opportunity will be given to synthesize several alternative systems, select the preferred design variables, precedence order the design computations, and estimate the gross profit for tentative operating conditions. This is the kind of activity we can engage in independently once the methods of cost estimation in Chapter 5 are established: here we have prepared the cost estimates as part of the problem statement.

A processing system is to be designed to process 40,000 lb/hr of a solvent S which contains 0.20 pounds of a solute s per pound of S. The solvent S can be sold at \$0.10/lb regardless of its solute content. However, the solute s is worth \$1.00/lb when in a solution with a wash solvent W. The required concentration of the solution is 0.80 pounds of s per pound of W. The solvent W costs \$0.70/lb and is available in unlimited amounts. All materials are available at 70°F.

The following items of processing equipment are available for the synthesis of a system to best accomplish the processing task. After describing the characteristics of the equipment, several system alternatives will be suggested.

Extractors. The transfer of solute s from solvent S to solvent W can be achieved in equilibrium stage extractors which can operate up to a maximum capacity of 10,000 pounds of S per hour. (See Fig. 3.M-1.)

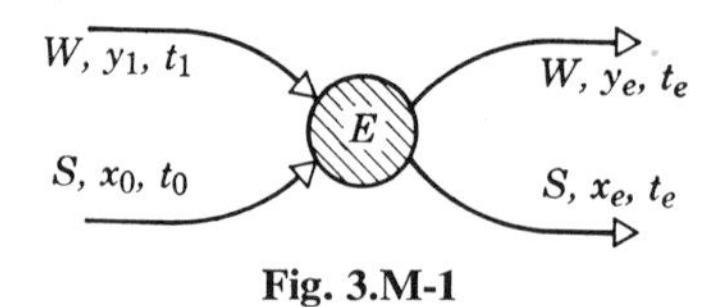

Fig. 3.M-1

Solute distribution between phases

$$y_e = 0.1 \frac{(t_e - 60)}{60} x_e$$

where y_e and x_e are the pounds of s per pound of W and S respectively in equilibrium at t_e°F. This expression is valid up to $t_e = 200$°F at which point the extractor is inoperable.

Material balances

$$Wy_1 + Sx_0 = Wy_e + Sx_e$$

Energy balance

$$Wt_1 + St_0 = (W + S)t_e$$

Cost of extraction. The total cost of extraction, including the amortization of equipment and operating costs, is

$$C = 0.50 + 0.10S \qquad \$/\text{hr}$$

Separators. A solution of s in W can be concentrated by a solvent separator (see Fig. 3.M-2). The separator can produce a maximum of 5,000 lb/hr of

W_3, t_3 = 250°F

W_1, y_1, t_1 — S — W_2, y_2 = 0.80 lb s/W

Fig. 3.M-2

pure W at a temperature of 250°F. The design equations are:

Material balances

$$W_1 = W_3 + W_2$$

$$W_1 y_1 = W_2\, 0.80$$

Cost of separation

$$C_S = 0.20 + \left(\frac{250 - t_1}{250}\right) 0.10 W_1 \qquad \$/\text{hr}$$

Heaters. Any stream of F pounds per hour can be heated at a rate less than $Q_H = 1{,}000{,}000$ Btu/hr by the heater shown in Fig. 3.M-3.

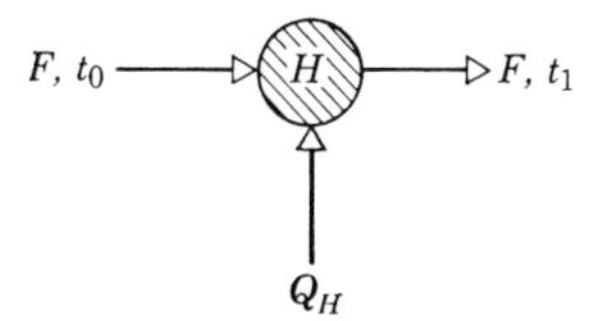

Fig. 3.M-3

Energy balance

$$Ft_0 C_p + Q_H = Ft_1 C_p$$
$$C_p = 1$$

Cost of heating

$$C_H = \$1.50/10^6 \text{ Btu of } Q_H$$

Coolers. Any stream of F pounds per hour can be cooled at a rate of less than $Q_c = 40{,}000$ Btu/hr. (See Fig. 3.M-4.)

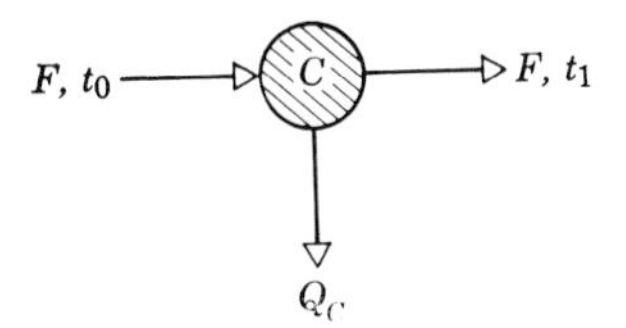

Fig. 3.M-4

Energy balance

$$Ft_0 C_p - Q_c = Ft_1 C_p$$
$$C_p = 1$$

Cooling cost

$$C_c = \$0.50/10^6 \text{ Btu}$$

Heat exchangers. Heat exchangers can be used to transfer heat from any hot fluid which is available at a rate F_1 to any cooler fluid available at a rate F_2. (See Fig. 3.M-5.)

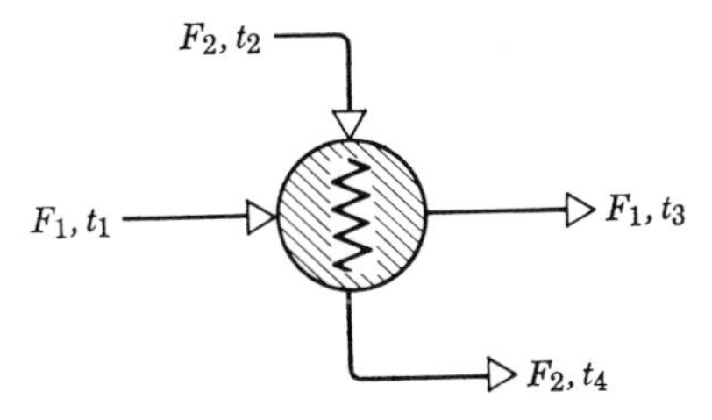

Fig. 3.M-5

Energy balances

$$Q = F_1(t_1 - t_3)C_p \qquad \text{Btu/hr}$$
$$C_p = 1$$

$$F_1 t_1 + F_2 t_2 = F_1 t_3 + F_2 t_4$$

Surface area design equation

$$Q = 10A(t_1 - t_2) \qquad \text{where } A \text{ is surface area in square feet}$$

Cost of heat exchange. The cost of the heat exchanger is correlated to the area of the exchanger by the following.

$$C_x = 0.02A^{1/2} \qquad \$/\text{hr}$$

Assuming that any number of these standard items of processing equipment are available, develop design alternatives, assemble the design equations for the alternative, select the design variables, and select reasonable values for the design variables. Then estimate the gross profit from the proposed processes.

(a) Synthesize a system consisting of parallel banks of extractors and separators with no heating or cooling.
(b) Synthesize a system with multiple counter-current extractor systems followed by separators.
(c) Introduce the heaters, coolers, and heat exchangers to take advantage of the improved extraction at higher temperatures.

3.N. A feed of given composition may be divided into two phases, provided the temperature of the flash, T, lies between the bubble point, T_{BP}, and the dew point, T_{DP}, temperatures of the feed, where the latter are evaluated at the pressure of the flash, P. For a binary feed we consider the system shown in Fig. 3.N-1.

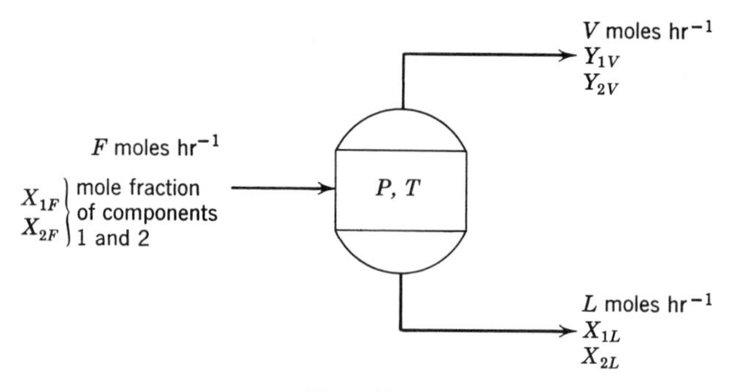

Fig. 3.N-1

Material balance and equilibrium relationships are given below:

$$F = V + L$$
$$X_{1F}F = Y_{1V}V + X_{1L}L$$
$$X_{2F}F = Y_{2V}V + X_{2L}L \qquad K_1 = Y_{1V}/X_{1L}$$

$$X_{1F} + X_{2F} = 1 \qquad K_2 = Y_{2V}/X_{2L}$$
$$X_{1L} + X_{2L} = 1 \qquad K_1 = K_1(T, P)$$
$$Y_{1V} + Y_{2V} = 1 \qquad K_2 = K_2(T, P)$$

A. Suppose that in some situation F, X_{1F}, T, and P are given.

(a) Are there any degrees of freedom in the calculation of the remaining variables, and, if there are, how many?

(b) Suppose that you found in part (a) that there are 3 degrees of freedom. What does this mean?

B. Suppose that in a different situation only F, and X_{1F} are specified. Are there any degrees of freedom in the calculation of the remaining variables, and, if there are, how many?

C. Let us now consider the thermal effects associated with this separation process where F and X_{1F} are specified as well as the feed temperature, T_F, and pressure, P_F. The enthalpies of the feed, overhead vapor, and bottoms liquid are H_F, H_V, and H_L, respectively. Here

$$H_F = H_F(T_F, P_F, X_{1F})$$
$$H_V = H_V(T, P, Y_{1V})$$
$$H_L = H_L(T, P, X_{1L})$$
$$P_F = \text{Pressure of the feed}$$
$$T_F = \text{Temperature of the feed}$$

and the enthalpies are on a per mole basis. The energy balance is

$$Q + FH_F = VH_V + LH_L$$

where Q is the amount of heat added to the system per unit of time.

(a) Why may the system require a supply of heat?

(b) How many degrees of freedom are available now?

(c) If component 2 is relatively nonvolatile, we may assume that $Y_{2V} = 0$, reducing the material balance, equilibrium, and energy balance relationships to

$$F = V + L \qquad K_1 = 1/X_{1L}$$
$$X_{1F}F = V + X_{1L}L \qquad K_1 = K_1(T, P)$$
$$X_{2F}F = X_{2L}L \qquad Q + FH_F = VH_V + LH_L$$
$$X_{1F} + X_{2F} = 1 \qquad H_F = H_F(T_F, P_F, X_{1F})$$
$$X_{1L} + X_{2L} = 1 \qquad H_V = H_V(T, P)$$
$$H_L = H_L(T, P, X_{1L})$$

Determine the number of degrees of freedom, and, if the number is greater than zero, select the design variable(s) that precedence order calculations. Show the order of calculations as well as the procedure used to determine this order.

(d) Suppose that in part (c) you found that there were 3 degrees of freedom and that selecting T, P, and X_{1L} precedence ordered the calculations. Indicate what ranges you would expect these variables to fall within.

4

ECONOMIC DESIGN CRITERIA

Having established that economic degrees of freedom exist in a process design, we shall now establish design criteria to guide the adjustment of these free variables and to aid in selection among processing alternatives. Most often the criterion takes the form of the question: Which of the alternatives would be expected to increase the financial health of the firm by the greatest amount in the reasonably near future? To answer the question requires an understanding of basic economic principles, a topic to which this chapter is devoted.

Typical Problem

Ten thousand dollars is available to purchase insulation for certain high-temperature components in a chemical process, and this available capital is insufficient to satisfy all the demands for insulation. How should the components be insulated best to utilize the limited capital?

4.1 INTRODUCTION

The preliminary screening of process alternatives is often based on an approximate analysis of the alternatives, seeking to detect weaknesses which are dominant. However, as attention is focused on the more plausible alternatives, precision is required: and the simple principles of preliminary screening are too coarse for the task. The engineer must define the economic environment in which the process is to function and establish a criterion which, when used during process design, will lead to the economically optimal process.

A fundamental dilemma arises. On the one hand, it is recognized that the project in question will require a commitment of resources in the form of investment capital and engineering talent and in this way will influence all the other activities of the firm. Thus, the engineering of a given process cannot be performed with complete accuracy unless all other projects are considered simultaneously. On the other hand, it is recognized that this extremely large and overly detailed problem cannot be handled and must be decomposed into a number of smaller and simpler approximate problems which can be solved in a reasonable length of time and with a reasonable expenditure of effort. The resolution of this dilemma has required the development of economic design criteria to guide a given project.

However, before we can undertake to examine the principles which lead to design criteria, it is necessary to lay a foundation of concepts and terms used in profitability studies. This is done in the next two sections. Then we show how the design criteria arise as approximate solutions to an overly detailed capital allocation problem. It is thereafter shown how a limit on available capital alters the design criteria, and how other criteria arise as the long-range goals of the firm change.

4.2 DEFINITION OF TERMS USED IN PROFITABILITY STUDIES

In this section we define the terms which enter into the estimation of the economic value of a proposed system. Figure 4.2-1 illustrates the symbols to be defined.

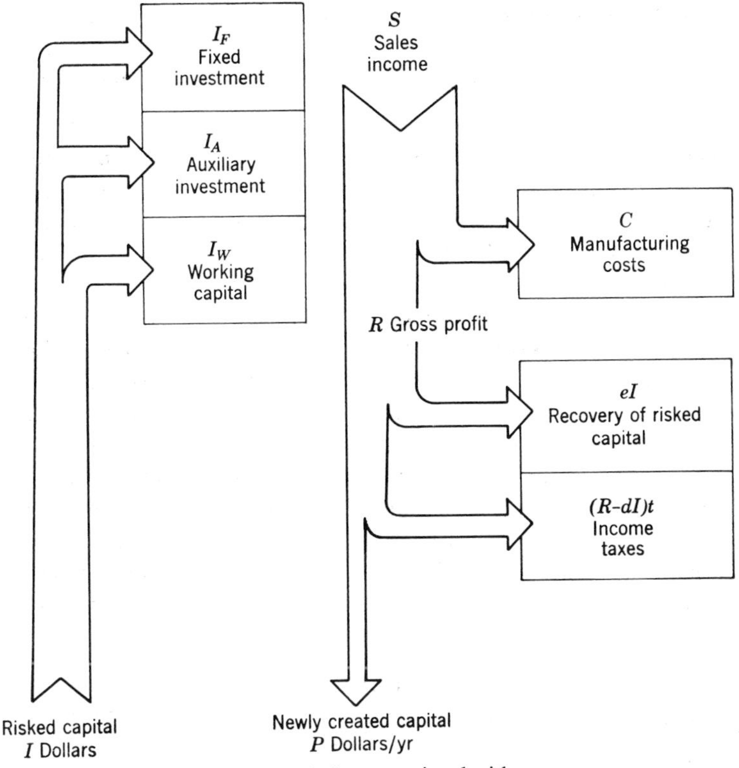

Fig. 4.2-1 The cash flow associated with a process.

THE TOTAL INVESTMENT, I. The total investment I can be broken down according to degree of risk into three parts, the fixed investment in the process area I_F, the investment in auxiliary services I_A, and the investment in working capital I_W.

$$I = I_F + I_A + I_W \tag{4.2-1}$$

decreasing risk →

These are fixed amounts of money that will be *tied up* and *risked* in the interest of the proposed system. The investment I has the dimension, dollars, and should not be confused with operating costs which have the dimension, dollars per year.

FIXED INVESTMENT, I_F. This is the investment in all processing equipment within the *processing area* or *battery limits*. The processing area is generally an area in the plant in which the specific operations are carried out, as distinguished from such general purpose areas as storage, loading, shop, steam generation, and so forth. Quite commonly the term refers to a specific room or area in a building or a plot of ground set aside for the process in question. Investment in equipment within the process area carries the greatest degree of risk, since this capital can only be partially recovered as salvage value in case the system is suddenly terminated, for example, when market changes make further operation unprofitable.

AUXILIARY INVESTMENT, I_A. Such items as steam generators, fuel stations, and fire protection facilities are commonly stationed outside the process area and serve the system under consideration as well as other processing systems within a large complex. Each system to be served with these auxiliary services must pay a portion of the cost.

The assessment of cost may be handled in a number of ways. In many cases these auxiliary services will have continuing use for other processes. The system presently being designed may then be charged a yearly fee or internal price for the services it will require in proportion to the specific amounts to be consumed, and in this case, the amount invested in the equipment for producing these auxiliary services does not enter into consideration as capital risked for this project. Where the specific consumption is not readily determined, as in fire protection, the system might be charged a fixed fee for the auxiliary services. The fixed fee might be assessed as a fraction of the value of fixed equipment. Such use of an annual fee or an internal price we consider as part of the manufacturing cost, discussed below.

WORKING CAPITAL, I_W. This is the capital tied up in the interests of the system in the form of ready cash to meet operating expenses, real estate, inventories of raw materials and products. It is common to require that a system have in stock raw materials, products and cash for one month's operation as a hedge against uncertainty. In an extreme example, the taconite mills in Northern Minnesota might stockpile a full winter's production of enriched iron ore until Lake Superior opens to traffic in the spring. The system must be charged a fee for the use of this working capital, even though it might be completely recoverable should the process become uneconomical and be terminated.

THE MANUFACTURING COSTS, C. The manufacturing costs C incurred in keeping the system running from day to day have the dimension, dollars per year, and can be divided roughly into terms proportional to the fixed investment, F_I, the production rate, F_Q, and the labor, F_L.

$$C = F_I + F_Q + F_L \tag{4.2-2}$$

FIXED INVESTMENT TERM, $F_I = aI_F$. The factors which are commonly independent of the production rate and may be taken as proportional to the fixed plant investment are as follows.

Maintenance—labor and materials (this is that part of the maintenance expense which is independent of production rate).
Property taxes.
Insurance.
Safety expenses—fire protection, police, first aid, etc.
General services—laboratory, roads, transportation, etc.
Administrative services—office, accounting, legal, etc.

Some of these costs may be allocated by the accountants for an existing process on various bases, such as direct labor, total production costs, investment, and sales. The use of labor cost as a basis for expressing these overhead costs appears especially risky with the advent of automation. It appears simple and more useful for the engineer to express all these costs as fractions of fixed plant investment in dealing with a proposed process in design and development stages. These then contribute to the factor, a.

PRODUCTION RATE TERMS, $F_Q = bQ$. A number of costs depend directly on the rate at which product is to be generated by the system, Q units per year. If Q_d = the *designed* capacity, then $\Phi = Q/Q_d$ may be called the *load factor*, which expresses the fraction of the expected capacity actually used. After the plant has been built, it is often found that the actual capacity is substantially different, either more or less than the designed capacity, Q_d. We may call this *actual*, experimentally determined, capacity, Q_a. During the design and development period, Q_d, is, of course, the only available value for plant capacity. The factors proportional to production rate are:

Raw material costs.
Utilities costs—power, fuel, cooling water, process water, steam, etc.

Maintenance costs incurred by operation.

Chemicals, catalysts, and materials consumed in manufacturing other than raw materials.

Warehouse and shipping expenses, quality control and routine analysis, royalties, and license fees.

These contribute to the factor, b.

LABOR TERMS, $F_L = cL$. We shall take only those costs directly incurred in maintaining the operating labor force as proportional to labor, L, man hours per year:

Direct operating labor.
Superintendence.
Payroll overhead—vacation pay, social security, etc.

Clearly, some factors could be included in one or the other of these categories with equal accuracy, but the point is that the factors must be included somewhere.

PROFIT RATES. The *gross profit rate*, R, \$/yr, is the difference between the net income from the annual sales, S (sales receipts less distribution, sales, and promotion costs), and the annual manufacturing costs, C

$$R = S - C \tag{4.2-3}$$

where $C = aI_F + bQ + cL$. The *net profit rate*, P \$/year, defined as the expected annual return on the investment after deducting depreciation and taxes, is

$$P = R - eI_F - (R - dI_F)t \tag{4.2-4}$$

where

e = a yearly fractional assessment calculated to recover the investment in fixed equipment (depreciation)
d = a yearly fractional loss of value of the fixed equipment allowed by tax authorities for computing taxable income
t = income tax rate, \$/\$ earned

It is assumed here that the cost of the auxiliary services is included in appropriate charges comprised by factors a and b, so that the term, I_A, investment in auxiliaries, does not appear explicitly in this equation.

For *straight line* depreciation $e = 1/n$, where n = expected project life. This procedure in effect withholds, from the project earnings, payments in a non-interest bearing fund sufficient to recover the fixed investment. This set aside capital is, in fact, reinvested in company operations, but the project in question is not credited with the earnings, which are merely absorbed in the overall company profit. To be fair to the proposed project, in the *sinking fund* method of depreciation, a savings fund is envisioned, bearing interest at a rate i, \$/\$-yr, the actual current earning rate of all of the company operations. Lower equal payments of

$$e = \frac{i}{\exp(in) - 1} < \frac{1}{n} \tag{4.2-5}$$

are then required to recover the investment. A short discussion of continuous compounding is presented in Section 4.3 to show how Eq. 4.2-5 is obtained.

The government provides a number of complex depreciation schedules which account for the obsolescence of equipment and the write-off of fixed investments. This is not the place to enter into a discussion of such accounting detail, and we prefer to use the constant depreciation factor d in the illustrative examples for simplicity, with the full understanding that this is an approximation to reality. The fact that the tax rate t is on the order of 50 per cent of adjusted profit provides the incentive for seeking the optimum allowable depreciation schedule.

4.3 THE PRESENT VALUE OF FUTURE DOLLARS

The life of a processing system necessarily extends into the future with investments planned for expansions, changing profit rates, and a generally variable economic future. To compare properly processing alternatives which exhibit different cash flows over their life times, the future dollars must be brought to some standard base. The present time is a commonly used base for such comparisons, the value of such monies then being called *present value*.

Let $S(\theta)$ be the amount in a fund at time θ. Let interest be credited to that fund at a rate i, \$/\$-yr (continuously compounded). Then in the increment of time $d\theta$, the change in the amount in the fund will be

$$dS = iS\,d\theta \tag{4.3-1}$$

The amount of money at time $\theta = 0$, $S(0)$, which will accumulate to $S(\theta)$ at time θ is obtained by integrating Eq. 4.3-1 from $\theta = 0$ to $\theta = \theta$.

$$S(0) = S(\theta)e^{-i\theta} \tag{4.3-2}$$

The term $S(0)$ is called the *present value* of the $S(\theta)$ dollars at time θ years in the future. This present value may be thought of as the net amount of cash $S(0)$ which must be committed to a banker at time zero to finance the future need for cash $S(\theta)$. It is appropriate to use the rate of return a firm has experienced on invested capital as a measure of i, defining the value of money in a risk free investment within the firm, Table 4.3-1.

Table 4.3-1 The Experienced Rate of Return on Investment for Several Processing Industries

Industry	Experienced Rate of Return i (\$/yr-\$)
Pulp and paper Rubber	0.08–0.10
Synthetic fiber Chemical and petroleum	0.11–0.13
Drugs and pharmaceutical Extractive and mining	0.16–0.18

If payments are flowing continuously into the fund at a rate of M dollars per year the change in the fund is

$$dS = Md\theta + iSd\theta \tag{4.3-3}$$

If $S(0) = 0$ we find that the fund will have accumulated in θ years to

$$S(\theta) = \frac{M[\exp(i\theta) - 1]}{i}$$

Hence, if we wish to accumulate I dollars in n years by assessing equal payments of $M = eI$ dollars per year, the payments must be in an amount

$$M = eI = \frac{iI}{\exp(in) - 1} \tag{4.3-4}$$

This accounts for the depreciation rate in Eq. 4.2-5.

Example:
Suppose we have to choose between a plain steel reactor, which will cost \$10,000, will last 2 years, and require maintenance averaging 5 per cent of the initial investment per year, and a stainless steel reactor, which will cost \$30,000, will last indefinitely, will require no maintenance, and will have a net salvage value of \$6,000 after 6 years' service. The different schedules of costs for these two alternatives complicate the comparison as does the element of risk in the parent project. In this example we assume that the parent project is running successfully and the element of risk is absent. Using the quite realistic concept that the remainder of the company's operations function in effect like a nonprofit "bank," fixing the value of money within the firm at i, the current earning rate, we can simply compute the present value for 6 years' service for both of the alternatives. This can be regarded as the amount of money, at time zero, which would purchase a service contract for 6 years' service of either type of reactor. Suppose the company earns 15 per cent on invested capital. (See Table 4.3-2.)

Table 4.3-2

	Reactor	
	Plain	Stainless Steel
Present cost of investments		
$S(0)=S(\theta)e^{-0.15\theta}$, \$		
$\theta=0$	10,000	30,000
$\theta=2$	7,400	0
$\theta=4$	5,500	0
Salvage		
$\theta=6$	0	−2,400
Present cost of maintenance		
$S(0)=\int_0^6 (0.05)(10{,}000)e^{-0.15\theta}\,d\theta$	1,980	0
Total present cost of the purchase of 6 years' service, \$	24,880	27,600

4.4 COMPETITION FOR CAPITAL

Having defined the terms which are used in a process profitability study, we now put these definitions to work by showing how a competition for investment capital determines the criterion which the engineer must employ to design a useful system.

For ease in comprehending the principles involved, we begin by discussing the competition for investment capital among a group of distinct proposed projects, each of which can assume its optimum technical design for the given amount of capital invested in it. Later we shall extend this thinking to the detailed design of a single project, showing how the optimum design is found.

Normally a firm will have an abundance of investment opportunities and will have only a limited amount of investment capital free to allocate among them. Now, assuming that the goal of this firm is to maximize the total net rate of profit earned by its various interests, we ask the question: *what is a proper design criterion for any one of the abundance of projects to assure cooperation in achieving the master goal of the firm?*

The firm can be considered as a source of investment capital, much like a bank, which allocates money to a number of competing projects. In return for the receipt of capital the individual projects agree to transfer their net yearly profits back to the bank. We now define the capital allocation problem and indicate why it cannot be solved exactly in any practical situation.

Figure 4.4-1 shows the firm as a bank allocating capital to a number of projects. The number of projects is limitless since capital may be invested in stocks, bonds, and the world of finance, as well as in projects within the strict confines of the firm, such as improving and expanding existing company projects and creating new processing systems. Each of these projects is assumed to have a known net rate of profit as a function of invested capital. This is identified as $P_K(I_K)$ for the Kth project.

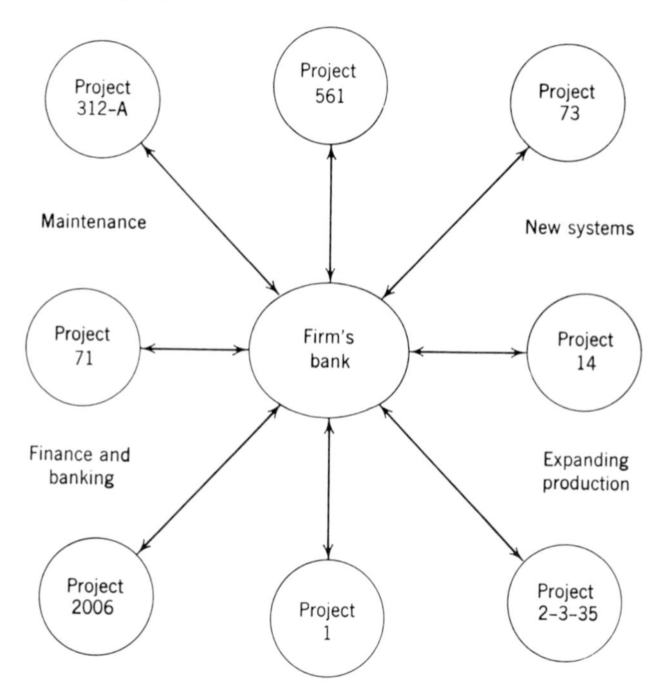

Fig. 4.4-1 The firm as a bank allocating capital.

It is assumed that the firm strives to allocate its capital to maximize the total rate of profit.[1]

$$\max_{\{I_K\}} \sum_K P_K(I_K) \tag{4.4-1}$$

Now, in a practical sense, this general capital allocation problem is beyond exact solution, since it requires a detailed knowledge of all facets of a large and involved enterprise encompassing and interacting with the worlds of finance as

[1] We might assume that the firm is striving to maximize the present value of all profits earned in the next five, ten, or twenty years. The developments in this chapter can be reproduced for any goal the firm might select to guide its investment policies, leading to different economic design criteria for individual projects.

well as process engineering, and is beyond solution because the parameter K ranges to exceedingly high values.

In a typical example, it is impossible for a large company situated on the Texas gulf coast to consider such details as whether part of the capital to be invested in a new multimillion dollar refinery could be better spent insulating a liquid methane storage tank in the company's Chicago facility. The converse is also true; it is impossible for an engineer assigned to a relatively small project, such as insulating a storage tank, to be aware of all the other investment opportunities available to the company. Yet, such awareness is essential and cooperation among projects is necessary.

A practical solution for any problem requires that the area of study be sufficiently small. But the mere act of isolation does violence to and distorts the actual picture in which there should be unrestricted competition for capital among *all* parts of *all* projects. This means that we *cannot* completely optimize any one portion of a process or any one segment of a firm's interest; this optimization can only be approached by the gradual inclusion of larger and larger segments of the firm's domain of interest. Clearly, the allocation of capital to projects (and even more so the optimization of processing systems) involves compromise between the easy solution of small but necessarily not completely realistic problems and the solution of larger, realistic but practically insoluble problems. The methods of optimization discussed in chapters 6 through 10 further attempt to resolve this dilemma.

4.5 THE EVOLUTION OF A DESIGN CRITERION

An approximation to the detailed capital allocation problem illustrated in Fig. 4.4-1 is shown in Fig. 4.5-1. Here, only the projects of immediate and special interest have been left unaltered, and the remaining abundance of projects has been combined into one large, limitless, and amorphous project called the other interests of the firm. It is assumed that the earning power of this large project determines the value of money within the firm, and that the projects of special interest must compete with it for capital.

This reduces the original impossibly large and complicated capital allocation problem from

$$\max_{\{I_K\}} \left[\sum_K P_K(I_K)\right] \tag{4.5-1}$$

where *K includes all the firm's opportunities*, to the far simpler problem

$$\max_{\{I_j\}} \left[\sum_j P_j(I_j) + i\left(I_{\max} - \sum_j I_j\right)\right] \tag{4.5-2}$$

where *j includes only those few projects of special interest.*

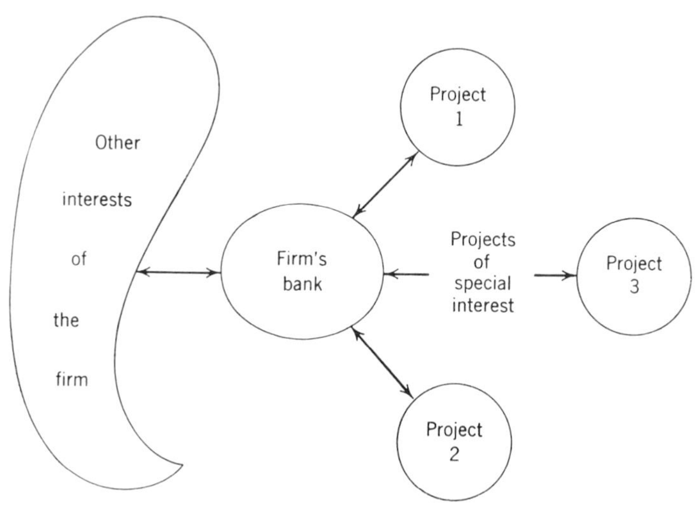

Fig. 4.5-1 Tearing several projects from the firm's other interests.

In Eq. 4.5-2 the terms within the left summation sign represent the profit to be achieved by the allocation of capital to the projects of special interest, and the term on the right is the profit to be made by allocating the remaining capital to the other interests of the firm at a rate of return i dollars per year per dollar allocated. The term I_{max} represents the total capital which is available for allocation.

Under certain conditions this approximate optimization problem admits of a particularly simple solution. Equation 4.5-2 can be written as

$$\max_{\{I_j\}} \left\{ \sum_j [P_j(I_j) - iI_j] \right\} + iI_{max} \tag{4.5-3}$$

subject to the constraint that

$$\sum_j I_j \leq I_{max} \tag{4.5-4}$$

Now, in the above expression the maximization can be performed inside the summation sign, term by term.

$$\sum_j \left\{ \max_{\{I_j\}} \left[P_j(I_j) - iI_j \right] \right\} + iI_{max} \tag{4.5-5}$$

as long as the I_j obtained by the individual maximizations (denoted here by I_j^*) do not violate the constraint of limited capital, Eq. 4.5-4.

Thus, if a given design project is to participate in the overall economic goals of the firm as defined by Eq. 4.4-1, the design criterion for that project must involve the maximization of the net profit to be earned minus a charge for capital risked. This simple design criterion is stated in Eq. 4.5-6

$$\max (P - iI) \tag{4.5-6}$$

Now, clearly if the maximand in Eq. 4.5-6 is negative for all investments I, the maximum is reached when $I = 0$. Thus, the criterion is established that money should not be risked in a project for which the *rate of return* P/I is less than the current earning rate i of the firm. Equation 4.5-7 must hold for a design which is a candidate for the enonomically optimal design

$$\frac{P}{I} > i \tag{4.5-7}$$

The maximum of Eq. 4.5-6 is reached when the derivative of the maximand with respect to the investment is zero. This leads to the *marginal investment design criterion* in Eq. 4.5-8.

$$\frac{dP}{dI} = i \tag{4.5-8}$$

Equation 4.5-8 states that capital should be invested in the design up to the point where the last increment of capital invested earns profit at a rate equal to the current earning rate of the firm.

In summary, these three design criteria are based on the assumptions that:

The firm is striving for maximum profit rate.
There exists an abundance of opportunities available to the firm which tend to determine the value of money.
Sufficient capital is on hand for each project.
The investments are risk free.

Thus, we see the origin and limitations of the several goals towards which the engineer might strive. Unfortunately, the restrictions listed above are too limiting to correspond to actual industrial situations, and as these restrictions are relaxed the design criteria change to force the accommodation of the process design to the more realistic environment. Next we shall relax the restriction that the investments be risk free.

4.6 ACCOUNTING FOR RISK

It is well recognized that certain industrial activities are economically riskier than others, and that correctives for uncertainty must appear in any decision-making criterion. Suppose that each project is required to purchase an "insurance policy" to guarantee the anticipated profits, and that the premium to be paid is proportional to the amount of capital to be risked and dependent upon the degree of risk associated with the project.

$$\text{“Insurance premium”} = hI_F \ \$/\text{yr} \tag{4.6-1}$$

The working and auxiliary capital I_W and I_A are here assumed risk free for simplicity.

The risk rate, h\$/\$-yr, would depend upon the fraction of the invested capital that can be recovered should the project terminate unexpectedly, the uncertainty of technical and economic forecasts, and general degree of risk associated with the investment.

This premium is then subtracted from the net profit to be earned by the projects, giving the following design criterion, the *venture profit*, first defined by Happel.

$$\max\,[P - (i + h)I_F - i(I_W + I_A)] \tag{4.6-2}$$

where $i_m = i + h$ is the *minimum acceptable rate of return* on invested capital for the particular project.

Let us remark on the use of the risk rate h as a corrective for uncertainty. Insurance policies can be purchased to cover certain physical risks such as fire, explosion and windstorm; however, most economic risks cannot be insured against, and the company is forced to cover its own losses. The concept of an insurance policy then becomes an artifact justifying the common business requirement that certain activities must offer more apparent profit than others before they can be undertaken. The minimum acceptable rate of return on invested capital $i_m = i + h$ then becomes the corrective for uncertainty, and such corrections are made intuitively. Until we delve deeper into the theory of engineering in the presence of uncertainty in Chapters 11 through 15, we shall use h as a practical corrective. Typical values of the minumum acceptable rate of return on invested capital employed by industry in a variety of risk environments are shown in Table 4.6-1.

Table 4.6-1 Typical Values of the Factor For Risk Compensation

Type of Project	$h = i_m - i$ \$/\$-yr
High risk. Projects involving considerable novelty or based on somewhat unproven sales data, products, or raw materials	0.20–1.00 or more
Fair risk. Projects somewhat outside the present field of activity, or novel projects and processes that have not been thoroughly investigated	0.10–0.20
Average risk. Projects in present field of operations but with some novel features or indefinite market information	0.05–0.10
Good risk. Expansion of existing operations in a known market	0.01–0.05
Excellent risk. Cost reduction in operating processes in stable environment	0.00–0.01

The design criteria corresponding to Eqs. 4.5-7 and 4.5-8 in the presence of risk are shown in Eqs. 4.6-3 and 4.6-4.

$$\frac{P - i(I_W + I_A)}{I_F} > i_m \quad (4.6\text{-}3)$$

$$\frac{dP}{dI_F} = i_m \quad (4.6\text{-}4)$$

Equation 4.6-3 is interesting in that attention is focused on the risk by charging the process for the benefits received from the working and auxiliary capital at rates sufficient to finance these completely. Thus the modified rate of return $[P - i(I_W + I_A)]/I_F$ is a more accurate measure of the profits to be earned by the risked capital than is a direct analog of Eq. 4.5-7, namely, the term $P/(I_F + I_W + I_A)$.

4.7 THE DESIGN OF A HEAT-EXCHANGE SYSTEM

In this section we now demonstrate that the principles developed so far can be used to guide the design of a heat-exchange system for a still involved in the recovery of ethanol. We also demonstrate the profit-investment field concept in interpreting the extensive tabular information accumulated during design computations.

The system under consideration is shown in Fig. 4.7-1. The entire system is in the planning stage and no equipment has been installed. Thus, for each arrangement of the exchangers the required heat-exchange capacity and, hence, the capital investment can be calculated. We present here only the results of such calculations in order to illustrate how the principles presented in this chapter can be used to estimate the optimal system. We also assume that this system must integrate into an environment which demands at least 20 per cent per year return on invested capital ($i_m = 0.20$).

The relative profit δP for this system can be computed with respect to a base case, as

$$\delta P_j = \sum_i [(C_s + C_c)_0 - (C_s + C_c)_j + f(I_{i_0} - I_{i_j})](1 - t)$$

where

C_s = steam cost \$/yr
C_c = cooling water cost \$/yr
$f = 0.30$ = sum of factors for maintenance, property taxes, insurance, depreciation etc, \$/yr-\$
I = investment in exchangers \$
t = tax rate, 0.50 \$/\$

with subscripts

j = case number
i = exchanger number
0 = base case

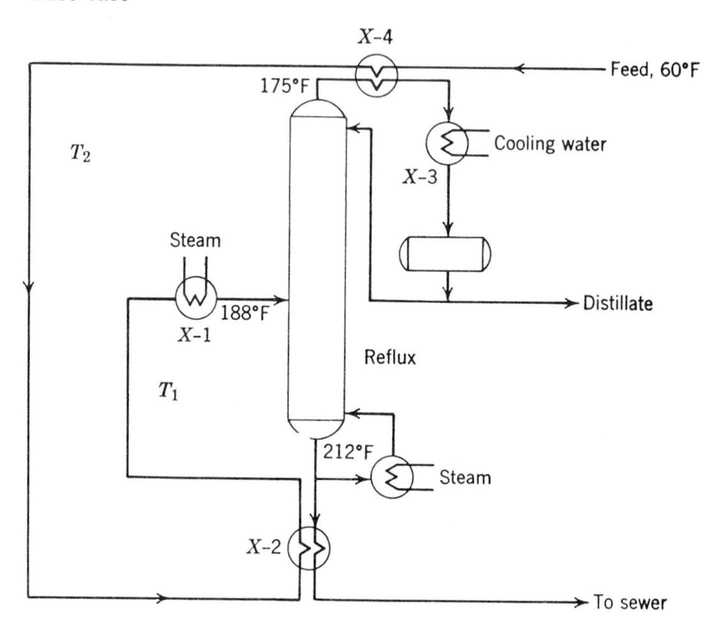

Fig. 4.7-1 The distillation tower and its associated heat-exchanger system. *Note.* The route of the feed from overhead vapor to bottoms to tower was selected from experience. In general, the routing of the flow constitutes a design parameter which takes on as many discrete values as possible routings.

Table 4.7-1 summarizes several design situations obtained by assuming values for the intermediate temperatures T_1 and T_2 and computing the required heat-exchanger investments and operating costs in the form of cooling water and steam. Case A is taken as the "base case," no extra heat exchangers being used in that case to save steam and cooling water.

A plot of the relative profit against the relative investment reveals the "field of design strategies" shown in Fig. 4.7-2. This plot indicates that there exists an *envelope of superior designs*, defined by the designs which exhibit the greatest profit rate for a given investment. Any design which falls below this envelope cannot possibly be part of an optimal processing plan. Focusing attention on the several designs which define the envelope, the best of the designs considered is found by invoking the design criterion.

$$\frac{dP}{dI} = i_m$$

Table 4.7-1 Summary of Design Computations

Case	A Base case	B	C	D	E	F	G	H	I	J
Intermediate Temperature °F										
T_2 feed after exchange with vapor	60	140	140	120	120	120	160	170	60	100
T_1 feed after exchange with bottoms	60	140	170	120	140	170	160	170	140	140
Exchanger investment, $										
I_2	0	0	3300	0	1900	4900	0	0	4700	3000
I_4	0	2150	2150	1600	1600	1600	3000	7000	0	1200
I_1	1350	800	500	950	800	500	650	500	800	800
I_3	3500	2750	2750	3100	3100	3100	2600	2400	3500	3100
Relative investment										
δI	0	850	3850	800	2550	5250	1400	5050	4150	3250
Relative profit										
δP	0	1125	985	825	775	675	1355	860	230	675
Relative venture profit										
$\delta V = \delta P - i_m \delta I$	0	955	215	665	265	−375	1075	−150	−600	25

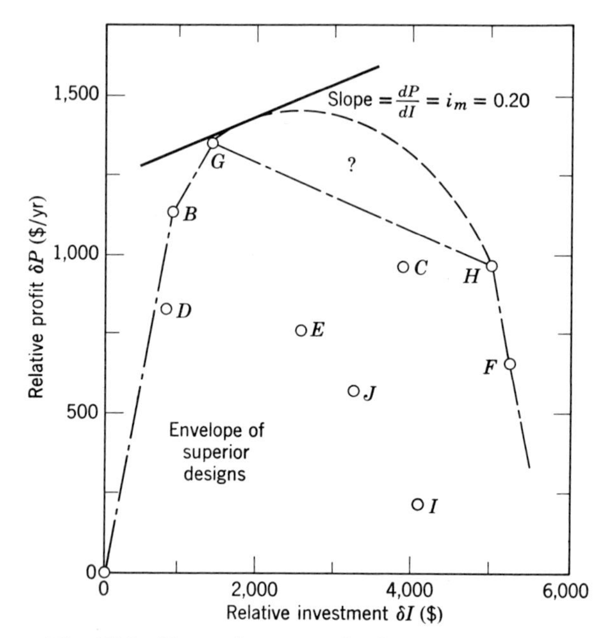

Fig. 4.7-2 Heat exchanger profit—investment field.

For clarity in presentation, only a few of the cases from a continuum of possible cases are presented in the form of Table 4.7-1. It will generally be necessary in this type of analysis to make use of a limited number of discrete cases, as analytical solutions tend to be inaccessible. The coverage of the cases is most clearly shown by giving the intermediate temperatures T_1 and T_2, and the investment in the exchangers, I_1, I_2, I_3, and I_4, as well as the total exchanger investment relative to the base case A, where exchangers *X*-2 and *X*-4 are eliminated. The profit, which is calculated from the heat costs and the appropriate fixed charges on the investment, is also presented relative to the base case A.

It can be seen that cases *C*, *E*, *J*, and *I* are obviously below the optimal design envelope and should not be considered. The maximum venture profit is achieved when $dP/dI = i_m$. For the cases studied, case *G* looks best, but the gap between cases *G* and *H* suggests the desirability of an intermediate case study. We have clearly shown the elimination of the inferior cases and the selection of the best case of those computed and have suggested a change in design which might lead to even more economy.

The plot in Fig. 4.7-3 of relative venture profit versus investment illustrates the alternative presentation of the economic data, invoking the equivalent criterion below

$$\max\,(\delta P - i_m\,\delta I)$$

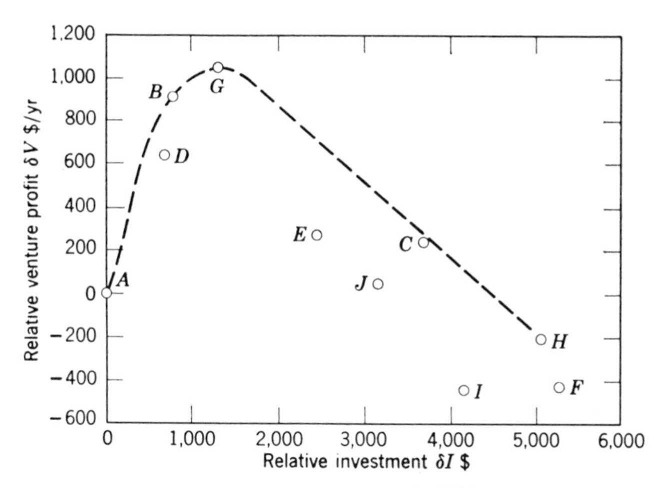

Fig. 4.7-3 Venture profit field.

This again results in the selection of case G as the best of the alternatives considered.

4.8 THE EFFECTS OF LIMITED CAPITAL

Frequently the demand for capital is greater than the supply, and if each project is allowed to strive for maximum venture profit the constraint of limited capital will be violated. As a matter of fact, we might remark that this excess demand for capital ought to be the normal situation in a field in which engineering talent is continually opening new economic possibilities. When this happy situation occurs, the allocation strategy must be modified, cutting back various projects from their points of maximum venture profit and eliminating certain of the elective projects. The criterion of maximum venture profit for each individual project is no longer valid, and its use as a guide for the design of a system will lead to a nonoptimal design.

Again the firm strives to allocate capital for maximum profit

$$\max_{\{I_j\}} \left\{ \sum_j [P_j(I_j) - i_{mj} I_j] + iI_{\max} \right\} \tag{4.8-1}$$

subject to

$$\sum I_j \leq I_{\max} \tag{4.8-2}$$

But, now the demands for capital I_j^* derived from the solution of the previously valid allocation criterion, Eq. 4.8-3,

$$\max_{\{I_j\}} \{[P_j(I_j) - i_{mj} I_j]\} \tag{4.8-3}$$

exceed the supply of capital. Some compromise allocation must be achieved; the projects are interfering with each other economically.

One approach is to assign a scarcity value λ to the value of money within the firm, and the original optimization problem is modified to include this factor.

$$\max_{\{I_j\}} \left\{\sum_j [P_j(I_j) - (i_{mj} + \lambda)I_j] + (i + \lambda)I_{\max}\right\} \tag{4.8-4}$$

We then search for the value of λ which results in the demand for capital within the constraints. That is, the individual projects are allowed to strive for the maximum of a modified venture profit; should that maximum occur for any project at $I_j = 0$, then project j is eliminated.

$$\max_{\{I_j\}} [P_j(I_j) - (i_{mj} + \lambda)I_j] \tag{4.8-5}$$

The demands for capital will then depend on λ

$$I_j^*(\lambda)$$

and we search for the value of λ which satisfies the constraint of limited capital.

$$\sum_j I_j^*(\lambda) \leq I_{\max} \tag{4.8-6}$$

The scarcity value λ is a *Lagrange multiplier*, an important tool in the theory of optimization. This is not the place to introduce that theory, since the interpretation of λ as a scarcity value provides an intuitive justification for the strategy, sufficient for our present needs.[2]

The treatment for the case of mandatory projects is similar to that above, except that there will be a *minimum investment* that must be made in the mandatory project. That is, a mandatory project cannot be allocated zero capital even if any finite allocation would result in a negative venture profit. The minimum investment allowed is that amount of capital required to provide the mandatory service.

The backing off from the investment in the case of limitless capital to that corresponding to the optimal use of limited capital can be illustrated graphically. As the scarcity value λ increases, the demands for capital drop off until the constraint of limited capital is satisfied. However, if for some project P/I becomes less than $i_m + \lambda$, that project should be eliminated, providing more capital for the other projects, as shown in Fig. 4.8-1.

[2] The valid use of the scarcity value concept requires that the problem satisfy the Kuhn–Tucker conditions. See W. H. Kuhn and A. W. Tucker, "Nonlinear Programming," *Proc. of the Second Berkeley Symposium on Mathematical Statistics and Probability*, Univ. of California Press, 1951.

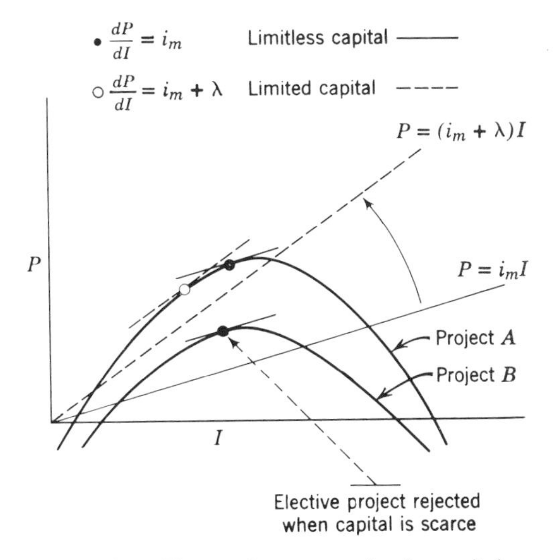

Fig. 4.8-1 Two projects competing for capital.

4.9 THE ALLOCATION OF INSULATION TO A REFINERY

We now illustrate the use of a scarcity value as a vehicle for allocating limited capital to provide insulation for a refinery.

Insulation is to be applied to certain portions of an oil refinery to reduce costly losses of heat. Suppose that:

The heat loss from the refinery can be calculated for any given arrangement of insulation.

The cost of the insulation, installation, and maintenance can be estimated.

The cost of the heat loss can be estimated by equating the value of the heat in the refinery to the cost of the purchase of steam for heating purposes.

The net profit achieved by the investment of I \$ in insulation can be then approximated by P \$/yr = yearly savings in heat − yearly charge for insulation investment. The results of such calculations are presented in Table 4.9-1.

The installation of insulation in this refinery is a relatively secure investment and we might estimate the minimum acceptable rate of return on investments to be 20 per cent.

$$i_m = 0.20$$

How much capital should be spent on insulation and how should it be allocated in the case of (1) adequate available capital, and (2) a limit of \$10,000 on capital?

Table 4.9-1 Net Profits To Be Earned by Insulating Certain Parts of a Refinery

	Net Savings P \$/yr		
Investment I \$	Reactors	Distillation Towers	Piping
2,000	2,200	500	1,000
4,000	2,300	1,500	2,000
6,000	1,000	1,800	3,000
8,000	—	2,000	4,000
10,000	—	1,800	5,000
14,000	—	—	7,000
16,000	—	—	6,000

SOLUTION. In the case of adequate capital we allow each of the projects to strive for maximum venture profit, as summarized in Table 4.9-2.

Thus, the investments of \$2,000, \$4,000, and \$14,000 in insulation for reactors, distillation towers, and piping are optimal, yielding net profits of \$2,200, \$1,500, and \$7,000 per year. A total investment of \$20,000 yields a total net profit of \$10,700 per year.

$$\frac{P}{I} = \frac{10,700}{20,000} = 0.51 > i_m = 0.20$$

The above allocation of capital is not optimal in the case of a limit of \$10,000

Table 4.9-2 Venture Profits

	Venture Profits, $V = P - 0.2I$ \$/yr		
Investment I \$	Reactors	Distillation Towers	Piping
2,000	1,800[a]	100	600
4,000	1,500	700[a]	1,200
6,000	−200	600	1,800
8,000	—	400	2,400
10,000	—	−200	3,000
14,000	—	—	4,200[a]
16,000	—	—	2,800
18,000	—	—	—

[a] Denotes maximum venture profit.

on available capital. It is necessary to assign a scarcity value to money increasing the minimum acceptable rate of return on invested capital to the point where \$10,000 or less is requested. This is sketched out in Table 4.9-3.

Table 4.9-3 Venture Profits in the Presence of Scarce Money

	Venture Profits, $V = P - (0.20 + \lambda)I$					
	Reactors		Distillation Towers		Piping	
Investments	$\lambda = 0.2$	$\lambda = 0.4$	$\lambda = 0.2$	$\lambda = 0.4$	$\lambda = 0.2$	$\lambda = 0.4$
0	0	0	0[a]	0[a]	0	0[a]
2,000	1,200[a]	1,000[a]	−300	−700	200	−200
4,000	700	−100	−100	−600	400	−400
6,000	—	—	−600	—	600	−600
8,000	—	—	—	—	800	—
10,000	—	—	—	—	1,000	—
14,000	—	—	—	—	1,400[a]	—
16,000	—	—	—	—	−400	—

[a] Denotes maximum venture profit.

$\sum I^*$	16,000	2,000
λ	0.2	0.4

Inspecting Table 4.9-3 we observe that a change in the scarcity value of money, from $\lambda = 0.2$ to $\lambda = 0.4$ changes the capital demands from \$16,000 to \$2,000. Since a limit of \$10,000 has been imposed, the proper value of λ has been bracketed. Notice that the only change in the individual requirements occurred in the piping insulation project. The demands for capital by the reactor and distillation insulation projects remain at \$2,000 and \$0, respectively. It, therefore, appears that the allocation of \$8,000 to the piping project would satisfy the constraints of limited capital. Thus, we recommend that \$2,000 be allocated to insulate the reactors and \$8,000 to insulate the piping, for a net profit of \$2,200 and \$4,000 per year, respectively.

$$\frac{P}{I} = \frac{6{,}200}{10{,}000} = 0.62 > i_m = 0.20$$

The constraint of limited capital has forced the consideration of the projects with higher P/I ratio at the expense of others. In the case of adequate capital, a lower quality investment was optimal, $P/I = 0.51$.

4.10 A SUMMARY OF INDUSTRIAL DESIGN CRITERIA

In the previous sections we illustrated how the following design criteria evolve from a *simplified* analysis of a capital allocation problem encountered by the firm.

Venture profit to be maximized

$$V = P - i(I_W + I_A) - i_m I_F \tag{4.10-1}$$

Return on investment

$$\frac{P - i(I_W + I_A)}{I_F} \geq i_m \tag{4.10-2}$$

Marginal return on investment

$$\frac{dP}{dI_F} = i_m \tag{4.10-3}$$

Also we indicated how these criteria might be modified to account for other factors such as limited capital. This analysis has defined the basic principles which lead towards valid and useful design criteria.

However, the engineer will encounter other criteria which differ in detail from those developed here. Rather than illustrate the principles which may lead to these other criteria, we merely describe their form.

The criteria may have arisen from valid attempts to simplify other capital allocation problems or may represent gross approximations to reality and may only be useful in very limited economic applications.

Methods based on total worth. Venture worth is the most general and sophisticated design criterion which is commonly used and is obtained by computing the present value of the venture profits to be accumulated by a project over its life time.

$$W = \int_{\substack{\text{Project}\\ \text{life}}} V e^{-\theta i}\, d\theta$$

Maximization of the venture worth for a project will tend to maximize the future worth of the firm in an uncertain environment.

The present worth is similar in concept without an allowance for income tax or hazard and is commonly useful for selection of design details in which tax and risk are not important factors.

Methods based on return from investment. A number of modifications of the basic criterion, Eq. 4.10-2 are used in industry. These are summarized in Table 4.10-1. They can only be used as indicators of project quality, and not as design criteria to be maximized.

Table 4.10-1 Some of the Economic Criteria Encountered in Industry[a]

	Comments
METHODS BASED ON TOTAL PROFITABILITY ($)	
Present worth	Present value of all future profits minus investments. Maximization of present worth will maximize future worth of company in an environment not plagued by uncertainty.
Venture worth	Present value of venture profits. Same as present worth with correction for uncertainty. Most suitable criterion for design purposes.
METHODS BASED ON RATIO OF PROFIT RATE TO INVESTMENT ($/$-YEAR)	
Return on original investment (also called, engineers method, duPont method, operators method, capitalized earning rate)	Ratio of average yearly profit during earning life to original fixed investment plus working capital.
Return on average investment (also called, return on book investment, accountants method)	Ratio of average yearly profit during earning life to average fixed investment plus working capital.
Interest rate of return (Also called, profitability index, discounted cash flow, internal rate of return, investors method)	Calculates the equivalent interest rate at which the investments are paid off by earnings from the project.
METHODS BASED ON TIME REQUIRED TO RECOVER INVESTMENT (YEARS)	
Payout time (Also called, payoff period, cash recovery period, payback period, payout period)	Ratio of original investment to profit rate plus yearly average depreciation.
Payout time including interest (Also called, cash recovery period)	Ratio of original investment plus interest to profit rate plus yearly average depreciation.

[a] Based on Table 26–27 *Perry's Chemical Engineers Handbook*, 4th ed., McGraw-Hill, New York, 1963, pp. 26–35.

Methods based on payout time. The inverse of a return on investment has the dimension, years, and is frequently used as a measure of the time to recover the fixed investment through the project earnings. The commonest of these is $T = I_F/R$, which is called payout time. Other similar criteria of time go by the same name and are mentioned in Table 4.10-1. These are useful as an indication of overall project quality but cannot be used as detailed design criteria.

4.11 ESTIMATING THE ECONOMIC LIFE OF A PROCESS

In this section the simple economic principles discussed so far are applied to the problem of estimating the economic life of a process when the process is in the planning stage. The reason for discussing this example is to point to the limitations of the general design criteria developed and to reinforce the idea of the necessity for a new economic analysis of each new design situation. A direct and thoughtless application, say, of the maximum venture profit criterion may not be appropriate, although the kinds of thinking which led to that criteria may still be valid.

A process with a fixed capital investment of I_F dollars faces the following future, as well as we now can predict.

The physical life of the equipment, p years, is greater than the economic life of the project, n years.

The earning power of the company is i \$/\$-yr.

The sales income from the process will decline as competition enters according to $S = S_0 \exp(-k\theta)$.

Manufacturing costs will remain constant over the economic life of the process, except for economic inflation which occurs at the rate j for all monies except the tax credits which are imposed by government edict.

The salvage value of the process will be zero.

An investment of I_W dollars in working capital must be committed to the process but is recoverable.

What is the optimal or economic life, n^* years, for this process in the case where the commitment of I_F dollars has not been made and the whole project is in the planning stage? This is an important question, since the economic life of a project enters into the calculations upon which the design criteria are based. It is a very simple example of the more complex economic problems which arise in practice, and which are discussed in Chapter 11.

Equation 4.11-1 is the present worth to the firm which can be attributed to the activities of this project over its economic life of n years.

$$W = \int_0^n (S_0 e^{-k\theta} - C_0)e^{j\theta}(1-t)e^{-i\theta}\,d\theta + \int_0^r \frac{I_F t}{r} e^{-i\theta}\,d\theta$$
$$- \int_0^n \frac{(i-j)I_F e^{-(i-j)\theta}}{e^{(i-j)n} - 1}\,d\theta - \int_0^n [I_F + I_W](i-j)e^{-(i-j)\theta}\,d\theta$$
$$- \int_0^n hI_F e^{-(i-j)\theta}\,d\theta \qquad (4.11\text{-}1)$$

The thinking which led to the formulation of each term will now be discussed. The first integral is the present value of all gross profits derived from the process, after tentatively removing taxes. The second term is the tax credits allowed by the government for depreciation of equipment. Notice that the tax term is allowed over r years, a period of time stated by the government as allowable and not related to the economic life of the process. An alternative formulation might be to include such a tax credit term only during the economic life of the project, and then claim a "loss on disposal" term of $(1 - n/r)I_F te^{-in}$ upon termination of the process. The third term in Eq. 4.11-1 is the accumulation of the depreciation fund for the replacement of the equipment after n years. While the equipment will not be replaced after the economic life of the process is over, it is necessary to replace the investment I_F in the banks of the firm. The fourth term is the earning the investment of $I_F + I_W$ dollars would have made in the other activities of the firm had they not been tied up in the interests of the project. Finally the last term is the present value of the hazard factor payments which must be subtracted from the earnings of the project, empirically to account for the uncertainties inherent in any project in the planning stage. These payments cover the risk involved in the commitment of capital to equipment which has no salvage value.

To find the economic life of the project, we merely compute the economic life which results in a maximum of the venture worth outlined in Eq. 4.11-1. Differentiating the present worth with respect to n and setting the result to zero, yields the following necessary condition for the optimal economic life

$$n^* = \frac{1}{k} \ln\left[\frac{S_0(1-t)}{C_0(1-t) + (i_m - j)I_F + (i - j)I_W}\right] \tag{4.11-2}$$

For example if

$$\begin{aligned}
I_F &= 500{,}000 && \$ \\
I_W &= 100{,}000 && \$ \\
S_0 &= 2{,}000{,}000 && \$/\text{yr} \\
k &= 0.20 && 1/\text{yr} \\
C &= 1{,}000{,}000 && \$/\text{yr} \\
t &= 0.50 && \$/\$ \\
i_m &= 0.35 && \$/\$\text{-yr} \\
i &= 0.15 && \$/\$\text{-yr} \\
j &= 0.05 && \$/\$\text{-yr}
\end{aligned}$$

then

$$n^* = 10 \text{ yr}$$

That is, ten years seems to be the life of the process, as best we now can predict. This bit of information might now be fed back into the economic design criterion to be used in the development of the optimal process to operate over this anticipated time period.

We have outlined in this example, a very simple deviation from the basic profitability study to be conducted on a proposed project. Our purpose in presenting this example is merely to point to the need for a more detailed and exacting study of the economics of a proposed process and to focus attention on the fact that any economic method of analysis is only suggestive and must be altered to fit the special conditions which arise in each new process.

4.12 CONCLUDING REMARKS

There are several points to be fixed in mind concerning the principles which lead to the evolution of economic design criteria. First, the criteria arise as guides for the approximate solution of excessively detailed and impossibly complicated capital allocation problems which the firm encounters in every phase of its activities. The engineer cannot be bothered by the bulk of these details, and therefore the attempt is made to condense the essential features of the vast activities of the firm into a single parameter i, the expected earning rate of capital invested in these activities. This simplification enables the decomposition of the impossibly detailed problem into approximate and more easily solved problems. This leads directly to the elementary design criteria used in driving a proposed process design towards the larger goals of the firm.

Second, the specter of an uncertain future confounds the process designer, and a second parameter is introduced, h the hazard factor. This leads to the definition of a minimum acceptable rate of return i_m to be used as a corrective, modifying the design criteria approximately to account for the risk of an uncertain economic and technical future. While the first parameter i can be estimated from the past earning rate of capital invested in the firm, the estimation of the minimum acceptable rate of return must necessarily include more speculation and hence less accuracy. Thus, this concept is on less secure ground than the concepts which led to the decomposition of the original capital allocation problem, but a correction for risk must necessarily be included in a valid design criterion.

Third, the design criteria must be modified further to account for the constraint of limited investment capital, a constraint which frequently arises in a technically active field of engineering. We saw how the scarcity value λ entered as a parameter to aid in the solution to such problems.

We must remark that only the most elementary of concepts have been developed here and that the engineer will encounter many other design criteria which either reflect a more detailed analysis of special economic situations or erroneous thinking which may have been passed down through the years without proper critical analysis.

The economic design criteria thus enable the engineer to get a foothold in the field of process design by defining computable criteria upon which to judge design alternatives. In specific situations these simple criteria will not be sufficiently detailed to guide the proper design of a system, and it will be necessary to delve deeper into the fundamental phenomena which may be confounding the evolution of the optimum design. This leads to the study of some of the most fascinating topics of process engineering which fall into the broad areas of (1) the analysis of the effects of uncertainty on the design of systems, and (2) the analysis of problems which involve an overwhelming number of alternatives and detail. These two topics comprise the bulk of material in the later chapters of this text.

References

For further study in the principles of economic analysis see:

W. J. Baumol, *Economic Theory and Operations Analysis*, Prentice-Hall, Englewood Cliffs, N.J., 1961.

H. Bierman and S. Smidt, *The Capital Budgeting Decision*, Macmillan, New York, 1966.

C. R. Carr and C. W. Howe, *Quantitative Decision Procedures in Management and Economics*, McGraw-Hill, New York, 1964.

The following articles offer a variety of industrial views on the subject of design criteria.

W. R. Earley, "How Process Companies Evaluate Capital Investments," *Chem. Eng.*, March 30, 1964.

D. P. Herron, "Comparing Investment Evaluation Methods," *Chem. Eng.*, January 30, 1967.

M. Souders, "Engineering Economy," *Chem. Eng. Progr.*, March 1966.

R. M. Adelson, "Criteria for Capital Investment, an Approach Through Decision Theory," *Operational Research Quarterly*, March 1965.

See also:

J. Happel, *Chemical Process Economics*, Wiley, New York, 1958.

M. S. Peters and K. Timmerhaus, *Plant Design and Economics for Chemical Engineers*, McGraw-Hill, New York, 1968.

PROBLEMS

Problems 4.A through 4.L are designed to familiarize the student with the contents of this chapter and involve no new ideas. Frequently, a problem statement will contain insufficient data, and it will be necessary to refer to the text for reasonable values, for instance, of the minimum acceptable rate to demand of a project.

Thus, the selection of reasonable data to use in the solution of a problem becomes part of the familiarization process.

The remaining problems involve extensions beyond the contents of the chapter.

4.A The following cost data are based on a process for the manufacture of 20,000 tons per year of cyclohexane by the hydrogenation of benzene. Would this project be a profitable outlet for the engineering and business talent of a firm?

Sales income	1,900,000 \$/yr
Manufacturing costs	1,700,000 \$/yr
Total investment	450,000 \$
Economic life of project	6 yrs

4.B A new plant is to manufacture 130 million pounds per year of acetylene and 200 million pounds per year of ethylene from light naphtha feedstock. On the basis of the following cost data:

(a) Estimate the cost of the ethylene-acetylene product in cents per pound when no allowance is made for risk or profit.

(b) Estimate the cost of the product when the project must compete for capital within a firm and when an allowance must be made for risk.

Investment data	
I_F \$	27,000,000
I_A \$	4,000,000
I_W \$	500,000
Manufacturing cost data	
C \$/yr	14,000,000
Business environment	

The economic life of the project is estimated as ten years, and the project is to be part of the business activities of a typical chemicals manufacturer

4.C Can you choose between the following exclusive projects? Does the magnitude of the proposed investments suggest further study of the company's investment situation?

Project		*A*	*B*
Total investment	I \$	1,000,000	6,000,000
Gross profit	R \$/yr	2,000,000	5,000,000
Minimum acceptable rate of return on investment	i_m \$/\$-yr	0.20	0.20
Tax rate	t \$/\$	0.50	0.50
Depreciation factors	d,e \$/\$-yr	0.10	0.10

4.D The following data are available on a proposed process for the manufacture of ethylene. What process capacity would you suggest as most economical?

Production rate (10^6 lb/yr)	100	200	600	1000
Total investment (10^6 \$)	7.5	11.6	23.2	35.0
Total manufacturing costs (10^6 \$/yr)	4.9	9.1	26.0	42.5
Credit from the sale of by-products such as heating fuel, butane, and propane (10^6 \$/yr)	3.7	7.4	22.1	36.9

The ethylene can be sold at 3¢/lb and the following cost parameters prevail

$$e = d = 0.10$$
$$t = 0.50$$
$$i_m = 0.20$$

4.E A plain steel pump has been found on the basis of past experience to fail after one year's service on the average. A stainless steel pump will last indefinitely but will cost three times as much as the plain steel pump and will have a net salvage value estimated at 30 per cent of its initial cost. If the economic life of the project in which the pump is to be used is three years and if the firm is earning 10 per cent per year on invested capital, which kind of pump should be bought? If the project life is six years? If the project life is three years and inflation of 5 per cent per year is expected to occur? Discuss how the degree of risk of the project affects these conclusions.

4.F In order to eliminate ice plugging of a distillation tower it has become necessary to install a set of feed gas driers. The drying system required to remove the moisture in the feed will cost \$9,200 and will result in a saving of \$4,800 per year in methanol consumption which is presently being used as a deicer. However, if the size of the dryer is increased significantly, requiring an investment of \$16,500, hydrogen sulfide and carbon dioxide in the feed will be removed in addition to the moisture. If these two components can be removed from the feed, it will be possible to shut down a caustic wash unit which presently removes these components after the distillation. The additional savings expected from the removal of the carbon dioxide and hydrogen sulfide amount to some \$6,000 per year. If the firm demands a minimum acceptable rate of return of 15 per cent per year on such cost saving projects, the firm is earning 10 per cent per year on the bulk of its activities and the economic life of the distillation project is five years, what do you suggest be done about drying the feed?

4.G Additional instrumentation has been suggested to determine more accurately the octane numbers of gasolines being blended in an in-line gas blending system. It is required that any investments return at least 15 per cent per year. The capital expenditures necessary to provide the instrumentation amount to some \$70,000: it will cost \$6,000 per year to operate the system: but a saving of \$30,000 per year is expected in tetraethyl lead consumption during blending due to the more accurate measurement of octane numbers. The depreciation factors d and e are 10 per cent per year and the tax rate is 50 per cent. Should the instrumentation be installed?

4.H When a new product is to be marketed, the decision must be made as to whether terminal storage facilities should be built immediately or whether the construction of the storage facilities should be postponed until the market for the product has grown sufficiently. Determine the optimum year for the construction of storage facilities for the following project by minimizing the terminalling cost over the 15 year life of the product.

The markets starts at zero and grows by 8,000 barrels per year until the maximum demand of 40,000 barrels per year is reached.

If storage facilities have not been constructed they may be rented at a cost of $1,250 plus $0.10 per year per barrel stored.

A storage tank to hold the 40,000 barrels will cost $25,000, will have no value at the end of the 15 year project, and will have negligible operating costs.

The firm earns 16 per cent per year on investments, the cost of the tank can be depreciated for tax purposes over 20 years, by straight line depreciation, and the corporate taxes are 50 per cent.

4.I During the construction of a storage tank a decision must be made regarding whether the tank should have a cone top or a floating roof. The cone top costs $20,000, but a vapor loss of $5,000 per year will occur. The floating roof costs $40,000 and will eliminate the vapor losses. Which roof should be installed if the tank will last for 15 years, the firm is earning 10 per cent per year on investments, and the corporate tax is 50 per cent.

4.J A process, for which the original investment was $\$(1.0)(10^6)$, has run successfully for one year, and the prospect of increased sales now justifies doubling capacity. We estimate an additional life of $n = 4$ years for the project; the income tax rate is 0.5; minimum acceptable return on the new capital invested is 0.15.

We should examine three cases:

(A) Retain original plant (I), $I_{F,\text{I}} = \$(1.0)(10^6)$.

(B) Retain original plant, and build a new one, II, of the same capacity, at an additional investment of $I_{F,\text{II}} = \$(0.8)(10^6)$.

(C) Scrap (I) (no salvage) and replace with a double size new unit, at $I_{F,\text{III}} = \$(1.2)(10^6)$.

Assume straight-line depreciation at 25 per cent per year for both tax and profit calculation.
Sales income for (A): $S_A = \$(1.00)(10^6)/\text{yr}$.
Sales income for (B) and (C): $S_B = S_C = \$(2.00)(10^6)/\text{yr}$.
Operating expense for (A), $\$(0.65)(10^6)/\text{yr}$; (B), $\$(1.07)(10^6)/\text{yr}$; (C), $\$(0.84)(10^6)/\text{yr}$.

Compare venture profits, charging only the *new* investment with depreciation and minimum acceptable return. This is because the $(1.0)(10^6)$ already invested in Unit I is to be regarded as a "sunk cost," with no further need for recovering or earning a profit on this investment.

In a more general comparison of investment strategies, in which some of the cases are in part made up of existing fixed capital investment, we would make use of the procedure illustrated by Fig. 4.7-3 (which is there a graphical, but in general a numerical analysis). The I_{Fi} in each case would be only the *new* investment, and the V_i would be the *increase* in venture profit resulting from the new investment.

4.K The U.S. Bureau of Mines has studied the Bayer process for producing pure Al_2O_3 from hydrated alumina ores. For a plant producing 1,000 tons/day of alumina, the following cost data are estimated:

Capital costs	
Grinding and Digestion	\$ 2,208,700
Clarification	4,614,000
Precipitation and calcination	13,628,100
Spent liquor recovery and lime calcination	2,565,000
Steam plant	3,093,600
Plant facilities, utilities	5,744,000
Construction costs, including contingency, interest on construction loan, etc.	13,862,600
Total fixed investment I_F =	\$ 45,716,000
Working capital	
Raw materials and supplies	706,900
Product inventory	1,398,800
Accounts receivable	1,398,800
Available cash	1,048,400
Total I_W =	\$ 4,552,900

The estimated annual operating costs may be summarized as follows:

Manufacturing costs	
Raw material	\$/yr 8,310,400
Utilities	2,016,600
Labor	929,800
Maintenance	1,150,800
Operating supplies	172,600
Indirect costs	1,040,300
Fixed costs (property, insurance)	879,200
Total C =	\$/yr 14,499,700

Sales price of alumina 0.033 \$/lb

Assuming the life of the project is 20 years, with no sales during the first year but constant sales during the following 19 years, calculate the venture worth. Assume that the process is technologically sound and that the parent company has a diversity of interests and capital earns about 10 per cent before taxes. Let the tax rate = 50 per cent and the depreciation for tax purposes be straight-line over 10 years.

4.L A captive mine is to be developed which potentially may be credited with sales of ore valued at \$500,000/yr. The cost of development is \$2,000,000 and the working capital tied up is \$200,000. Its lifetime is estimated at 20 years. A mine depletion allowance of one half of the sales can be included as a tax credit.

Suppose, on the other hand, that the cost to the parent company of outside purchased ore (identical with respect to total metallics) would be \$600,000/yr and a contract can be signed, guaranteeing delivery for 20 years. If the parent company's invested capital earns 10 per cent per year, is it worthwhile to go ahead with the development or to sign the contract?

4.M A firm has decided to allocate capital to projects in a way so that the firm's growth is maximized. A criterion for maximum growth is

$$\max\left[\sum_j (I_j + S_j')\right]$$

where I_j is the investment to be made immediately in project j, and S_j' is the present value of all the earnings this investment will make. The earnings will be reinvested into the activities of the firm and will be compounded continuously at the anticipated rate of return for the firm i \$/\$-yr. The maximand in the above criterion may then be interpreted as the capital invested in all projects plus the total earnings resulting from these investments.

(a) Show that the amount of money compounded by the year T is

$$S = \frac{P_j}{i}(e^{iT} - 1)$$

where P_j is the net profit rate for project j.

(b) S_j' is the present value of S_j. Show that this equals

$$S' = \frac{P_j}{i}(1 - e^{-iT})$$

(c) Demonstrate that the firm's capital allocation criterion reduces to the following in the case of long-range planning (i.e. T is large)

$$\max_{\text{all projects}}\left[\sum_j\left(I_j + \frac{P_j}{i}\right)\right]$$

(d) The firm's capital allocation criterion is impossible to invoke since it necessarily involves the simultaneous analysis of large numbers of projects j. Tear one project, say project k, from this complex allocation problem and consider all other projects as the other activities of the firm which earn at the rate of i \$/\$-yr. What design criterion for this project forces cooperation with the other activities of the firm?

Answer: For project k

$$\max P$$

$$P \geq 0$$

$$\frac{dP}{dI} = 0$$

4.N In an uncertain field of manufacture a firm wishes to allocate capital so as to pay off its investments in equipment as rapidly as possible. The following capital allocation criterion is thus established.

$$\min_{\text{all projects}} \left(\sum_j T_j \right)$$

where T_j is the years required to pay off the investment I_j through the earnings after taxes from project j.

(a) If the gross earnings in a project after taxes are reinvested into the activities of the firm, derive an expression for the payout time for project j

$$\textit{Answer}: T_j = \frac{1}{i} \ln \left[1 + \frac{iI_j}{R_j(1-t) + dI_j t} \right]$$

(b) What economic design criterion for project K will force that project to accommodate to the firm's capital allocation criterion?

Hint: Show that the payoff time for capital invested in the overall activities of the firm is

$$T_{\text{firm}} = \frac{1}{i} \ln \left[1 + \frac{1}{(1-t)} \right]$$

Answer: Min T_K subject to the condition

$$T_K \leq \frac{1}{i} \ln \left(1 + \frac{1}{1-t} \right)$$

4.O In an excellent article Westbrook and Aris demonstrate how a chemical reactor can be designed using a high speed digital computer.[3] Peruse this article and determine the goal of the firm which a reactor designed by their computer program will accommodate to.

[3] G. T. Westbrook and R. Aris, "Chemical Reactor Design," *Ind. Eng. Chem.*, No. 3, **53** (1961).

5

COST ESTIMATION

In this chapter we bridge the remaining gap between the synthesis of a processing concept and the beginning of the optimization of the process based on that concept. The bridge is made by means of cost equations which relate the design variables to the economic parameters which appear in the design criterion. In this way, proposed changes in the design can be assessed immediately and the optimization of the process can begin.

Typical Problem

A study of the information flow structure of the design equations for a heat exchanger reveals one economic degree of freedom. This economic degree of freedom will be consumed by assigning as the design variable the rate at which cooling water is to be pumped through the exchanger. How are changes in the cooling water rate reflected into changes in the fixed investment cost and the operating cost for the heat exchanger?

5.1 INTRODUCTION

In Chapter 2, *The Synthesis of Plausible Alternatives*, methods were outlined for transforming a primitive problem into a specific process design problem. In Chapter 3, *The Structure of Systems*, it was observed that a rather wide latitude exists in the specification of the technical details of a process, and that this latitude or freedom expresses itself in the existence of design variables over which the engineer has control. In Chapter 4, *Economic Design Criteria*, it was established that the usefulness of a proposed process can be assessed once certain parameters are estimated, such as investment requirements, sales income, and manufacturing costs. What remains is to trace the effects of proposed changes in the design variables to the design criteria, for if this trace can be made with sufficient accuracy the design variables can be adjusted to yield a more useful process. The remaining gap is now closed by the development of cost equations.

Figure 5.1-1 shows the flow of information which results from the change of the design variables. The change is first felt by the design equations, as the conditions under which the process is to be operated change. The change in the design variables and the resulting changes in the state variables then reflect themselves in the cost equations from which the process investment requirements, sales income, manufacturing costs, and so forth are obtained. This information then passes on to the economic design criteria and the usefulness of the proposed change in the design variables is assessed. Based on the results of this assessment, further changes in the design may be in order: this leads to the methods of process optimization to be dealt with in Part II of this text.

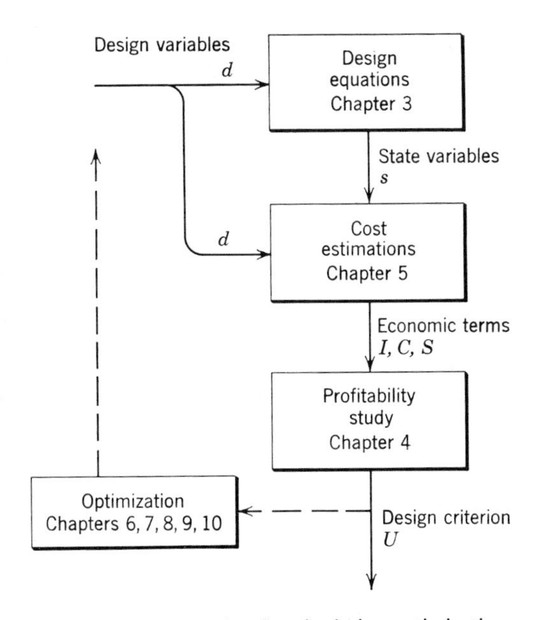

Fig. 5.1-1 Information flow in design optimization.

We distinguish between the *detailed* cost estimates for construction and the development of *approximate* cost estimates for design optimization. The former are carried out by specialists usually in a cost estimating department. This group takes the drawings and specifications of a completely designed system and carries out cost estimations for the purposes of making or judging bids for construction. Such estimates are usually relatively accurate, being based on the detailed design in which the vessels have been sized, the piping laid out, the buildings designed, the instrumentation completely specified, and so forth. In essence the process engineering must be completed before a detailed cost estimate can be made.

To be useful in optimizing a nonexistent system, design cost estimation must not rely on details of the proposed design. Rather, estimation methods must be based only on major processing concepts, gross features of the system, and important design variables. For example, given the location of a proposed system, a rough sketch of the process flow sheet, a list and approximate sizes of the major items of processing equipment, and roughly estimated requirements for steam, water, electricity, and so forth, the competent engineer can, in favorable cases, estimate the investment and operating costs with a probable error of less than 15 per cent.[1] This accuracy may be sufficient to guide the adjustment of major design variables during the preliminary optimizations. In fact, there are many situations in which further resolution of cost data is obtained only with an inordinate expenditure of time, and the designer must be content with this level of accuracy.

There are several points to keep in mind during the study of this chapter.

1. The rapid estimation of the capital investment and operating costs for a process in the early stages of design is a critical area of process engineering.
2. Uncertainty in cost estimation results in uncertainty in the optimization of the process.
3. In order to reduce the labor and time of assembling such estimates, it is often possible to approximate the cost data by correlations and factored estimates; and these may be of sufficient accuracy in the initial stages of design.
4. There is still great need for improvements in the principles of cost estimation tailored to the needs of the various stages of design development.

5.2 ESTIMATING THE INVESTMENT IN MAJOR ITEMS OF EQUIPMENT

A useful approach when building up an estimate of the investment required in a proposed process is first to list the major items of processing equipment needed along with their estimated capacity. This list would include vessels, heat exchangers, pumps, filters, dryers, and so forth, but would not include such items as insulation, electrical work, fencing, piping, and other items which do not depend directly on the mode of processing. Once this list is available, the delivered cost of each item on the list is estimated, and this estimate forms the starting point for the *factored estimate* method of system cost estimation. In this section we discuss methods for obtaining rapid and accurate estimates of the costs of major items of equipment, and then in Section 5.3 we show how this information can be expanded into an estimate of the complete cost of a system.

[1] J. W. Hackney, *Control and Management of Capital Projects*, Wiley, New York, 1965.

The process designer makes extensive use of *correlations* of process equipment investment costs, in which the costs are usually plotted on log-log coordinates, against one or more capacity parameters. For example, the installed cost of stainless steel centrifugal pumps with discharge pressures up to, say, 100 lb/sq in., with a particular type of mechanical seal, tends to correlate with maximum pumping capacity, in gallons per minute. The cost of pumps with other types of seal, other discharge pressures, and made of other materials will plot as similar lines on the same graph. However, a suitable correlation variable for high pressure compressors is shaft horsepower, while for low pressure fans the correlating variable is the maximum volumetric flow rate. It may be necessary to resort to more than one correlating variable to produce a satisfactory correlation.

Figure 5.2-1 shows typical correlations which the process engineer might use to get a quick estimate of the cost of major items of process equipment. Such correlations are compiled from the past experience of an engineering firm and are often regarded as part of the firm's private information. Similar correlations are published frequently in the public literature cited at the end of this chapter and we can accumulate a fairly complete cost estimation notebook from these sources.

Since such cost data tend to correlate against a capacity parameter on log-log plots, it has become common practice to present cost data in the form of a power-law estimating equation.

$$I = I_B\left(\frac{Q}{Q_B}\right)^M \tag{5.2-1}$$

Where I is the estimate required for a piece of equipment of capacity Q, I_B is the known base investment for equipment with capacity Q_B, and M is the slope of the correlating curve on the log-log cost plot.

The range of validity of these and similar equipment cost correlations must be carefully noted. At the lower limit, units of smallest practical size may be approached. If extended to smaller sizes, the correlation line would tend to become horizontal, at a minimum cost. At the upper limit, the maximum commercially available size of unit may be approached, after which further capacity must be obtained with multiple units. The curve then becomes a 45° line on the usual log-log plot.

Bauman[2] reports cost correlations for a large number of items of processing equipment, a portion of that tabulation appearing as Table 5.2-1. Tabulated are the correlating parameter Q, the base size Q_B, the base investment I_B, the range of validity of the estimate and the power factor M. Notice that the factor M varies widely, generally within the range of $0 \leq M \leq 1$.

[2] H. C. Bauman, *Fundamentals of Cost Engineering in the Chemical Industry*, Reinhold, New York, 1964. See also M. D. Winfield and C. E. Dryden, *Chem. Eng.*, December 24, 1962.

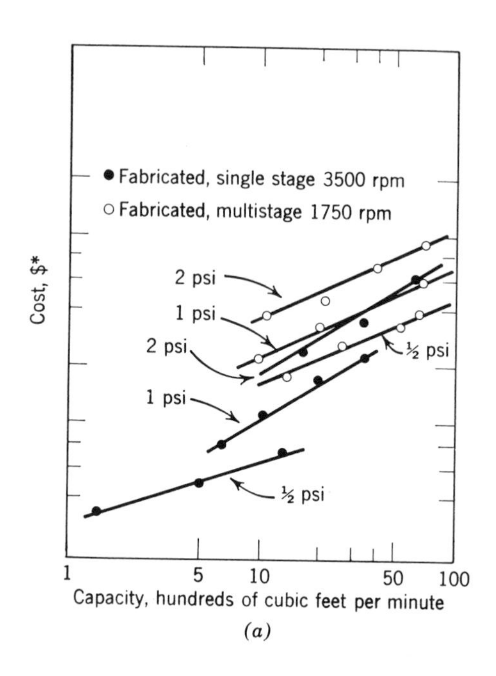

(*a*)

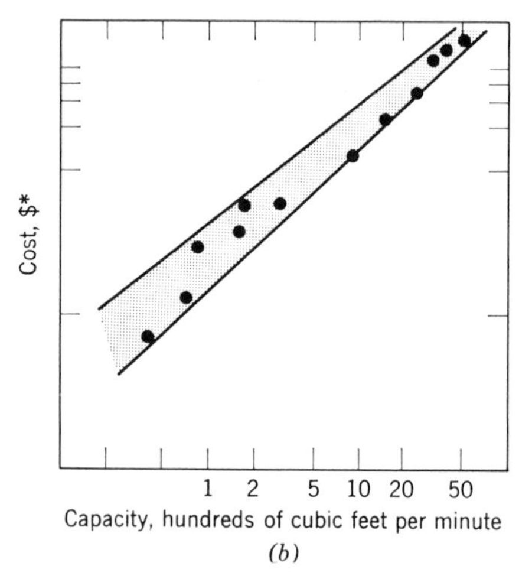

(*b*)

* See Table 5.2-1 for correlations corresponding to this kind of cost information.

Fig. 5.2-1 A page from a cost estimation notebook. (*a*) Radial flow high pressure blowers for three pressures. (*b*) Positive rotary blowers and gas pumps for pressures between 3 and 10 psi.

Example:
Estimate the investment cost of a growth type crystallizer to process 20 tons of material per day.

Table 5.2-1 contains all the basic data needed to assemble such an estimate.

$$I = I_B\left(\frac{Q}{Q_B}\right)^M = \$22{,}100\left(\frac{20}{10}\right)^{0.63} = \$33{,}500\ (1961)$$

While we must be satisfied with approximate costs, for process design studies, we do not want these correlations to contain avoidable errors. Thus, it is necessary to examine the sources of the cost information upon which the correlations are based.

As a rule, prices are F.O.B., that is, without transportation. It is desirable to convert such data to a current basis, including transportation, duties and instalation. These items, for a distillation tower in a chemical process in Taiwan, accumulated to nearly the F.O.B. cost of the component in the United States, where it was manufactured. A costly mistake could have been made had these factors been neglected during the analysis which led to the selection of that component. In any case, it is important to note the basis of the estimate of equipment costs, a fact which frequently is not stated clearly in published cost correlations.

The prices of new equipment are usually found in one of two ways:

The items are of standard design. The prices of small items that might be carried in stock, which might exist as unassembled parts, or of larger standard items, which might exist only as drawings, may be listed in catalogs or in special price lists. However, these prices are commonly subject to a number of discount factors, which are generally not known to the casual observer. These discounts may collectively reach half of the listed price and depend upon the amount of past and future business between the vendor and purchaser, among other things. This again is a source of uncertainty which confounds decision making.

The items are of special design. When the item is of special design and is not available in the market place an inquiry must be made of a manufacturer or supplier, who will furnish a "quotation." This may be a letter or a more formal document, specifying the cost of the exact items and services to be furnished, the time of delivery, guarantees, and all conditions of the proposed sale. The price quoted is usually valid for 30 days and is referred to as a "firm price." The quotation *may* contain an "escalation" clause which permits specific, automatic price increases because of increases in the costs of material or labor occurring prior to shipment.

In summary, while cost correlations are a practical compromise, providing approximate cost estimates quickly, the engineer should be aware of the sources

Table 5.2-1 Typical Equipment Cost Correlations[a]

Item	Size, Q_B	Cost, I_B 1961 $	Size Range	M
Blowers				
1 psi	70 cfm	360	70–1,400	0.46
7 psi	1,400	6,900	1,400–6,000	0.35
Waste heat	1,000 lb steam/hr	2,800	1,000–4,000	0.90
Boiler	4,000	9,800	4,000–20,000	0.67
Centrifuge				
Carbon steel	40 in. basket	28,700	40–66	0.81
Stainless	40	43,000	40–66	0.63
Air Compressor				
Steam turbine drive	240 hp	80,000	240–2,000	0.29
Belt Conveyor	200 ft^2	7,000	200–1,200	0.81
Crystallizer (growth type)	10 tons/day	22,100	10–1,000	0.63
Dryers				
Drum (atm)	60 ft^2	11,600	60–400	0.63
Drum (vacuum)	40	31,000	40–180	0.86
Screw conveyor	16	6,400	16–265	0.53
Evaporators				
Falling film	4 ft^2	9,200	4–9	0.24
	9	11,200	9–33	0.36
	33	17,900	33–66	0.55
	66	26,300	66–145	0.67
Plate and frame filter	10 ft^2	800	10–300	0.85
Heat exchangers				
Shell and tube	50 ft^2	1,350	50–300	0.48
Fin tube	700	5,400	700–3,000	0.58
Plate coil	15	45	15–40	0.52
Reboiler	400	4,070	400–600	0.25
Kettles (jacketed)				
Cast iron	250 gal	5,800	250–800	0.24
Glass lined	800	9,900	800–2,500	0.65
Mixer (propeller type)	15 hp	3,900	15–25	0.19
Unfired pressure vessels				
Carbon steel	3,000 lb	1,060	3,000–6,000	0.60
	30,000	4,830	30,000–100,000	0.80
Rubber lined	800 gal	1,730	800–20,000	0.54
Centrifugal pumps				
Alloy	10 hp	1,300	10–25	0.68
	25	2,480	25–100	0.86
Reactors				
Glass lined	50 gal	1,000	50–300	0.41
and jacketed	300 gal	2,100	300–2,800	0.69

Table 5.2-1 (continued)

Item	Size, Q_B	Cost, I_B 1961 $	Size Range	M
Refrigeration units				
40°F	7 ton	4,200	7–4,000	0.72
20°F	7	5,100	7–4,000	0.72
0°F	7	7,000	7–4,000	0.72
−20°F	7	9,600	7–4,000	0.72
Tanks				
Carbon steel	300 gal	240	300–1,400	0.66
Stainless	150	730	150–500	0.69
Process towers	6,000 lb	2,900	6,000–20,000	0.79
Carbon steel	20,000	7,600	20,000–300,000	0.71
Tower trays				
Sieve	3 ft diam	50	3–7	1.63

[a] A limited selection from H. C. Bauman, *Fundamentals of Cost Engineering*, Reinhold Book Corp., a subsidiary of Chapman-Reinhold, Inc., New York, 1964.

of cost data going into the construction of these correlations. He may have to secure firm quotations to reduce the uncertainty in the cost of major items of special design, despite the time and trouble required to obtain such figures. The economics of improving the accuracy of cost estimates falls in a developing area, which is to be discussed in Chapter 12, *Accounting for Uncertainty in Data*.

In addition to the correlations of the investment cost of major items of processing equipment, correlations have been made of the investment cost of completed processing systems, such as those shown in Table 5.2-2. While these correlations are probably in error by more than ± 30 per cent, they may be useful during the preliminary economic screening of alternatives. Further, these correlations give rise to what has been called the *six-tenths rule*. Notice in Table 5.2-1 how the power factor M for individual items of equipment varies widely, and how in Table 5.2-2 the power factor M seems to vary more or less about $M = 0.60$. This empirical observation is useful in the preliminary design phases of process engineering and has an interesting explanation.

As mentioned earlier, the costs of much individual equipment, such as tanks, reactors, pumps, filters, etc. vary as power functions of capacity, with M ranging from about 0.30 to about 0.90. As plant capacity increases, one after another of these major components reaches maximum commercial size, and thereafter its contribution to the total plant cost will be directly proportional to its capacity. However, the attendant costs for piping, wiring, foundations,

Table 5.2-2 Complete Plants Battery Limits Costs[a]

Plant	Size	Unit	1962 Cost ($1,000)	Size Range
Acetylene (from natural gas)	10	tons/day	2,000	10 –100
Aluminum sulfate (alum)	25	tons/day	375	25 –100
Ammonia (from natural gas)	100	tons/day	3,000	100 –300
Ammonia (from blast furnace gas)	100	tons/day	4,000	100 –300
Ammonium nitrate	50	tons/day	630	50 –300
Ammonium sulfate	100	tons/day	1,100	100 –300
Butadiene	5M	tons/yr	370	5M –400M
Butyl alcohol (*n*-butanol)	4M	tons/yr	600	4M –350M
Carbon dioxide	50	tons/day	450	50 –300
Caustic	4M	tons/yr	3,500	4M –300M
Chlorine (electrolytic)	5M	tons/yr	4,000	5M –300M
Dicyandiamide	20	tons/day	1,500	20 –100
Ethanol (synthetic)	1.3M	tons/yr	1,300	1.3M –100M
Ethylene	11M	tons/yr	2,600	11M –370M
Ethylene oxide (including glycol mfg. facilities)	1.5M	tons/yr	830	1.5M –100M
Formaldehyde (100 per cent)	25	tons/day	900	25 –100
Hydrochloric acid (anhydrous 99.5 per cent)	10	tons/day	90	10 – 20
Hydrochloric acid (water white)	50	tons/day	130	50 –130
Hydrocyanic acid	10	tons/day	1,800	10 – 50
Hydrofluoric acid (anhydrous)	10	tons/day	900	10 – 50
Hydrogen (from natural gas)	1MM	scfd	410	1MM – 10MM
Hydrogen (from natural gas)	10MM	scf/day	1,400	10MM – 50MM
Isopropyl alcohol	3.8MM	gal./yr	2,400	3.8MM –120MM
Liquid alum	20	tons/day	320	20 –100
Maleic anhydride	10	tons/day	4,000	10 – 25
Melamine (from dicyandiamide)	15	tons/day	2,100	15 –150
Methanol	50MM	tons/day	2,000	50MM –300MM
Nitric acid	50	tons/day	820	50 –300
Oxygen	20	tons/day	430	20 – 1M
Phosphoric acid (54 per cent including concentration)	50	tons/day	1,100	50 –300
Polyethylene (high pressure)	2.5M	tons/yr	3,000	2.5M – 90M
Polyethylene (low pressure)	3M	tons/yr	2,200	3M – 70M
Styrene	2M	tons/yr	1,150	2M –100M
Sulfuric acid (contact)	50	tons/day	300	50 – 1M
Titanium dioxide	10	tons/day	3,400	10 –100
Urea	150	tons/day	3,700	100 –300

[a] From H. C. Bauman, *Fundamentals of Cost Engineering in the Chemical Industry*, Reinhold, N York, 1964. In the table, M means (10^3), MM means (10^6); "scfd" means standard cubic feet day.

roads, buildings, etc. will continue to increase less than proportionally to capacity. Thus it is possible for the total plant cost to vary as the 0.6 power of capacity over a wide range, for many processes. Examining cost data in great detail will reveal, however, that the best exponent for variation of plant investment with capacity will depend upon the process being considered, see Table 5.2-2.

The six-tenths rule explains why there are economic advantages to designing very large processing systems, the investment per unit capacity decreases as capacity increases as long as an expansion in capacity can be achieved without complete replication of components. We must remark that this apparent economic advantage of large processing units may be opposed by the large losses which attend process shutdown due to component failure. This topic is discussed in Chapter 13, *Failure Tolerance*.

Example:
What investment is required to obtain a process for the manufacture of 30 tons a day of hydrocyanic acid.

Table 5.2-2 gives the data for the following cost estimating equation

$$I = \$1{,}800{,}000\left(\frac{30}{10}\right)^{0.52} \cong 3{,}100{,}000 \text{ (1962)}$$

with a probable error of more than 30 per cent. This accuracy may be sufficient for preliminary screening of alternatives.

Example:
What increase in investment would be anticipated should the design capacity of a proposed commercial system be doubled.

$$\frac{I_2}{I_1} = \left(\frac{2}{1}\right)^{0.6} = 1.5$$

Statements of costs are based on the value of the dollar at a given year, and a *cost index* is needed to compare costs which are presented on differing bases.

The use of a *cost index* to express the changing purchasing power of the dollar dates back to 1913 when the *Engineering News Record Magazine* began to publish the ENR Cost Index which is a weighted average of the costs of steel, lumber, cement and common labor. Since then a number of other indices have been established, among them the Marshall and Stevens Index, Nelson's Refinery Index, Chemical Engineering Magazine Cost Index, Boeck's Index, and the Bureau of Labor Statistics Index. Of these, the Marshall-Stevens equipment cost indices, Nelson's index and the Chemical Engineering Plant Cost index seem most pertinent to process engineering. These indices give the ratio of the cost of various items at a given date to the cost at some base da te

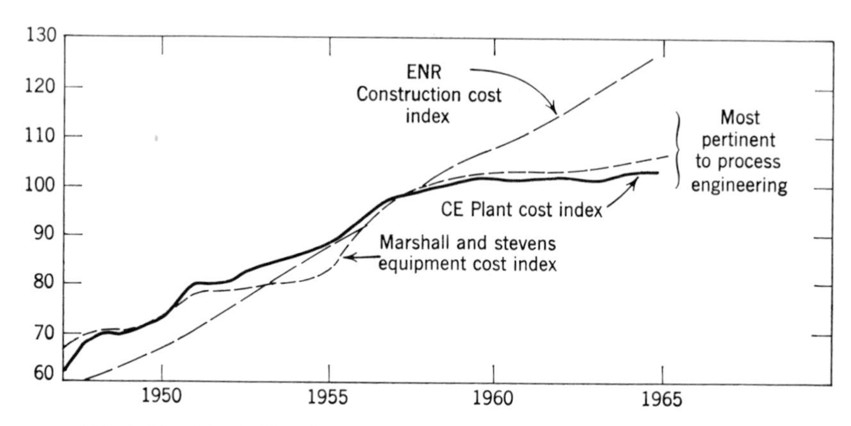

Fig. 5.2-2 Cost indices brought to the common base of 1957–1959 ≡ 100.

thereby allowing extrapolations and interpolations in time. Figure 5.2-2 shows the ENR Index, the Marshall Stevens Index for equipment costs, and the Chemical Engineering Plant Cost index brought back to the common base of 1957–59 ≡ 100 for comparison.

5.3 THE FACTORED ESTIMATE METHOD

The factored estimate method of systems cost estimation is a method by which the investment cost in a completed system can be extrapolated from the delivered cost of the major items of processing equipment. In addition to being useful in giving more accurate estimates of the investment cost of a system than can be obtained from tabulations such as Table 5.2-2, the data which compose the factored cost estimation method are useful in developing the cost equations needed to optimize the details of a proposed process.

The starting point in the factored estimate method is an estimate of the investment cost in the major items of processing equipment, denoted by I_E (see Table 5.2-3).

Table 5.2-3

Major Items of Processing Equipment	Estimated Installed Cost
Vessels	I_v
Pumps	I_p
Heat exchangers	I_h
Towers	I_t
Compressors	I_c
⋮	⋮
	Total I_E

It has been observed that the costs of other essential items needed to complete the process system can be correlated with the investment cost in major items of equipment, and that the total investment can be estimated by the application of experience factors to the base investment I_E. Thus the factored estimate equation 5.3-1 results, in which the experience factors f are obtained from a study of many similar processing systems.

$$I_F = \left[I_E + \left(\sum_i f_i I_E\right)\right] f_I \tag{5.3-1}$$

I_F = fixed investment in complete system
I_E = cost of major processing equipment
f_i = multiplying factors for the estimation of cost of piping, instrumentation, buildings, and so forth
f_I = multiplying factor for the estimation of indirect expenses such as engineering fee, contractors costs and profit, contingency, and so forth.

Table 5.3-1 is taken from an early study of Chilton[3] and is presented as typical of the data which can be accumulated from the analysis of existing processes. The factored estimate method was initiated by Lang[4] in 1947 by the analysis of over a dozen existing processes. This kind of activity has continued, one of the most extensive studies being that of Arnold and Chilton[5] in which 156 projects were analysed during the development of the *Chemical Engineering* plant cost index.

The proportionality factor between the installed cost (or often the delivered cost) of the major items of process equipment and the total fixed investment has come to be called the Lang factor, f_L.

$$\begin{aligned} I_F &= f_L I_E \\ f_L &= \left(1 + \sum f_i\right) f_I \end{aligned} \tag{5.3-2}$$

An inspection of the data on experience factors reveals that the Lang factor may be on the order of $f_L \cong 3$ for an ordinary process. It is interesting that the investment in the major items of equipment may amount to as little as one half, one third or even one quarter of the total fixed investment, depending on the nature of the process.

The experienced engineer can develop an estimate of the total fixed investment required of a new process to within a probable error of 10 to 15 per cent, by carefully selecting the appropriate experience factors from the ranges given. Table 5.3-2 is a similar breakdown of the costs of existing processes, and this tabulation will be useful in the development of cost estimates for optimization, Section 5.6.

[3] C. H. Chilton, *Chem. Eng.*, June 1949.
[4] H. J. Lang, *Chem. Eng.*, October 1947.
[5] T. H. Arnold and C. H. Chilton, *Chem. Eng.*, February 1963.

Table 5.3-1 Factors for Estimating Total Plant Cost

INSTALLED COST OF PROCESS EQUIPMENT	I_E
EXPERIENCE FACTORS AS FRACTION OF I_E	
Process piping	f_1
Solids processing	0.07–0.10
Mixed processing	0.10–0.30
Fluids processing	0.30–0.60
Instrumentation	f_2
Little automatic control	0.02–0.05
Some automatic control	0.05–0.10
Complex, centralized control	0.10–0.15
Manufacturing buildings	f_3
Outdoor construction	0.05–0.20
Indoor-outdoor construction	0.20–0.60
Indoor construction	0.60–1.00
Auxiliary facilities	f_4
Minor additions at existing site	0.00–0.05
Major additions at existing site	0.05–0.25
Complete at new site	0.25–1.00
Outside lines	f_5
Among existing facilities	0.00–0.05
Separate processing units	0.05–0.15
Scattered processing units	0.15–0.25
TOTAL PHYSICAL COST	$I_E\left(1 + \sum_i f_i\right)$
EXPERIENCE FACTORS AS FRACTION OF PHYSICAL COST	
Engineering and construction	f_{I_1}
Straightforward engineering	0.20–0.35
Complex engineering	0.35–0.50
Size factor	f_{I_2}
Large commercial unit	0.00–0.05
Small commercial unit	0.05–0.15
Experimental unit	0.15–0.35
Contingencies	f_{I_3}
Firm process	0.10–0.20
Subject to change	0.20–0.30
Tentative process	0.30–0.50
Indirect costs factor $f_I = 1 + \sum_i f_{I_i}$	
TOTAL PLANT COST $I_F = I_E\left(1 + \sum_i f_i\right) f_I$	

Table 5.3-2 Typical Experience Factors for Fluids Processing Systems[a]

DELIVERED COSTS OF MAJOR EQUIPMENT		I_E
ADDITIONAL DIRECT COSTS AS FRACTION OF I_E		
Labor for installing major equipment		0.10–0.20
Insulation		0.10–0.25
Piping (carbon steel)		0.50–1.00
Foundations		0.03–0.13
Buildings		0.07
Structures		0.05
Fireproofing		0.06–0.10
Electrical		0.07–0.15
Painting and clean-up		0.06–0.10
	$\sum f_i$	1.09–2.05
TOTAL DIRECT COST $(1+\sum f_i)I_E$		
INDIRECT COSTS AS FRACTION OF DIRECT COSTS		
Overhead, contractors costs, and profit		0.30
Engineering fee		0.13
Contingency		0.13
	$f_I = 1 + 0.56 = 1.56$	
TOTAL COST $I_F = (1+\sum f_i)f_I I_E = (3.1\text{–}4.8)I_E$		

[a] Adapted from J. Happel, *Chemical Process Economics*, Wiley, New York, 1958.

It is reasonable that the accuracy of factored estimates may be increased significantly by including more detail in the breakdown of the process into its major components. For example, H. C. Bauman[6] of the American Cyanamid Company has developed special equations to estimate from the cost of a given plant the cost of:

1. A *new* plant of the *same* capacity at a *new* location.
2. A *new* similar plant of a *different* capacity with the *same number* of process units at a *new* location.
3. A *new* similar plant of a *different* capacity with *multiples* of the *original* process units at a *new* location.

Hirsch and Glazier[7] modified the original Lang factor method to include

[6] H. C. Bauman, *Fundamentals of Cost Engineering in the Chemical Industry*, Reinhold, New York, 1964.
[7] J. H. Hirsch and E. M. Glazier, *Chem. Eng.*, No. 12, **56** (1960).

special factors for fractionating columns, pressure vessels, heat exchangers, fired heaters, and so forth to obtain a more accurate cost estimate. Further, R. M. Waddell[8] reported a method called expansion estimating which is an elaboration of the factored estimate method.

In summary, Eq. 5.3-1 is sufficient for the limited purposes of this text for a rapid estimate of the required investment in a process system. However, we must be aware that more elaborate special purpose methods of estimation have been developed and may be necessary in a given process design problem.

Example:
As a simple example of the factored estimate method, we shall present an approximate plant cost analysis for the ammonia synthesis unit shown in Fig. 5.3-1 for a 30 ton per day ammonia plant. Details are summarized in Table 5.3-3.

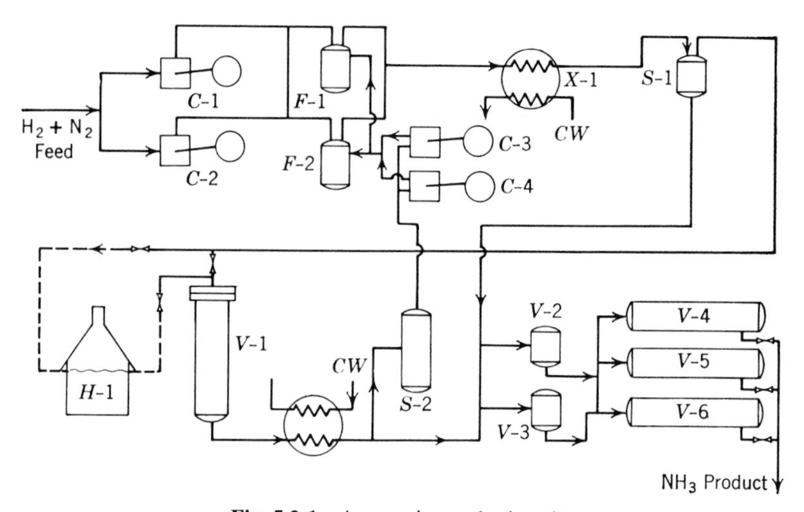

Fig. 5.3-1 Ammonia synthesis unit.

Thus, the fixed investment in the converter section, Fig. 5.3-1, for a 30 ton per day ammonia synthesis plant is estimated at one million dollars (1961).

It is interesting to compare this factored estimate to that obtained for a *total* ammonia plant from Table 5.2-2.

$$I_F = \left(\frac{30}{100}\right)^{0.63} 3{,}000{,}000 = \$1{,}500{,}000$$

This is very reasonable, as it leaves \$500,000 for the synthesis (N_2 and H_2) gas preparation.

[8] R. M. Waddell, *Chem. Eng. Progr.*, **57** (1961).

Table 5.3-3

Symbol	Equipment Item	Quantity	Size	Delivered Cost (1961 \$)
F-1,2	Filters	2	2×10 ft, 5,300 psi	\$ 6,400
S-1,2	Separators	2	2×8 ft, 5,300 psi	5,200
V-1	Reactor	1	30 in. $\times$ 25 ft, 5,300 psi	12,700
V-2,3	Weigh tanks	2	5×15 ft, 300 psi	4,000
V-4,5,6	Storage tanks	3	13,000 gal	16,500
X-1	Primary cooler	1	1,530 ft^2	6,400
X-2	Secondary cooler	1	1,230 ft^2	5,200
H-1	Start-up heater	1	10^6 Btu/hr	4,000
C-1,2	Compressors	2	700 hp	150,000
C-3,4	Circulators	2	100 hp	24,000
P-1,2	Ammonia pumps	2	10 gpm	2,700
C-5,6	Recovery compressors	2	5 hp	2,000
	Instruments			10,000
I_E	Total delivered cost of major equipment items			249,100
	Rounded to			(250,000)
$\sum f_i$	f_1, Installation at 15%			
	f_2, Insulation at 15%			
	f_3, Piping at 75%			
	f_4, Foundations at 10%			
	f_5, Buildings at 7%			
	f_6, Structures at 6%			
	f_7, Fireproofing at 6%			
	f_8, Electrical at 10%			
	f_9, Painting and cleanup 6%			
$\sum f_i$		1.50		
Total direct cost $= (1 + \sum f_i) I_E$				625,000
Indirect costs				
		Contractors overhead and profit 30%		
		Engineering fee at 13%		
		Contingency at 13%		
		$f_I = 1 + 0.56 = 1.56$		
Total fixed capital investment		$= I_F = (1 + \sum f_i) f_I I_E =$		975,000
Rounded to		\$(1961)		(1,000,000)

5.4 INVESTMENT IN AUXILIARY SERVICES

The cost estimating method described above produces an approximate value of the investment I_F within the process area. Often no account is taken of the auxiliary systems and services investment I_A such as in systems for the generation and distribution of utilities (steam, etc.), laboratories, machine shops, facilities for personnel, and so forth. These costs may be substantial and must be taken into account in evaluating any proposed project.

When these auxiliaries have general and continuing use for other processing systems within an industrial complex, account is best taken by charging the project under consideration an internal purchase price based on the amount of the service to be consumed. When the exact consumption cannot be measured, account may be made by the charge of an appropriate "overhead" factor, usually an annual fee proportional to I_F.

Table 5.4-1 Typical Variation in Per Cent of Total Installed Plant Cost of Auxiliary Facilities (Grass Roots and Large Additions)[a]

Auxiliary	Range	Median (Per Cent)
Auxiliary buildings	3.0–9.0	5.0
Steam generation	2.6–6.0	3.0
Refrigeration including distribution	1.0–3.0	2.0
Water supply cooling and pumping	0.4–3.7	1.8
Finished product storage	0.7–2.4	1.8
Electric main substation	0.9–2.6	1.5
Process waste systems	0.4–1.8	1.1
Raw material storage	0.3–3.2	1.1
Steam distribution	0.2–2.0	1.0
Electric distribution	0.4–2.1	1.0
Air compression and distribution	0.2–3.0	1.0
Water distribution	0.1–2.0	0.9
Fire protection system	0.3–1.0	0.7
Water treatment	0.2–1.1	0.6
Railroads	0.3–0.9	0.6
Roads and walks	0.2–1.2	0.6
Gas supply and distribution	0.2–0.4	0.3
Sanitary waste disposal	0.1–0.4	0.3
Communications	0.1–0.3	0.2
Yard and fence lighting	0.1–0.3	0.2

[a] From H. C. Bauman, *Chem. Eng. Progr.*, No. 1, **51** (1955).

At the other extreme is the case of the "grass roots" plant, isolated from any industrial complex, which must provide complete auxiliary services for its sole use. In this case, the "process area" should be expanded to include the complete auxiliary service systems, and their investment should be included as part of I_F.

For example, an ammonia plant might be under design to be included within a large chemical manufacturing complex where it can purchase hydrogen, nitrogen, electric power, steam, and cooling water at suitable inplant prices. In this case, that part of the fixed investment which is tied to the risk of the project includes only the investment in the ammonia plant. However, suppose this same plant is to be located in an isolated area where such auxiliary services are not available. Now the investment which shares the risk includes the investment in the ammonia plant plus the substantial investment in a hydrogen plant, a nitrogen plant, steam boilers, electric power generating station, a cooling water system, and perhaps even housing for the workers and their families.

Table 5.4-1 gives a rough idea of the fraction of the total cost of a grass roots plant or a large addition which is accounted for by the several auxiliaries cited.

5.5 THE ESTIMATION OF MANUFACTURING COSTS

The methods presented in the preceding sections enable the engineer to estimate the investment required to obtain an operating process. The engineer must also have methods for estimating the cost of the day-to-day operation of the proposed process. These are the manufacturing costs C, which in Chapter 4

Table 5.5-1 Indirect Costs Which Appear as Manufacturing Cost

Item	Cost as Fraction of Investment per Year
Maintenance and repair	
Simple processes	0.02–0.06
Highly corrosive, high pressure, or high temperature process	0.07–0.11
Property taxes	0.01–0.04
Insurance	0.01
General services	0.02–0.10
Reasonable total indirect cost factor	$a = 0.10$–0.20

were divided into terms proportional to the total investment in the system I, the anticipated production rate for the system Q, and the labor requirements L.

$$C = aI + bQ + cL$$

Attention is now focused on methods for estimating these factors.

Costs proportional to investment. A process is burdened with a number of annual costs which tend to correlate with the total investment in the process. These are listed in Table 5.5-1 with reasonable factors to account for these services.

Thus for the purposes of this text we shall charge processes under design an assessment of from 10 to 20 per cent per year of the total fixed investment to finance maintenance, insurance, property taxes, fire protection, police protection, general administrative services, and other indirect services, which do not depend strongly on the rate at which the process produces products.

In the literature cited at the end of this chapter, more detailed information will be found.

Table 5.5-2 Approximate Consumption of Electricity, Steam and Process Water during the Manufacture of Several Common Products[a]

Process	Kwhr/Lb Product	Lb Steam/Lb Product	Gal Water/Lb Product
Acetic acid from acetylene	0.210	3.1	43.0
Aluminum sulfate	0.015	3.4	0.5
Ammonium sulfate	0.014	0.2	2.5
Boric acid	0.025	2.2	0.1
Ascorbic acid	4.100	240.0	870.0
Butadiene from alcohol	0.055	22.0	190.0
Dry ice	0.156	8.5	9.0
Gasoline, catalytic	0.001	0.1	1.2
Hydrofluoric acid	0.140	0.5	43.0
Kraft pulp	0.170	4.2	25.0
Nitroglycerine	0.043	1.8	0.7
Oxygen, tonnage	0.220	2.2	13.0
Phosphoric acid, Dorr process	0.052	0.7	7.3
Rayon	2.600	70.0	170.0
Silica gel	0.030	14.0	10.0
Sodium	4.900	5.1	37.0
Sulfur dioxide, liquid	0.002	3.3	9.0
Sulfuric acid, contact process	0.015	0.1	2.2

[a] R. S. Aries and R. D. Newton, *Chemical Engineering Cost Estimation*, McGraw-Hill, New York, 1958.

Cost proportional to production rate. The major portion of the manufacturing costs is approximately proportional to the production rate of the system. The consumption of raw materials, electric power, steam, water, catalysts (and frequently the cost of license fees) is directly proportional to the process operation rate. Usually the annual requirement for these items can be estimated from material and energy balances over the process in question. The rates of consumption are then multiplied by the unit costs for the commodity consumed to estimate this total annual cost.

Table 5.5-3 Utility Rates[a]

Type of Utility	Rate
Steam	
400 psi	\$0.50–0.90/1,000 lb
100 psi	0.25–0.70/1,000 lb
Exhaust	0.15–0.30/1,000 lb
Electric power	
Purchased	0.01–0.02/kwhr
Self-generated	0.005–0.01/kwhr
Cooling water	
Well	0.02–0.10/1,000 gal
River or salt	0.01–0.04/1,000 gal
Tower	0.01–0.05/1,000 gal
Process water	
City	0.07–0.25/1,000 gal
Well	0.02–0.10/1,000 gal
Filtered and softened	0.10–0.20/1,000 gal
Distilled	0.60–1.00/1,000 gal
Compressed air	
Process air	0.015–0.03/1,000 cu ft
Filtered and dried for instruments	0.04–0.10/1,000 cu ft
Coal	6.00–10.00/ton
Fuel oil, No. 6	0.04–0.08/gal
Gas	
Natural	0.30–0.80/1,000 cu ft
Manufactured	0.60–1.30/1,000 cu ft
Refrigeration	
Steam-jet, 50°F	0.55/ton day
Ammonia, 34°F	0.50/ton day
Ammonia, 0°F	0.90/ton day
Ammonia, −17°F	1.20/ton day

[a] R. S. Aries and R. N. Newton, *Chemical Engineering Cost Estimation*, McGraw-Hill, New York, 1958.

Tabulations have appeared in the literature of steam, water, and electric power consumption by a number of common processes, and these tabulations are useful in obtaining a rough estimate of the demands of a system for these utilities (see Table 5.5-2). Table 5.5-3 contains approximate unit prices for these services.

The prices of chemicals for raw materials, catalysts, and other purposes in a process are reported periodically in *Chemical and Engineering News*, the *Oil Paint and Drug Reporter*, and in *European Chemical News*. Actually, the prices of chemicals used in large amounts are usually negotiated and the prices will differ from the reported prices, depending on the volume purchased, the location, the extent of past and future business between the supplier and purchaser, and the general economic climate.

The cost of transporting chemicals from the point of supply to the point of manufacture must be included in the manufacturing costs. Table 5.5-4 gives a rough indication of these costs.

Table 5.5-4 Average Shipping Distances per Day and Rates for Bulk Chemicals

Shipping Method	Average Distance (Miles/Day)	Long Haul[a] Rate Range (Dollars/Ton-Mile)	
		Large Volume	Small Volume
Pipeline	65	0.0020	0.0025
Barge	100	0.0025	0.0060
Rail	200	0.0100	0.0220
Tanker	300	0.0025	0.0060
Truck	700	0.0300	0.0450

[a] Over 250 miles. Short haul rates range from two to five times as high.

Costs proportional to labor. The chemical industry is becoming more and more automated and the requirements for operating labor are decreasing. In large scale liquid or gaseous processes the costs associated with this labor normally only amount to from 5 to 10 per cent of the total manufacturing costs, whereas in fields of manufacture which involve much handling, the labor cost may amount to 25 per cent of the manufacturing cost.

Example:
Estimate the cost in dollars per pound of manufacturing 30 tons per day of oxygen for use within a chemical manufacturing system by the liquefaction of air.

The oxygen plant is to service other facilities within a manufacturing center and must neither make a profit nor run at a loss. The cost of oxygen must be computed on the basis of a neutral contribution to the economics of the firm, i.e., on the basis of zero venture profit.

$$V = (S - C) - (S - C - dI)t - eI - i_m I = 0$$

The production rate required is

$$Q = (30 \text{ tons/day})(2{,}000 \text{ lb/ton})(365 \text{ days/yr}) = 2.2 \times 10^7 \text{ lb/yr}$$

The sales income is the production rate times the cost of oxygen.

$$S = pQ$$

Thus, the cost of the oxygen is computed as

$$p = \left[C + \frac{(e + i_m - dt)I}{(1 - t)}\right] \Big/ Q$$

Reasonable values for the several parameters are given below.

Capital recovery factor	$e = 0.10$ \$/\$-yr
Tax depreciation factor	$d = 0.10$ \$/\$-yr
Minimum acceptable rate of return	$i_m = 0.20$ \$/\$-yr
Tax rate	$t = 0.50$ \$/\$

In this economic situation

$$p = \frac{C + 0.5I}{Q}$$

The problem has been reduced to that of estimating the investment required in the oxygen plant I and the manufacturing costs C.

Investment. Table 5.2-2 gives data for the following order of magnitude estimate of the investment in a complete oxygen plant.

$$I = \$430{,}000\left(\frac{30}{20}\right)^{0.56} = \$540{,}000 \text{ (1962)}$$

Manufacturing costs. Table 5.5-1 indicates that the manufacturing costs which are proportional to investment tend to run about 15 per cent per year of the investment.

$$aI = (0.15 \text{ \$/\$-yr})(\$540{,}000) = \$81{,}000\text{/yr}$$

The terms proportional to production rate involve the consumption of electricity, steam, and water, factors which can be estimated from Tables 5.5-2 and 5.5-3.

$$\begin{aligned}
&\text{Electricity: } (0.220\ \text{kwhr/lb})(\$0.01/\text{kwhr}) &&= \$0.0022/\text{lb}\\
&\text{Steam: } (2.2\ \text{lb/lb})(\$0.50/1{,}000\ \text{lb}) &&= \$0.0011/\text{lb}\\
&\text{Water: } (13.0\ \text{gal/lb})(\$0.02/1{,}000\ \text{gal}) &&= \$0.00034/\text{lb}\\
& && \overline{\phantom{\$0.00034/\text{lb}}}\\
& b &&= \$0.004/\text{lb}
\end{aligned}$$

Thus

$$bQ = (\$0.003/\text{lb})(2.2 \times 10^7\ \text{lb/yr}) = \$88{,}000/\text{yr}$$

The manufacturing costs proportional to labor might amount to 10 per cent of the total manufacturing costs. Thus

$$cL = \frac{0.1}{0.9}(aI + bQ) = \$16{,}000/\text{yr}$$

Therefore the total manufacturing costs are estimated to be

$$C = aI + bQ + cL = \$185{,}000/\text{yr}$$

The price of oxygen is estimated at

$$p = \frac{\$185{,}000 + 0.5(540{,}000)}{2.2 \times 10^7} = \$0.02/\text{lb}$$

It should be noted that what is referred to in the literature and elsewhere as the "cost" of a product quite frequently includes only the "manufacturing cost," as we have defined it, and may not include an adequate assessment of the costs of risking capital.

5.6 COST EQUATIONS FOR OPTIMIZATION

In this section we shall develop in detail the cost equations for the design of the shell and tube heat exchanger shown in Fig. 5.6-1, which is to be used as a cooler to reduce the temperature of 1,000 pounds per hour of a fluid from 200°F to 100°F using cooling water available at 60°F. A direct tie will be made between the single economic design variable, the coolant flow rate, W pounds per hour, and the economic design criterion, venture profit, V dollars per year.

We shall see how the design equations and cost equations interact. This kind of analysis must be performed for each part of a processing system before process optimization can begin.

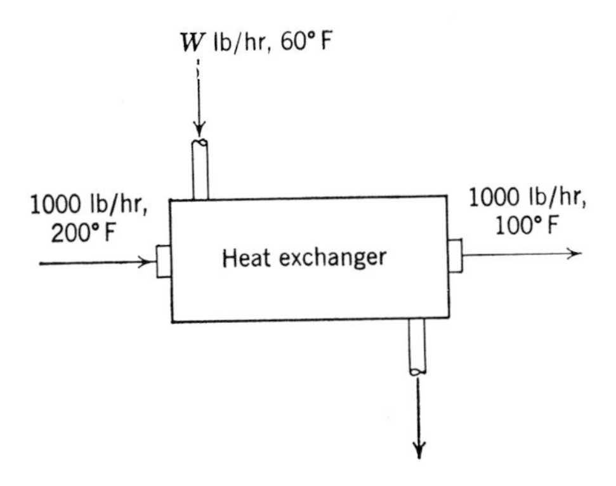

Fig. 5.6-1 The cooler.

Cost equations. The investment in a shell and tube heat exchanger has been shown to correlate with the heat-transfer surface area A ft^2, as shown in Table 5.2-1.

$$I_E = \$1{,}350\left(\frac{A}{50}\right)^{0.48} \tag{5.6-1}$$

for $50 \leq A \leq 300$ ft^2.

However, this is only part of the investment required to achieve an operating exchanger. Table 5.3-2 can be used to estimate the following experience factors:

Cost of installation	$f_1 = 0.15$
Insulation (none)	$f_2 = 0$
Piping	$f_3 = 0.50$
Foundations	$f_4 = 0.08$
Structures (none)	$f_5 = 0$
Fire proofing (none)	$f_6 = 0$
Electrical (none)	$f_7 = 0$
Painting and clean up	$f_8 = 0.08$
	$\sum f_i = 0.81$

Total direct cost factor $(1 + \sum f_i) = 1.81$

Indirect costs

Overhead, engineering and contractors profit	$f_1 = 0.20$
Engineering fee (standard design to be used)	$f_2 = 0.05$
Contingency (no unusual problems expected in standard item of equipment)	$f_3 = 0.05$
	$f_I = 1 + 0.30 = 1.30$

Total cost

$$I_F = (1 + \sum f_i) f_I I_E = 2.34 I_E$$

Thus a more complete estimate of the investment in the heat exchanger is given by Eq. 5.6-2.

$$I_F = \$3{,}200 \left(\frac{A}{50}\right)^{0.48} \tag{5.6-2}$$

It is interesting to notice at this point the multiplying factors that Hand[9] had developed to obtain total installed costs of equipment from delivered costs.

Item	Multiplying factor f_L
Fractionating columns	4
Pressure vessels	4
Heat exchangers	3.5
Fixed heaters	2
Pumps	4
Compressors	2.5
Instruments	4
Miscellaneous equipment	2.5

These general factors were arrived at by the same accumulation of factors which led to $f_L = 2.34$ for the exchanger in Fig. 5.6-1. We see now that the general factor suggested by Hand for heat exchangers, $f_L = 3.5$ is not sufficiently accurate for this particular problem in which insulation, structures, fire proofing, and electrical work are not required.

The operating cost of the heat exchanger will include pumping costs, both for the hot fluid and for the coolant, and the cost of supplying the cooling water. In addition there will be maintenance and repair costs.

[9] W. E. Hand, "From Flow Sheet to Cost Estimate," *Chem. Eng. Progr.*, **37**, 331 (1958).

Local cost of cooling water including pumping	5×10^{-6}	\$/lb
Maintenance and repair for exchanger at 2 per cent of investment per year	$0.02I_F$	\$/yr

Thus the total operating costs might be approximated by

$$C = (5 \times 10^{-6}\$/\text{lb})(W\ \text{lb/hr})(8{,}640\ \text{hr/yr}) + 0.02I_F$$

$$C = 4.32 \times 10^{-3}\ W + 0.02I_F \qquad \$/\text{yr}$$

Design criterion. The design criterion for the exchanger might involve the maximization of the venture profit.

$$\max_{\{W\}} [R - (R - dI_F)t - eI_F - i_m I_F] \tag{5.6-3}$$

Where $R = S - C$ and where S is constant since the outlet temperature and flow rate for the exchanger are specified.

This then reduces to a minimization of costs

$$\min_{\{W\}} [C + (-C - dI_F)t + eI_F + i_m I_F]$$

Now if

$$\begin{aligned} e &= 0.05 && \$/\$\text{-yr} \\ d &= 0.10 && \$/\$\text{-yr} \\ i_m &= 0.10 && \$/\$\text{-yr} \\ t &= 0.50 && \$/\$ \end{aligned}$$

The design criterion becomes

$$\min_{\{W\}} (0.5C + 0.10I_F) \tag{5.6-4}$$

Design equations. Having tied the design criterion to the major process variables by means of the cost equations, there remains the problem of relating the area of the heat exchanger to the single design variable, the coolant flow rate W. This is done by means of the precedence-ordered design equations Fig. 3.6-6, in which $C_p = 1$ Btu/lb-°F.

Figure 5.6-2 illustrates the flow of information which traces out the structure of the cooler design problem. A similar web of information flow will result from the detailed study of the design equations, the cost estimates, and the design criterion for each of the several components which compose an industrial process system. We can visualize the complete optimization problem for a

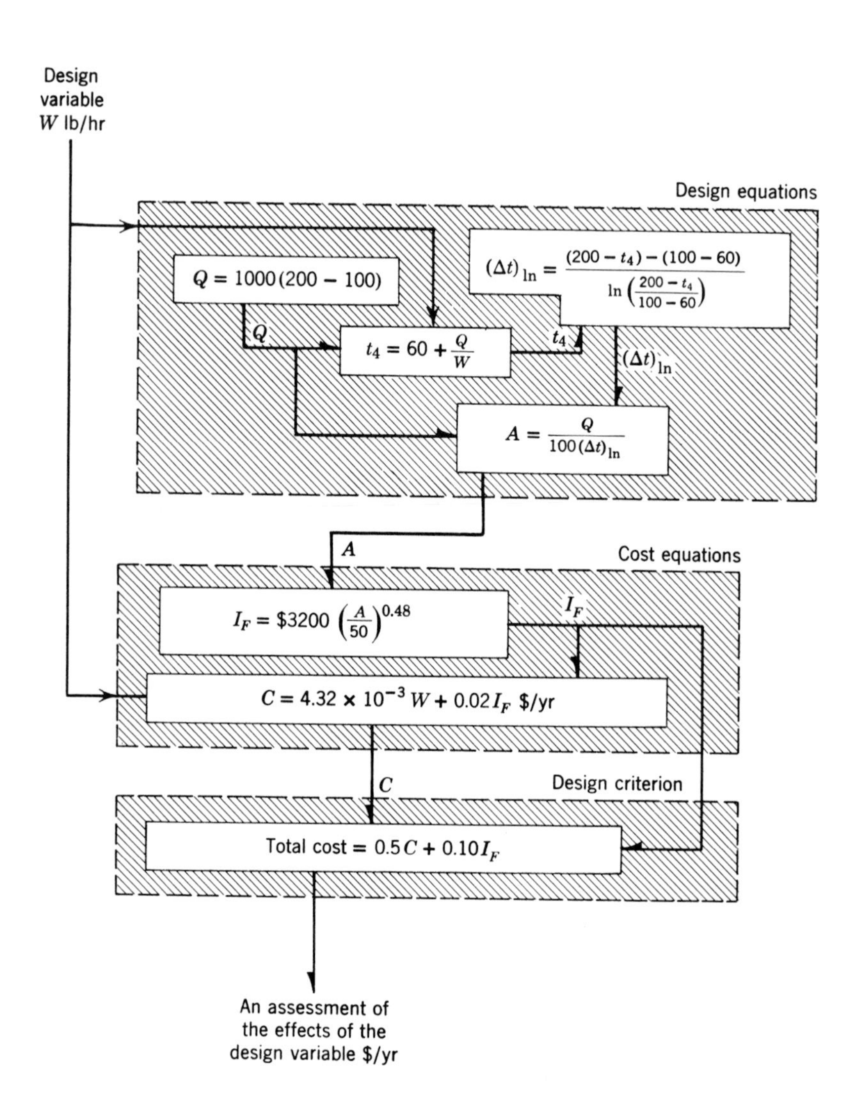

Fig. 5.6-2 The flow of information through the design.

large process, such as an oil refinery which involves hundreds of components, as a complex spider web which must be carefully teased apart so that the optimum design can be estimated without getting entangled in unnecessarily difficult problems. Part II of this text contains a series of chapters on optimization; the strategy of getting the most information out of a complex problem with the least amount of effort. Using this strategy to untangle the web of information flow can result in orders of magnitude reductions in the difficulty of process optimization.

5.7 CONCLUDING REMARKS

We have now concluded an elementary discussion of the kinds of thinking which transform a primitive problem into a specific and well defined engineering problem. The development of cost estimates is the last link in the definition of specific problems.

There are several observations that must be made at this point. The first is that cost data are alive and changing with time and location, in contrast to physical property data which are dead and constant. The general cost indices cited are not responsive enough completely to guide alterations in equipment costs; since, for example, competition in a given area of design may force the cost of a special item down while the general and average costs are experiencing an upward trend.

Further, there are other considerations which were not even mentioned in the elementary introduction to estimation, such as the use of *location factors* to account for changes in cost with geographic location.

The craft of cost estimation has become so specialized that large design firms will discover the need for specialists called *cost engineers* to work with the process engineers to provide a realistic economic base for design. We shall see some extensions of these simple ideas in Chapter 11, *Accommodating to Future Developments* and Chapter 12, *Accounting for Uncertainty in Data.*

References

Several books are available dealing with the problem of cost estimation.

H. Carl Bauman, *Fundamentals of Cost Engineering in the Chemical Industry*, Reinhold, New York, 1964.

Cecil H. Chilton, ed., *Cost Engineering in the Process Industries*, McGraw-Hill, New York, 1960.

O. T. Zimmerman and I. Lavine, *Cost Engineering*, Industrial Research Services, Dover, NH 1950.

R. Aries and R. Newton, *Chemical Engineering Cost Estimation*, McGraw-Hill, New York, 1955.

J. W. Hackney, *Control and Management of Capital Projects*, Wiley, New York, 1965.

C. E. Dryden and R. H. Furlow, *Chemical Engineering Costs*, Ohio State University, 1966.

Frequent plant and equipment cost data appear in journals such as Chemical Engineering, Hydrocarbon Processing, Cost Engineering Quarterly, and various trade journals and commercial publications. A general source is:

S. Katell, H. H. Faber, and J. W. Douglas, "Bibliography of Investment and Operating Costs for Chemical and Petroleum Plants," annual joint publication of Bureau of Mines and American Association of Cost Engineers. See also similar listings such as U.S. Bureau of Mines Circulars, 1949 to date.

To get some indication of the kinds of work published in the general area of cost estimation we suggest that the following be perused.

C. A. Miller, "New Cost Factors Give Quick Accurate Estimates," *Chem. Eng.*, Sept 13, 1965.

H. E. Mills, "Costs of Process Equipment," *Chem. Engr.*, March 16, 1964.

A. J. Weinberger, "How to Estimate Required Investment," *Chem. Eng.*, November 25, 1963.

———————, "Calculating Manufacturing Costs," *Chem. Eng.*, December 23, 1963.

F. C. Zevnik and R. L. Buchanan, "Generalized Correlation of Process Investment," *Chem. Eng. Progr.*, No. 2, **59** (1963)

Westbrook and Aris present an excellent example of the development of cost equations for the design of a reactor.

G. T. Westbrook and R. Aris, "Chemical Reactor Design," *Ind. Eng. Chem.*, No. 3, **53**, 182–183 (1961).

PROBLEMS

5.A. Estimate the costs at the present time of the following items of process equipment.

(a) A 500 hp air compressor.
(b) Screw conveyor dryer with 200 ft^2 of heat-transfer surface.
(c) A reboiler with 400 ft^2 of heat-transfer surface.
(d) A 1,000 ton refrigeration unit to operate at 0°F.
(e) A 100 gal glass-lined reactor.

5.B. In Table 5.B-1 Garrett and Rosenbaum report on the cost of crystallizers.[10] Develop a cost equation for quickly obtaining an approximate estimate of the cost of crystallizers during the preliminary stages of process design.

[10] D. E. Garrett and G. P. Rosenbaum, *Chem. Eng.*, August 1958.

Table 5.B-1 Costs of Forced-Circulation and Growth Crystallizers

General Type and Description	Capacity Tons/Day	Crystal Size, Mesh	Complete, FOB Price, $ (1958)	Instru-mentation, $ (1958)
Growth-type, Oslo				
Two-stage evaporator-crystallizer, mild steel	15	60% + 50 90% + 50	22,000	4,400
Single-stage evaporator-crystallizer, mild steel construction	10	Large	22,500	10,000
Two-stage cooler crystallizer, mild steel construction	85	50% + 40 95% + 100	100,000	—
Three-stage vacuum crystallizer, mild steel	235	95% + 50	100,000	—
Three-stage vacuum crystallizer, 316 stainless	235	95% + 50	210,000	—
Three-stage. Vacuum crystallizer first stage; ammonia cooling second and third stages; 304 stainless steel construction	520	90% + 40	343,000	—
Growth-type Pachuca				
Three-stage vacuum crystallizer, rubber-lined	390	98% + 80	143,000	—
Three-stage vacuum crystallizer, 316 stainless	390	98% + 80	200,000	—
Three-stage vacuum crystallizer, rubber-lined	390	98% + 30	159,000	—
Three-stage vacuum crystallizer, 316 stainless	390	98% + 30	205,000	—
Three-stage vacuum crystallizer, 316 stainless	130	98% + 80	68,000	—
Conventional forced-circulation				
Single-stage vacuum crystallizer, mild steel	30	—	31,000	—
Two-stage cooler-crystallizer, cast iron	80	95% + 100	50,000	1,500
Single-stage vacuum crystallizer, rubber-lined	120	90% + 100	43,000	—
Three-stage vacuum crystallizer, mild steel	550	—	154,000	—
Triple-effect evaporator-crystallizer, mild steel, stainless heaters	5,000	—	1,900,000	—
Quadruple-effect evaporator-crystallizer, steel with stainless heaters	5,000	—	2,700,000	—

What per cent accuracy would you assign to your cost estimating equation? How does it compare to the data in Table 5.2-2? How does the material of construction influence the cost? The capacity? The mesh size?

5.C. It has been observed that the cost of a certain type of process vessel is directly proportional to the weight of the vessel. If the vessel is thin walled and spherical, what power factor M will correlate the cost of the vessel to its volume? How does this compare with the observed power factor of Table 5.2-1?

5.D. The actual data used by Lang in 1947 to develop experience factors for process cost estimation are shown in Table 5.D-1.
Prepare a factored estimate table similar to Table 5.3-2 from these data. Carefully assess the accuracy of the experience factors. Within what range of confidence can you predict the Lang factor for a new process from these data? Is it reasonable to express the Lang factor to three significant figures, as is often done in the literature?

Table 5.D-1

		Percentages of Total Cost						
Plant Number	Approximate Cost of Plant	Total Other Costs[a]	Yard Improve-ments	Buildings In-cluding Services	Process Equip-ment	Process Piping	Elec-trical Instal-lations	Service Facil-ities
I: Solids processing plants								
7	$1,500,000	21.9	4.7	11.3	47.3	3.4	6.2	5.2
8	1,500,000	21.6	9.1	10.3	44.7	3.4	5.9	5.0
II: Solids and fluid processing plants								
6	1,000,000	23.4	1.6	9.7	46.7	6.5	6.7	5.4
10	2,250,000	20.3	3.7	7.8	39.1	11.4	5.6	12.1
11	2,500,000	22.2	2.6	12.0	37.4	13.0	4.0	8.8
12	4,000,000	24.2	0.7	6.7	47.0	7.2	6.2	8.2
13	8,500,000	17.2	1.4	7.6	49.4	12.8	3.1	8.5
14	15,000,000	17.4	2.3	8.3	44.3	11.6	3.2	12.9
III: Fluids processing plants								
1	100,000	35.7	—	—	41.3	8.7	8.1	6.2
2	150,000	32.7	—	—	43.0	9.4	8.1	6.8
3	400,000	28.6	—	15.8	36.4	14.1	5.1	—
4	400,000	27.8	—	17.4	35.0	8.6	8.3	2.9
5	500,000	26.5	—	14.1	35.2	18.0	4.7	1.5
9	2,000,000	26.0	1.4	6.1	32.3	21.3	3.9	9.0

[a] Total Other Costs include home and field office expenses, contingency and insurance and taxes.

5.E. The process shown in Fig. 5.E-1 is under design. Estimate the cost of the major items of equipment and the total fixed investment required of the system. It is a standard low temperature process which is not highly instrumented.

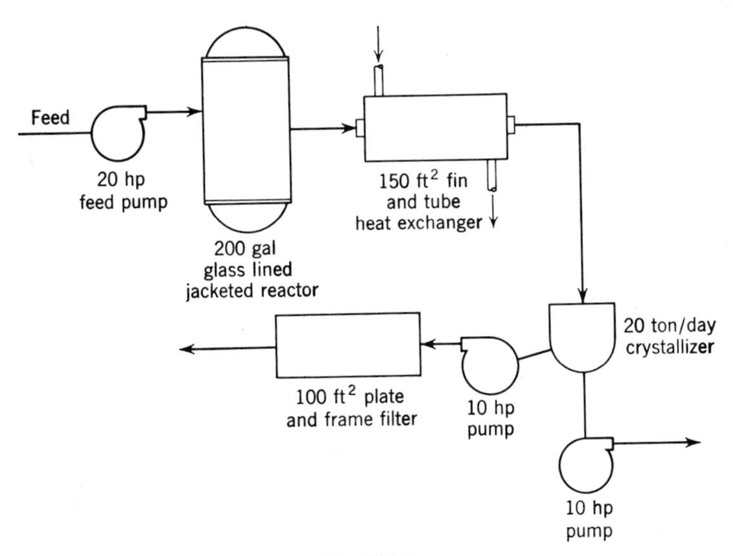

Fig. 5.E-1

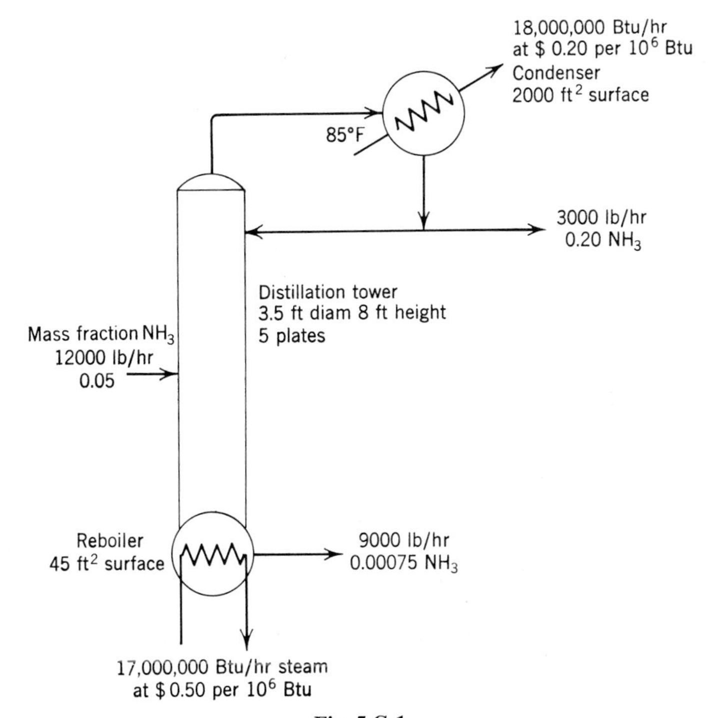

Fig. 5.G-1

5.F. From the data given in Tables 5.2-2 and 5.5-2 estimate the cost, in dollars per pound, of manufacturing sulfuric acid (1) in a 100 ton per day contact process, (2) in a 50 ton per day process. Sulfur is available at $24/ton and other data required will be found in the text. The market price of 100 per cent sulfuric acid is about $25/ton. How does your estimate compare?

5.G. The distillation process shown in Fig. 5.G-1 has been suggested as a solution to the primitive problem 2.N. 12,000 lb/hr of a 5 per cent ammonia in steam mixture is to be concentrated into a 20 per cent ammonia solution.

(a) Estimate the total investment required of this design.
(b) Estimate the manufacturing costs.
(c) If the product is worth $50/ton of NH_3, estimate the venture profit for the process.

5.H. Ten thousand lb/hr of an aqueous industrial waste contains 0.002 pounds of benzoic acid per pound of water. It is proposed to recover the benzoic acid by washing with benzene. The benzene is to be recovered by evaporation and condensation for reuse. The proposed process is shown in Fig. 5.H-1.

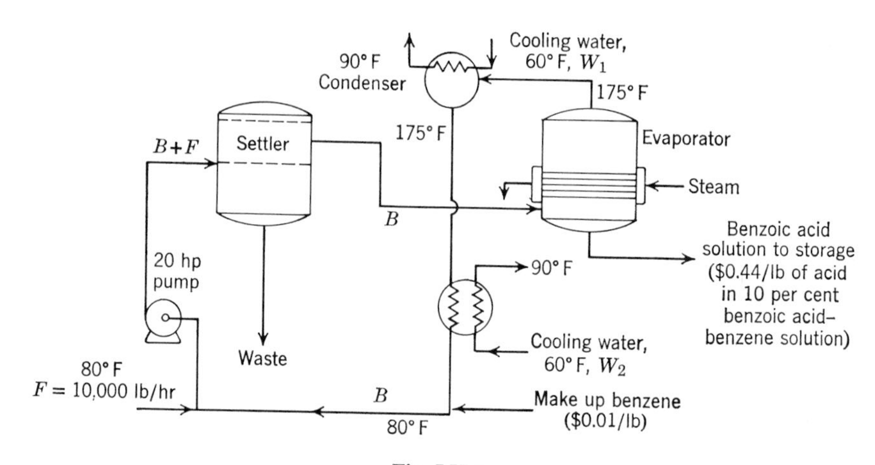

Fig. 5.H-1

Extractor. The extractor is a mixer-settler requiring a 20 hp mixer pump and sufficient volume in the settling vessel to insure a 5 min residence time based on the total fluid input. A carbon steel vessel should be used. The benzoic acid distributes itself between the aqueous and benzene phases according to the distribution $y = 4x$, where $y =$ pounds of benzoic acid per pound of benzene, and $x =$ pounds benzoic acid per pound of water. The benzene is soluble in the water phase in the amount of 0.07 lb benzene/100 lb water.

Evaporator. The benzene is boiled off (boiling point 80°C and heat of vaporization 100 cal/gm) in a falling film evaporator at 1 atm using steam at 300°F. The overall heat-transfer coefficient is 100 Btu/hr-ft^2-°F.

Condenser. The pure benzene vapors are condensed in a shell and tube heat exchanger using 60°F cooling water which may be heated to 90°F. The overall heat-transfer coefficient for the condenser is 100 Btu/hr-ft^2-°F. The condensate is cooled to 80°F in a shell and tube cooler with a heat-transfer coefficient of 20 Btu/hr-ft^2-°F.

(a) Develop the design equations for the process and determine the information flow structure.
(b) Estimate the manufacturing and investment cost in terms of the design and state variables.
(c) Estimate the venture profit for the process in the case of 1,000 lb/hr of benzene circulated. Repeat for 10,000 and 20,000 lb/hr of benzene.

5.I. A study is to be made of the optimum size of a heat exchanger to exchange heat between an outgoing reactor product stream and the incoming feed stream. The plant is already in existence and has adequate feed preheat and product cooling exchangers. The proposed exchanger is therefore elective rather than mandatory. Develop a cost equation which will contain the variable, A, heat-transfer area, of the proposed exchanger and will show realistically how the venture profit of the processing unit will vary as A is varied in a search for the optimum. Remember that the delivered cost of the exchanger must be augmented by the appropriate factor to produce the true change in the investment and operating costs of the unit. Remember also that the optimum may be no exchanger at all!

5.J. A vent gas is variable in amount and composition. It contains HCl, chiefly, at less than 2 per cent by volume. We now use HCl in our process, paying \$0.02/lb of muriatic acid. The concentrated acid is diluted to 20 per cent acid for use.

(a) Synthesize a processing system to make use of the acid in the vent gas.
(b) Assuming that the vent gas contains 1,000 lb/day of HCl, estimate the investment requirements and operating cost for the process.
(c) Would you risk the required capital in this venture?

5.K. The process flow sheet for the concentration plant at the Lac Jeannine iron ore plant is shown in Fig. 5.K-1. The plant is to take 8,800 tons/hr of feed ore and concentrate it to 2,600 tons/hr of high grade concentrate.

(a) Based on this information, the process flow sheet and data available in the text, and cost estimation texts, estimate the major equipment costs.
(b) Estimate the total investment cost in the system.
(c) Estimate the operating cost in dollars per ton of concentrate.
(d) Will this system be a worthwhile sink for a firm's investment capital?

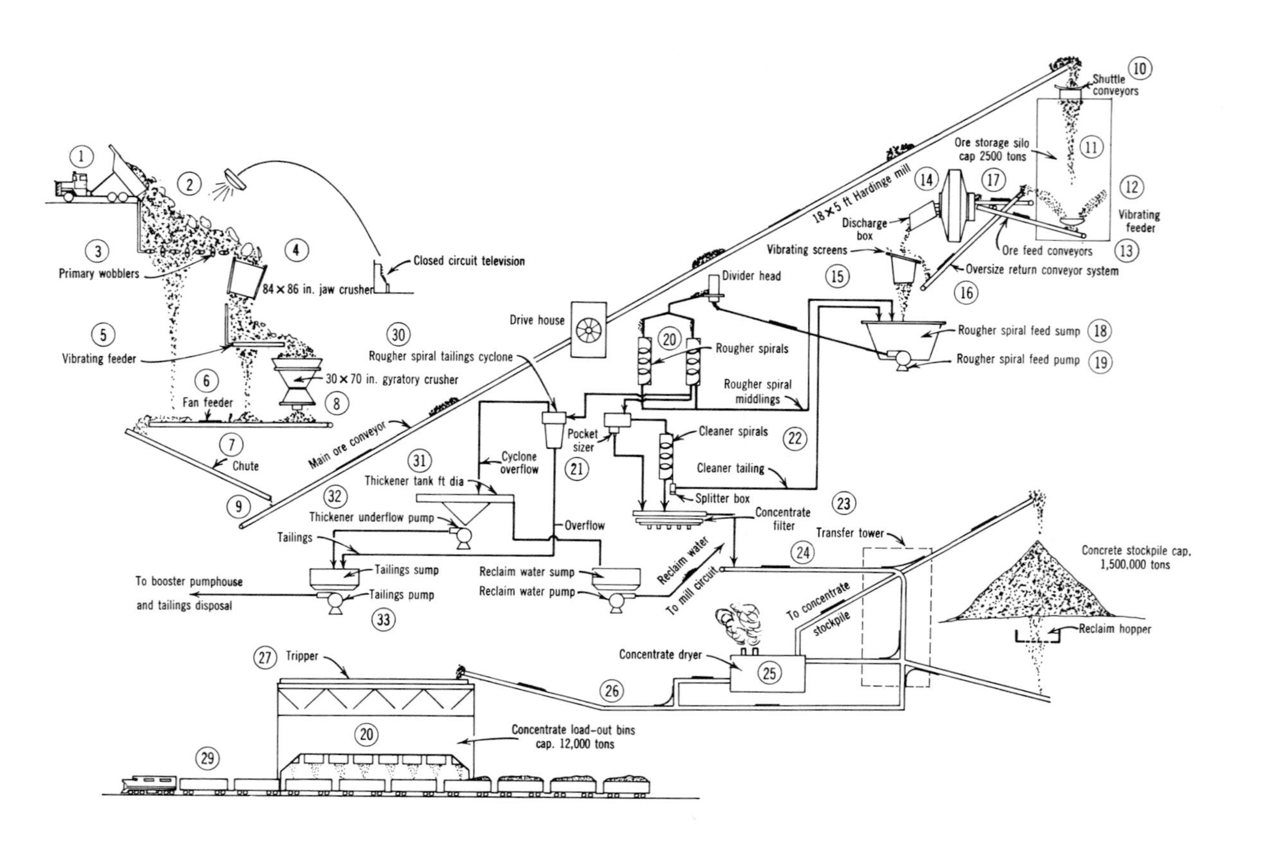
Shuttle conveyors
Ore storage silo cap 2500 tons
Vibrating feeder
18 × 5 ft Hardinge mill
Discharge box
Ore feed conveyors
Oversize return conveyor system
Vibrating screens
Divider head
Rougher spiral feed sump
Rougher spiral feed pump
Rougher spirals
Rougher spiral middlings
Cleaner spirals
Cleaner tailing
Splitter box
Concentrate filter
Transfer tower
Concrete stockpile cap. 1,500,000 tons
Reclaim hopper
Primary wobblers
84 × 86 in. jaw crusher
Closed circuit television
Vibrating feeder
Fan feeder
30 × 70 in. gyratory crusher
Chute
Main ore conveyor
Drive house
Rougher spiral tailings cyclone
Pocket sizer
Cyclone overflow
Thickener tank ft dia
Thickener underflow pump
Overflow
Tailings
Tailings sump
Tailings pump
To booster pumphouse and tailings disposal
Reclaim water sump
Reclaim water pump
Reclaim water
To mill circuit
To concentrate stockpile
Concentrate dryer
Tripper
Concentrate load-out bins cap. 12,000 tons

(1) 40- and 65-ton dump trucks.
(2) Two 38 × 26-ft feed hoppers, 300-ton capacity each. Four trucks can dump simultaneously.
(3) Two wobbler feeders in parallel, 72-in. width
(4) Two 84 × 66-in. jaw crushers in parallel, setting 10 in.
(5) Two vibrating feeders.
(6) Two 72-in. pan feeders.
(7) Two chutes.
(8) Two 30 × 70-in. gyratory crushers, 250-hp motors $6\frac{1}{2}$-in. setting.
(9) Two main conveyors, 1000-ft × 54-in., 15 per cent grade, three 400-hp drive motors each.
(10) Two shuttle conveyors.
(11) Twelve 2,500 ton silos.
(12) Twenty-four 48 × 96-in. vibrating feeders.
(13) Twelve 36-in. feed belts, 17° slope.
(14) Twelve 18 × 5-ft Cascade autogenous grinding mills, 600 hp motors. Power and sound controls regulate water and ore addition.
(15) Forty-eight 15 × 10-ft vibrating screens (four per mill).
(16) Twelve oversize belts, 24-in., 18° slope.
(17) Twelve conveyors, 24 in.
(18) Twelve rougher spiral feed sumps.
(19) Twelve rougher spiral feed pumps.
(20) One hundred twenty-eight rougher spirals for each mill. Size 24A, five turns.
(21) Pocket sizers.
(22) Sixty-four cleaner spirals per mill. Size 24A, five turns.
(23) Twelve rotary, horizontal concentrate filters 13-ft diam.
(24) One 36-in. concentrate collecting conveyor to transfer tower.
(25) Drying plant equipped with two Fluo-solids dryers, capacity 500 tph, Bunker C fuel oil.
(26) One 42-in. belt to load out bins, 15° slope.
(27) Tripper.
(28) Load-out bins, 12 chutes, three per car. Four cars loaded at one time.
(29) Ore train composed of 125 to 140 100-ton cars pulled by four 1750- and 1800-hp diesel locomotives to Port Cartier.
(30) Thirty-six 36-in. rubber-lined tailings cyclones.
(31) Three 160-ft thickeners.
(32) Pump, 6-in. centrifugal.
(33) Four two-stage tailings pumping systems in parallel, three operating, one spare. Each system includes three pumps first stage, two pumps second stage.

Fig. 5.K-1 Flow sheet of Lac Jeannine concentrator.

PART II

Optimization

Part II is a series of five chapters dealing with methods for adjusting a specific design to the most profitable level.

Chapter 6, *The Search for Optimum Conditions*, deals with methods for conducting the numerical search for the optimum values of the design variables in the case of a few degrees of freedom.

Chapter 7, *Linear Programming*, is concerned with a special method of direct search which is useful for the solution of linear optimization problems. Most oil refineries are scheduled by the solution of massive linear programming problems.

Chapter 8, *The Suboptimization of Systems with Acyclic Structure*, is an introduction to dynamic programming. Using this method it is possible to fracture a large acyclic optimization problem into smaller parts for suboptimization.

Chapter 9, *Macrosystem Optimization Strategies*, deals with methods for organizing the attack on complex design problems which contain persistent recycle. This class of problem arises quite often in chemical process design.

Chapter 10, *Multilevel Attack on Large Problems*, is a glimpse at a developing area of optimization, in which it is possible to approach the extremely large problems which arise, for example, in the design of integrated processing systems. These problems are too large to attack by ordinary methods.

6

THE SEARCH FOR OPTIMUM CONDITIONS

This is the first in a series of chapters on the strategy of optimization. We now limit attention primarily to systems or subsystems which exhibit one or a few degrees of freedom. A proper strategy for locating the best level of operation often involves a sequence of well planned numerical experiments, an approach that has come to be called *direct search.* The techniques of direct search are the basic tools of optimization theory and play an important role in the strategy of optimization of larger systems. Hence, this chapter, while important by itself, contains the background essential for subsequent chapters.

Typical Problem

A proposed reaction-separation system exhibits one major degree of freedom, namely, the conversion in the reactor. Should a cheap and simple reactor be elected with low conversion, an elaborate and costly product separator is required. On the other hand, the use of an elaborate and costly reactor with a high conversion reduces the load in the separator, thus reducing its cost. How can the compromise conversion which minimizes the cost of the total system be found with a *minimum of effort*?

6.1 CALCULUS

The formal procedure for locating an extremum of the design objective function using calculus follows.[1] We merely differentiate $U(d)$, the objective function, with respect to d, the design variable; set the derivative equal to zero;

[1] See H. Hancock, *Theory of Maxima and Minima*, Dover Pub., New York, 1960, for a definitive study of calculus in optimization.

and solve for d^*. In review, the necessary and sufficient conditions for locating an extremum in the region $a < d < b$ using calculus are:

$U(d)$ must be continuous for $a \leq d \leq b$.
$U(d)$ must be differentiable for $a < d < b$.
The first derivative must vanish at some point d^* where $a < d^* < b$.
At least one higher derivative must not vanish at d^*.
The first nonvanishing derivative must be an *even* one.
If this nonvanishing derivative is positive, the extremum is a minimum, and if it is negative, the extremum is a maximum.

Example:
To design a minimum cost cylindrical storage vessel which is to contain a volume V of material, where the sides cost c_S dollars per square foot, the top costs c_T dollars per square foot, and the bottom costs c_B dollars per square foot, we might proceed as follows.

The total cost is

$$C_T = c_S \pi D H + (c_B + c_T) \frac{\pi D^2}{4}$$

where

$$V = \frac{\pi D^2 H}{4}$$

$$D = \text{diameter}$$

$$H = \text{height}$$

Substituting for H in the total cost gives

$$C_T = c_S 4 \frac{V}{D} + (c_B + c_T) \frac{\pi D^2}{4}$$

Differentiation with respect to D, the free design variable, yields

$$\frac{dC_T}{dD} = -\frac{4 c_S V}{D^2} + (c_B + c_T) \frac{\pi D}{2} = 0$$

or

$$D^3 = \frac{c_S}{(c_B + c_T)} \frac{8V}{\pi}$$

Substituting for the volume in terms of D and H gives the following expression for the minimum cost diameter to height ratio.

$$\frac{D}{H} = \frac{2c_S}{c_B + c_T} \tag{6.1-1}$$

Thus, if the sides, top, and bottom cost the same,

$$c_S = c_T = c_B$$

$$\frac{D}{H} = 1$$

If there is no top, and the sides and bottom have the same cost,

$$c_S = c_B \qquad c_T = 0$$

$$\frac{D}{H} = 2$$

If an elaborate foundation is needed or land is expensive, and

$$c_S = c_T \qquad \text{but} \quad c_B = 2c_T$$

$$\frac{D}{H} = \frac{2}{3}$$

It appears that Eq. 6.1-1 partially explains the variety of shapes of storage vessels we see in an industrial complex.[2]

6.2 THE NEED FOR SEARCH METHODS

The classical method of locating the extreme of a function using calculus is of limited value in most design problems since the practical use of calculus is limited to functions which are easy to manipulate. The functions involved in process design are not the well behaved ones which appear in the mathematics texts but quite often are tabular, discontinuous, or exhibit other characteristics which prevent the use of calculus.

Fortunately, some rather efficient numerical *search methods* exist which can be used to obtain the optimum in problems which involve one or perhaps a few design variables. There are really two optimization problems to consider. The first is that of *optimizing the process*, and the second is that of *optimizing the method* for optimizing the process. We can eventually locate the optimum

[2] See L. E. Brownell and E. H. Young, *Process Equipment Design: Vessel Design*, Wiley, New York, 1959, for a detailed study of the design of vessels.

conditions by almost any reasonable scheme, but the important question that must be considered is "which method of optimization makes the most efficient use of the engineer's time?" This is a critical question in process engineering, since all but the simplest designs can easily involve inordinate amounts of computational effort.

There are two reasons for considering problems with only a few degrees of freedom, the first being that a number of industrially important processes have only a few major design variables. The second reason is that large processing systems are composed of a number of such smaller processing units, and the methods of large system optimization often involve the fracturing of systems into smaller components, which may then be attacked by the methods of this chapter.

The methods of search presented in the remainder of this chapter are practical for problems involving a few degrees of freedom.

6.3 SEARCH OVER A SINGLE DESIGN VARIABLE

The simplest problem, and, therefore, the proper place to begin, is the adjustment of a *single* design variable. The optimum value of this single design variable often can be isolated quite quickly, using a concept known as *region elimination.* There exists a best method for isolating an optimum, when only one relative optimum is known to exist, as well as a number of slightly less efficient but far simpler methods. We now lay the background for the development of these methods.

The region over which a design variable can be adjusted is nearly always limited in practical problems. The area of a heat exchanger, for example, cannot be less than 0 and cannot be exceedingly large. Mass fractions lie in range zero to 1, pressures must lie within the constraints of safety, conversions lie between zero and 100 per cent. The problem is to find the value of the design variable within these constraints which gives the best value of the objective function U.

To illustrate this, consider the problem of adjusting the diameter of a gas absorption tower. The tower must process a gas stream at a fixed rate to remove a soluble gas component by adsorption in a liquid phase. Increasing the diameter of the tower lowers the gas velocity in the bed reducing the pressure drop and, hence, lowering the pumping costs. But a larger diameter tower is more costly to construct. Some balance must be reached between the pumping costs and the construction costs to lower the costs of operation.

There are constraints on the tower diameter. When the tower diameter is too small, flooding occurs and liquid is carried up with the gas stream, making the tower inoperative. Also, it is not practical to construct a tower of extremely large diameter because of liquid distribution problems. These factors are shown in Fig. 6.3-1.

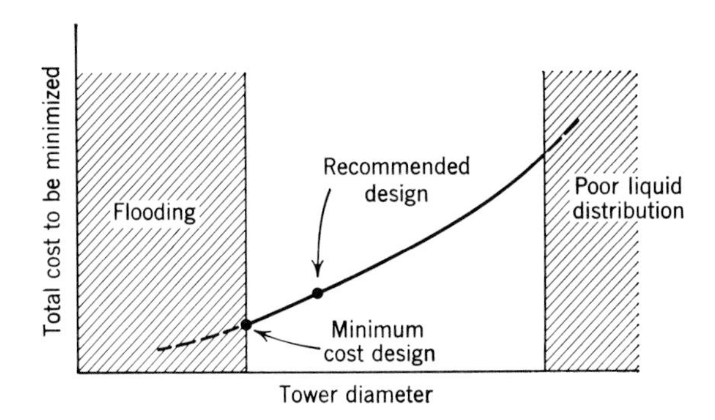

Fig. 6.3-1 Scrubbing tower optimization.

In this cost situation, we are led to the selection of the smallest possible tower diameter consistent with the flooding constraint. In actual practice, a larger diameter would be recommended to allow for changes in the absorption demands of the tower and to protect against the errors and uncertainties which are always present in the design, applying the concepts developed in Part III of this text.

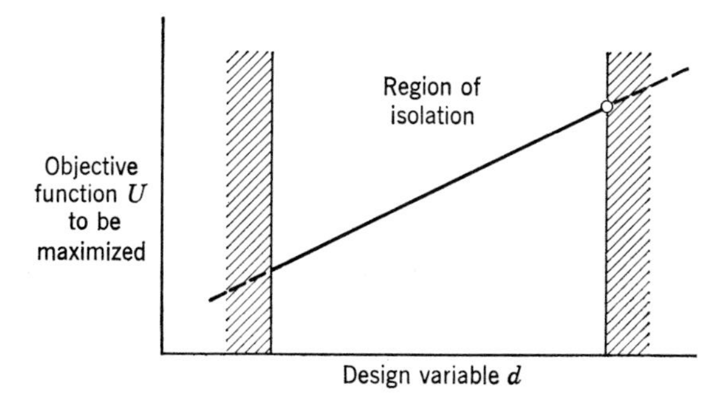

Fig. 6.3.2 Linear function.

It is worthwhile at this point to discuss the common shapes of the objective function; since when certain properties are thought to exist, the optimum can be isolated quite easily, without the need for sketching out the entire curve.

If the objective is a *linear* function of the design variables the optimum must *always* lie on a constraint (see Fig. 6.3-2). The optimum may lie on the constraint for a nonlinear function also, so the obvious strategy is to first investigate the conditions at the constraints. This simple observation concerning linear objective functions leads to linear programming, a powerful method of optimization to be discussed in Chapter 7.

There is an even more important reason for investigating these limiting conditions. The nature of the objective function at the constraint may be completely different from that away from the constraint.

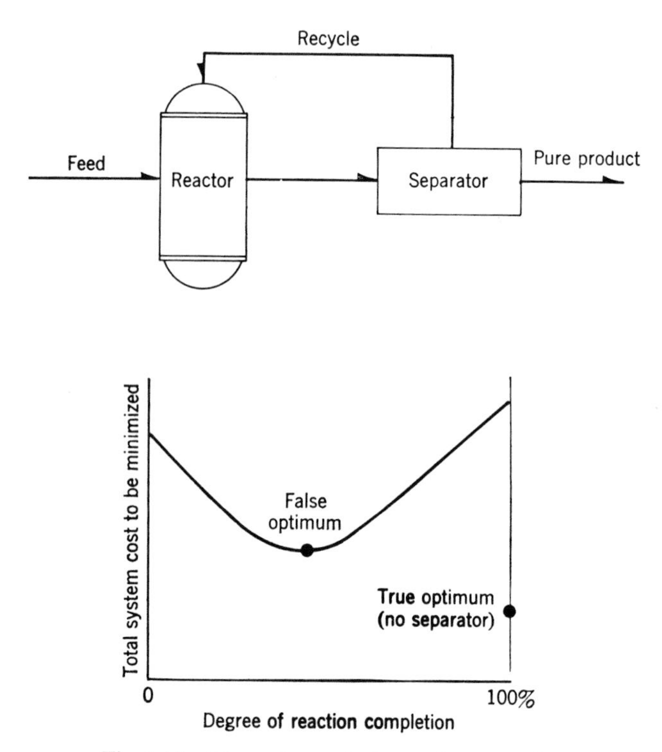

Fig. 6.3-3 The optimum design on the constraint.

For example, consider the process shown in Fig. 6.3-3 which consists of a reactor and separator. The design variable to be adjusted is the degree of reaction completion. In the case of complete reaction, no separator is needed and, hence, no construction costs for the separator are included. However, when any unreacted material is present in the reactor product, the separator must be used and its construction costs included. This causes the discontinuity in the cost and leads to the false optimum which might have been selected as the optimum design were the limiting conditions not investigated separately.

The objective function may be *unimodal* as is the case with a large number of engineering problems. A unimodal function is one for which the design becomes progressively worse as we move away from the optimum conditions. Such functions have only one relative optimum. Several unimodal and non-unimodal functions are shown in Fig. 6.3-4. When the objective is unimodal, the optimum can be isolated quite efficiently by region elimination, the topic of the next section.

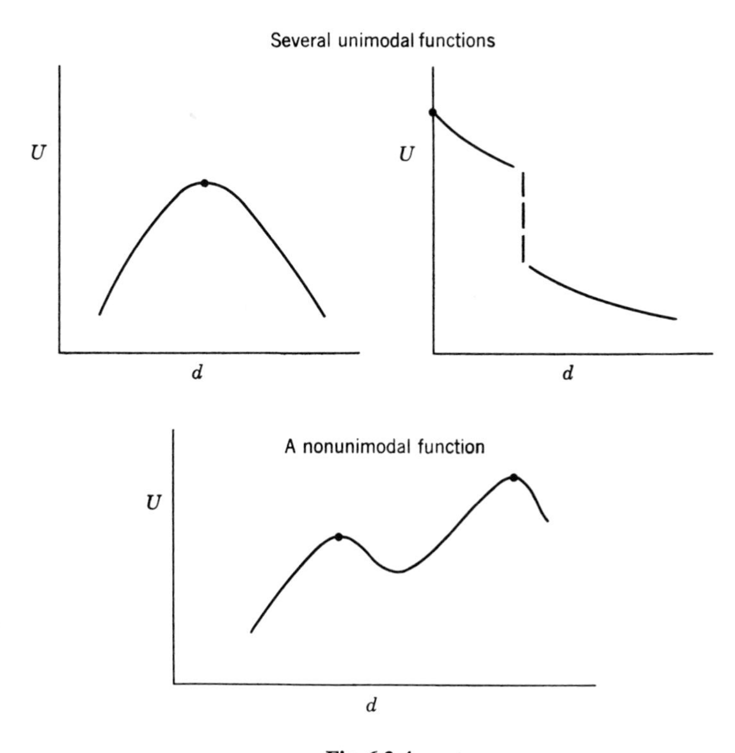

Fig. 6.3-4

6.4 REGION ELIMINATION

Region elimination is a very efficient way of isolating the optimum of a function that is known to be unimodal. Suppose we wish to isolate the optimum conditions to a region smaller than the original region defined by the constraints. We calculate the design at two values of the design variable and examine the results (see Fig. 6.4-1).

We see that the right-most value of the design variable gives a higher value. If the function is unimodal, the optimum must lie to the right of the left-most value of the design variable. On the basis of these two experiments, the left-most region can be eliminated from further consideration. This procedure is repeated in the smaller region until the optimum is isolated to a region of the desired size.

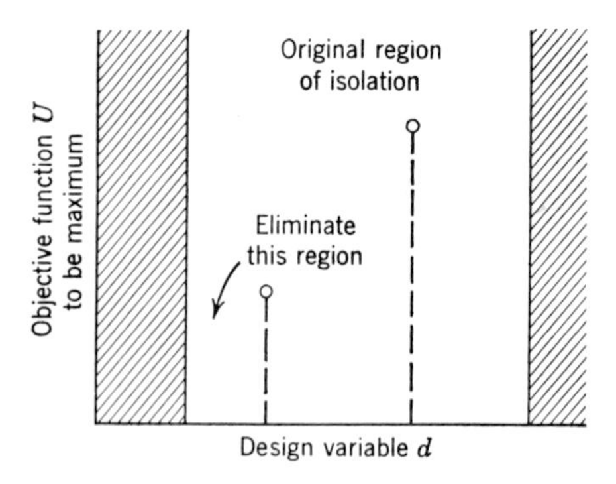

Fig. 6.4-1 Region elimination.

Within this smaller region there will be one point for which the objective function is known, the better of the two previous designs. We, therefore, need to calculate only one new design to further isolate the optimum.

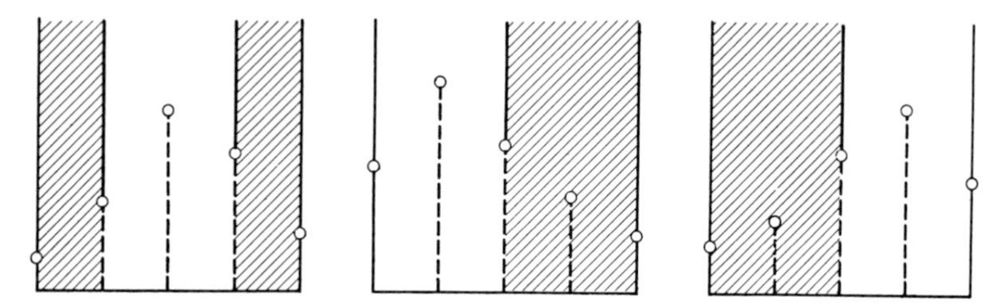

Fig. 6.4-2 A five point search plan.

We now present a simple and relatively efficient method of isolating the optimum. A pattern is started by determining the objective function at five values of the design variable, the two extreme values and three other values equally spaced in the interval. On the basis of these five calculations, the optimum can be isolated to a region one-half the size of the original. This is done by eliminating those intervals in which the optimum cannot lie because of its presumed unimodal property. Should the situation in Fig. 6.4-3 occur, even a greater region can be eliminated. The worst one can do with the first five calculations is to isolate the optimum to a region one-half of the original region.

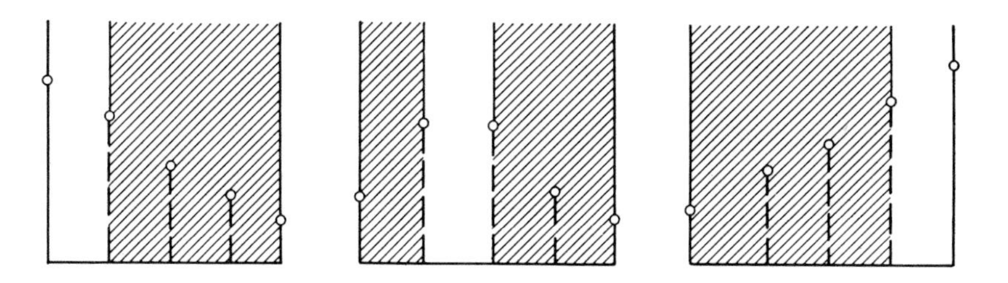

Fig. 6.4-3 A happy occurrence.

There remains the smaller region with the central point already calculated. Therefore, by calculating two new points (the one-quarter and three-quarter points) the same situation will be reached again, and the region can be further reduced by one-half. This search pattern is summarized in Table 6.4-1.

Table 6.4-1

Step Number	New Calculations	Total Calculations	Region of Isolation of Optimum
1	5	5	$\frac{1}{2}$
2	2	7	$(\frac{1}{2})^2$
3	2	9	$(\frac{1}{2})^3$
m	2	$N = 3 + 2m$	$(\frac{1}{2})^m$

Therefore, the total number of calculations N to isolate the optimum to a fraction Δ of the original is

$$N = 3 + 2m$$

$$\Delta = (\tfrac{1}{2})^m$$

$$-m = \frac{\log \Delta}{\log 2}$$

$$N = 3 - 2\frac{\log \Delta}{\log 2}$$

For example, if the optimum is to be isolated to one-tenth of a per cent of the original region of isolation, $\Delta = 10^{-3}$, a total of $N = 23$ search points will be needed. The search pattern to be discussed next is more efficient than this simple five point pattern.

6.5 SEARCH BY THE GOLDEN SECTION

The symmetrical placement of the search points (see Fig. 6.5-1) leads to a more efficient search method called search by the Golden Section.[3] This method is for most practical purposes the most efficient method. An optimal scheme known as *Fibonaccian Search*[4] exists, but it is more complicated and only slightly more efficient than the search by the Golden Section and is not discussed in this text.

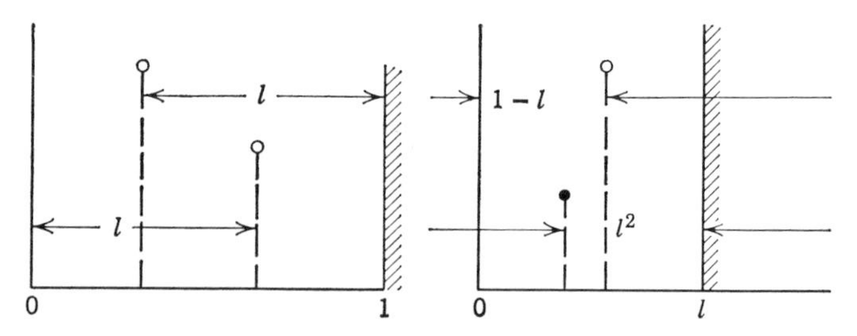

Fig. 6.5-1 The symmetrical placement of search points.

Suppose that the two search points are placed at a yet undetermined distance l from each side of the region of interest.

On the basis of these two calculations, a portion of the region can be eliminated. We wish to place these first points so that the point remaining after the first region elimination is a fraction l of the distance from the side of the remaining region.

This leads to the relationship

$$l^2 = (1 - l)$$

or

$$l = \frac{\sqrt{5} - 1}{2} = 0.618 \ldots$$

In this way, by placing the first two points a distance 0.618 of the original region of isolation from the sides, symmetry will be preserved. Each new point

[3] The Golden Section, a ratio $l = 0.618$ held mystical significance to the ancients. Greek temples were designed with that ratio of dimensions, since that was felt to be most pleasing to the eye. An interesting experiment for nature lovers is to measure the distance along a branch of a tree to its major subbranch and the distance along that subbranch to its major subbranch. The ratio of these distances has been said to average to the Golden Section—if we select the branches carefully.

[4] J. Kiefer, "Sequential Minimax Search for a Maximum," *Math. Soc.*, **4** (1953); S. M. Johnson, "Optimal Search for a Maximum is Fibonaccian," Rand Corp. Report, P-856 (1956).

reduces the region of isolation by 0.618 (see Table 6.5-1). The efficiency of the method is given in Table 6.5-2.

Table 6.5-1

Step Number	New Calculations	Total Calculations	Region of Isolation of Optimum
1	2	2	(0.618)
2	1	3	$(0.618)^2$
3	1	4	$(0.618)^3$
.			
.			
.			
m	1	$N = 1 + m$	$\Delta =$ $(0.618)^m$
	$\Delta = (0.618)^{N-1}$	$N = 1 + \dfrac{\log \Delta}{\log 0.618}$	

Table 6.5-2

Region of Isolation Δ	Number of Calculations, N By Five-Point Method	By Golden Section
10^{-1}	10	6
10^{-2}	17	11
10^{-3}	23	17
10^{-4}	30	20

6.6 DETERMINATION OF BEST BATCH TIME

We now demonstrate the application of region elimination by optimizing the batch process in Fig. 6.6-1 consisting of a reactor and a separator, which operate in sequence. A cycle begins when a charge is placed in the reactor. After a period of T days, the partially reacted charge is transferred to the separator where the product is removed.

The amount of product in the partially converted charge depends upon the reaction time, T, and is measured as X lb product/lb charge. The separation cost per batch depends on the mass fraction X of product in the partially

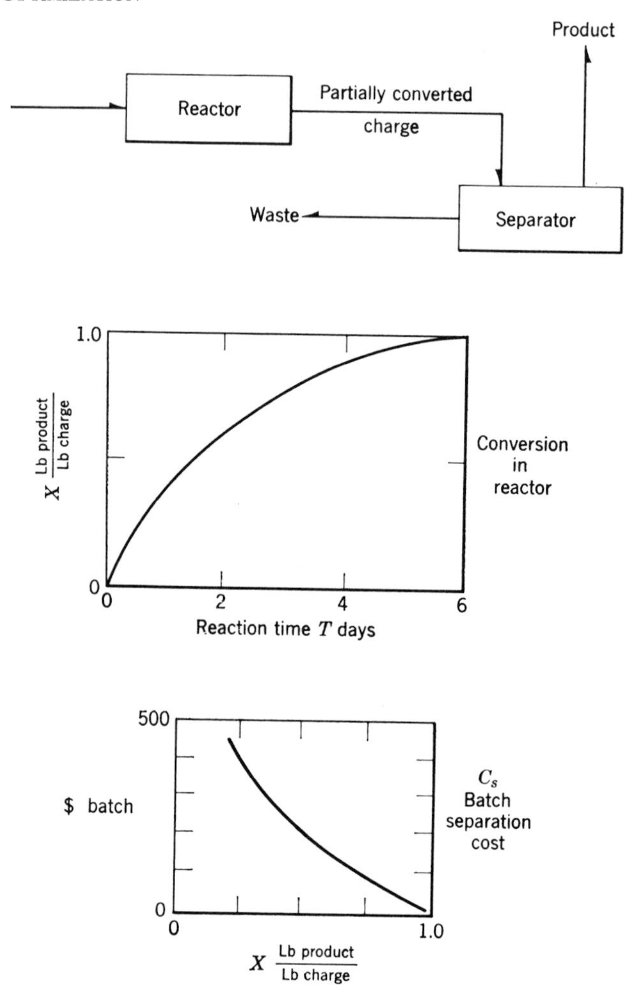

Fig. 6.6-1 A batch reaction-separation system.

converted charge. The cost of running the reactor is \$100/day. The value of the product is \$2.00/lb. The reactor holds 1,000 lb. of material. The costs and delay incurred by refilling the reactor are negligible. What is the batch time T which maximizes the gross profit R from the process?

$$R = \underbrace{\frac{1}{T}}_{\frac{\text{Batches}}{\text{Unit time}}} \left[\underbrace{(2.00)}_{\substack{\text{\$/lb} \\ \text{product}}} \underbrace{(1{,}000X)}_{\substack{\text{lb product} \\ \text{per batch}}} - \underbrace{100T}_{\substack{\text{Operation} \\ \text{cost of} \\ \text{reactor}}} - \underbrace{C_s}_{\substack{\text{Separation} \\ \text{cost}}} \right]$$

SOLUTION. We may search either T, the reaction time, or X, the conversion. The latter seems the most useful since X is bounded between 0 and 1. Using the Golden Section, we can isolate the best conversion to within 10 per cent in six trials or 1 per cent in eleven (see Table 6.5-2). Seven trials are summarized in Table 6.6-1.

Table 6.6-1

Search Number	X	\$/day	Eliminated	Remaining
1	0.382	450	0.618–1.00	0.000–0.618
2	0.618	380		
3	0.236	165	0.000–0.236	0.236–0.618
4	0.472	430	0.472–0.618	0.236–0.472
5	0.326	372	0.236–0.326	0.326–0.472
6	0.418	480	0.326–0.382	0.382–0.472
7	0.433	530	0.382–0.418	0.418–0.472

The peak has been isolated in 7 calculations to the region $0.418 \leq X \leq 0.472$ with a return of at least \$530 per day.

6.7 IN CASE OF MULTIMODAL FUNCTIONS

The methods of region elimination are based on the assumption that the functions being searched are unimodal. Fortunately, this seems to be generally the case, especially over limited regions in the design variable. However, we must be cautious, if the function is thought to possess more than one peak, as shown in Fig. 6.7-1.

The only really safe way for locating the true optimum is an exhaustive search, in which every single point is considered as a candidate for the optimum. For example, if we wish to isolate the peak of a function $U(d)$ to a fraction Δ of the original range of the design variable, $a < d < b$, calculations would be made at

$$d = a,\ a + \frac{\Delta}{2}(b - a), \qquad a + \frac{2\Delta}{2}(b - a) \ldots$$

using a total of $(2/\Delta) + 1$ search points to cover the region. This exhaustive search is extremely inefficient, requiring 20,001 calculations to isolate the peak to $\Delta = 10^{-4}$, for example, while the Golden Section Search, which is more dangerous, requires only 20 calculations.

When we suspect the presence of several peaks, a coarse grid exhaustive search might be employed to isolate the peak to some smaller region within which it might be safer to employ region elimination. For example, K preliminary exhaustive search points followed by N search points using the Golden Section method, would isolate the peak to

$$\Delta = \frac{2}{K-1}(0.618)^{N-1}$$

This compromise strategy would be safer but less efficient than pure region elimination and would be more dangerous but more efficient than a purely exhaustive search, the degree of compromise being determined by the relative values of K and N.

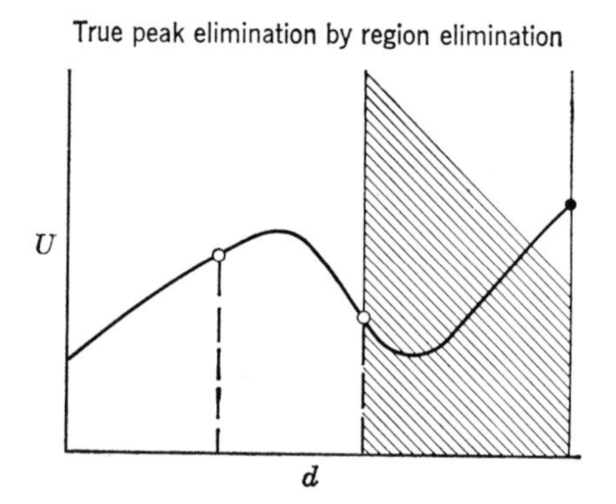

Fig. 6.7-1 A multimodal hazard.

6.8 SEARCH OVER SEVERAL DESIGN VARIABLES

Single variable search strategies have been well developed, and such problems offer no great difficulty. This happy situation does not occur, however, when more than one degree of freedom exists. Multivariable search is vastly more complicated than single variable search, principally because:

The unimodal assumption is less plausible.

There are no simple *a priori* measures of search efficiency.

The sheer size of the computational problems can be overwhelming.

Since there has been no best way invented for conducting a direct search in several variables and since the efficiency of any search plan depends on the particular characteristics of the surface being searched, a large number of multivariable search plans have been suggested, each with proponents. The engineer would hope for the simplest possible method to implement, requiring the minimum of computations of and manipulation with the design equations and objective function.

The simplest search plan to implement is the *sectioning* or *one-at-a-time* search plan.[5] In this strategy, all design variables except one are fixed at reasonable values, and this single variable is adjusted to maximize the objective function. Once this local maximum has been reached, the free variable is fixed at the value which achieved that maximum, and one of the other variables is selected as the search variable. This pattern of single variable searches is repeated in a cyclic fashion through all of the variables until no further improvement is observed.

If the contours of the objective function being searched are "circular," the sectioning method will converge to the optimum in F single variable search trials, where F is the number of design variables to be searched (see Fig. 6.8-1).

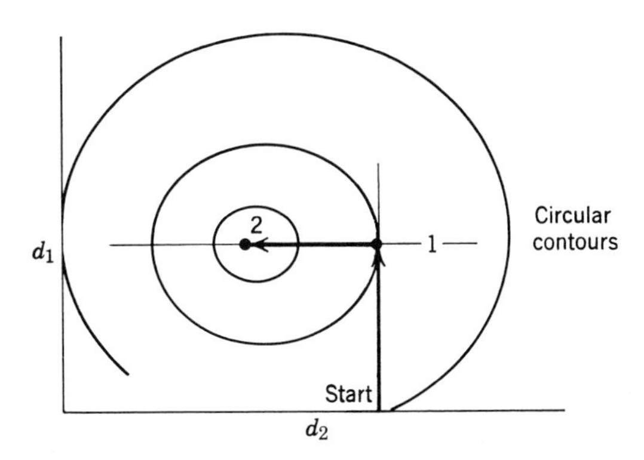

Fig. 6.8-1 Sectioning a two variable problem.

Since, a reasonably accurate single variable search might involve 10 search points, the sectioning search plan would involve $10F$ search points under the *ideal* search conditions mentioned above.

When the contours are not circular, convergence may take more than F single variable searches, as shown in Fig. 6.8-2. In engineering problems such distortions of the objective function contours exist to the degree that this simple one-at-a-time search plan converges extremely slowly and may not converge at all to the optimum.[6] In conclusion, the simplest of all multivariable search plans, the method of sectioning, is not suitable in practical engineering problems for it may lead to an inordinately large number of trials to reach the optimum.

[5] M. Friedman and L. S. Savage, *Selected Techniques of Statistical Analysis*, McGraw-Hill, New York, 1947.

[6] D. J. Wilde, *Optimum Seeking Methods*, Prentice Hall, New York, 1964, Sect. 5.01.

Rather than direct the search along lines perpendicular to the coordinates, (as in Fig. 6.8-2), it has been suggested that the search should proceed in the direction of *steepest ascent*, which is perpendicular to the local contour lines.[7]

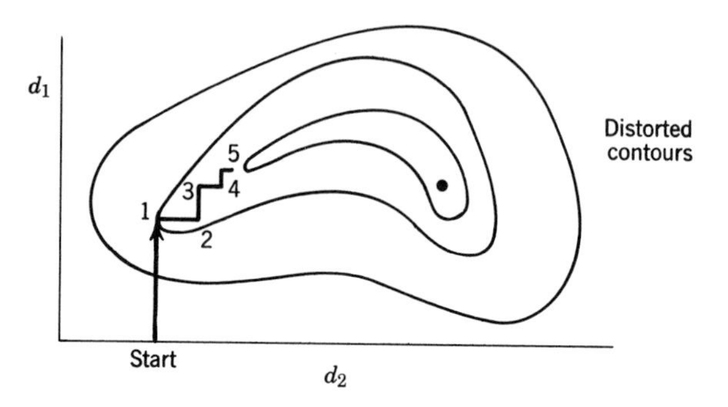

Fig. 6.8-2 Stepping towards the optimum.

Thus, we must make a few local explorations to obtain an estimate of this direction of steepest ascent. Once the direction is located, a search is made along that line until a local maximum is reached. Then the cycle is repeated by locating a new direction of steepest ascent (as shown in Fig. 6.8-3) for the case of two design variables.

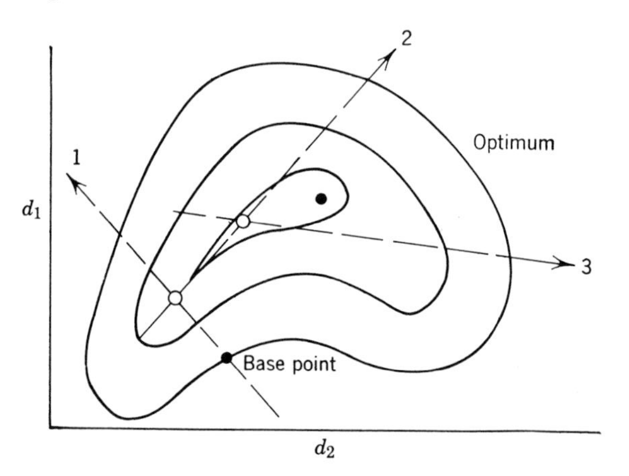

Fig. 6.8-3 Steepest ascent method.

The direction in which the objective function increases most rapidly from some initial base design can be estimated quite simply in the case of a few design variables by an analysis of the results of local explorations. Consider

[7] G. E. P. Box and K. B. Wilson, "On the Experimental Attainment of Optimum Conditions," *J. Royal Stat. Soc.*, **B13**, 1 (1951).

the case of a design with two design variables in which one has computed the objective function at a base point, with one variable perturbed, and with the second variable perturbed.

$$\begin{aligned} U_0 &= U(d_{1_0}, d_{2_0}) && \text{base design} \\ U_0 + \Delta U_1 &= U(d_{1_0} + \Delta d_1, d_{2_0}) && \text{variable } d_1 \text{ perturbed} \\ U_0 + \Delta U_2 &= U(d_{1_0}, d_{2_0} + \Delta d_2) && \text{variable } d_2 \text{ perturbed} \end{aligned}$$

If a plane is fitted through these three points, an approximate linear relation between the objective function and design variables is obtained as

$$U - U_0 = \frac{\Delta U_1}{\Delta d_1}(d_1 - d_{1_0}) + \frac{\Delta U_2}{\Delta d_2}(d_2 - d_{2_0})$$

Now, using this approximate relation, we ask: for what changes in d_1 and d_2 does U increase most rapidly? If we measure the "distance" moved by any design change which involves the simultaneous manipulation of the two variables by

$$r = \sqrt{(d_1 - d_{1_0})^2 + (d_2 - d_{2_0})^2}$$

we then wish to find the direction in which a change in r causes the greatest increase in the objective function. That direction is at right angles to the contour tangent obtained by differentiating the approximate objective function with respect to r and setting the derivative to zero.

$$\frac{d(U - U_0)}{dr} = \frac{\Delta U_1}{\Delta d_1}\frac{d(d_1 - d_{1_0})}{dr} + \frac{\Delta U_2}{\Delta d_2}\frac{d(d_2 - d_{2_0})}{dr} = 0$$

or, after some minor manipulations

$$\frac{d_1 - d_{1_0}}{d_2 - d_{2_0}} = \left(\frac{\Delta U_1}{\Delta d_1}\right) \Big/ \left(\frac{\Delta U_2}{\Delta d_2}\right) = \text{slope of line of steepest ascent}$$

Thus, by an analysis of the local explorations about a base design, a direction of steepest ascent has been found, and search can then be limited to a line along that direction until a new improved base design is found. The direction of steepest ascent is then found at this improved base design, and a new line of search is located. This procedure is iterated until the optimum design is found.

Any number of methods have been suggested over the years to improve the efficiency of direct search optimization, and several of the more popular methods have been compared in the literature.[8] Each may have certain features desirable to the process engineer, such as ease of implementation or speed of

[8] See for example, A. Leon, "A Comparison of Eight Known Optimizing Procedures," Space Sciences Laboratory, University of California, Berkeley, August 1964.

convergence on certain types of optimization problems; and each may possess certain undesirable features, such as an inability to cope with the discontinuities which seem to be fairly common in process optimization problems. There is no best method of direct search, but the logical search method to be discussed next will be taken as the standard method of direct search for use in this text, as it is a practical compromise between the methods which are extremely easy to implement and slow to converge and the methods which are quick to converge but troublesome to implement.

The several methods of multivariable search mentioned previously derive efficiency by reducing the multivariable search problem to a single variable search along a line which is thought to pass near the optimum. These methods attempt to take advantage of the *geometry* of the objective function by finding local slopes or directions of ascent. Hooke and Jeeves[9] describe a *logical* search plan which has many of the properties desired by the process engineer; namely, ease of implementation with no direct determination of slopes or direction and a relative quickness to converge on process optimization problems.

A logical search begins with an initial guess of the best design. A search increment is then established, at say 5 or 10 per cent of the range of variation of each design variable, and each design variable is changed in turn by its corresponding increment. Should such a *local exploration* result in an improved design, the improved design is called the best design and the initial guess is called the *next best design.*

Now rather than repeat these local explorations of the design variables, a pattern move is made. A pattern move is based on the conjecture that changes in the design which resulted in improvements during the exploration phase should result in still greater improvements. Each design variable is changed directly by the amount it was changed in the last exploration cycle. If this results in better design, the conjecture was correct; if not, the conjecture was not correct and more conservative local explorations should be undertaken until a valid pattern of design variable changes is established.

A strategy of using local explorations about a base design to establish pattern moves towards the optimum design is outlined in Fig. 6.8-4, the details of local exploration outlined in Fig. 6.8-5, and pattern moves in Fig. 6.8-6. Should the exploratory and pattern moves converge to a design, that design is the locally optimal design within the increment of design changes made during exploration; either the optimum has been found or the exploration increment should be reduced further to focus in on the optimum. The use of this kind of a logical search strategy is illustrated in the next section.

[9] R. Hooke and T. A. Jeeves, "Direct Search Solution of Numerical and Statistical Problems," *J. Assoc. Comp. Mach.*, No. 2, **8** (1961).

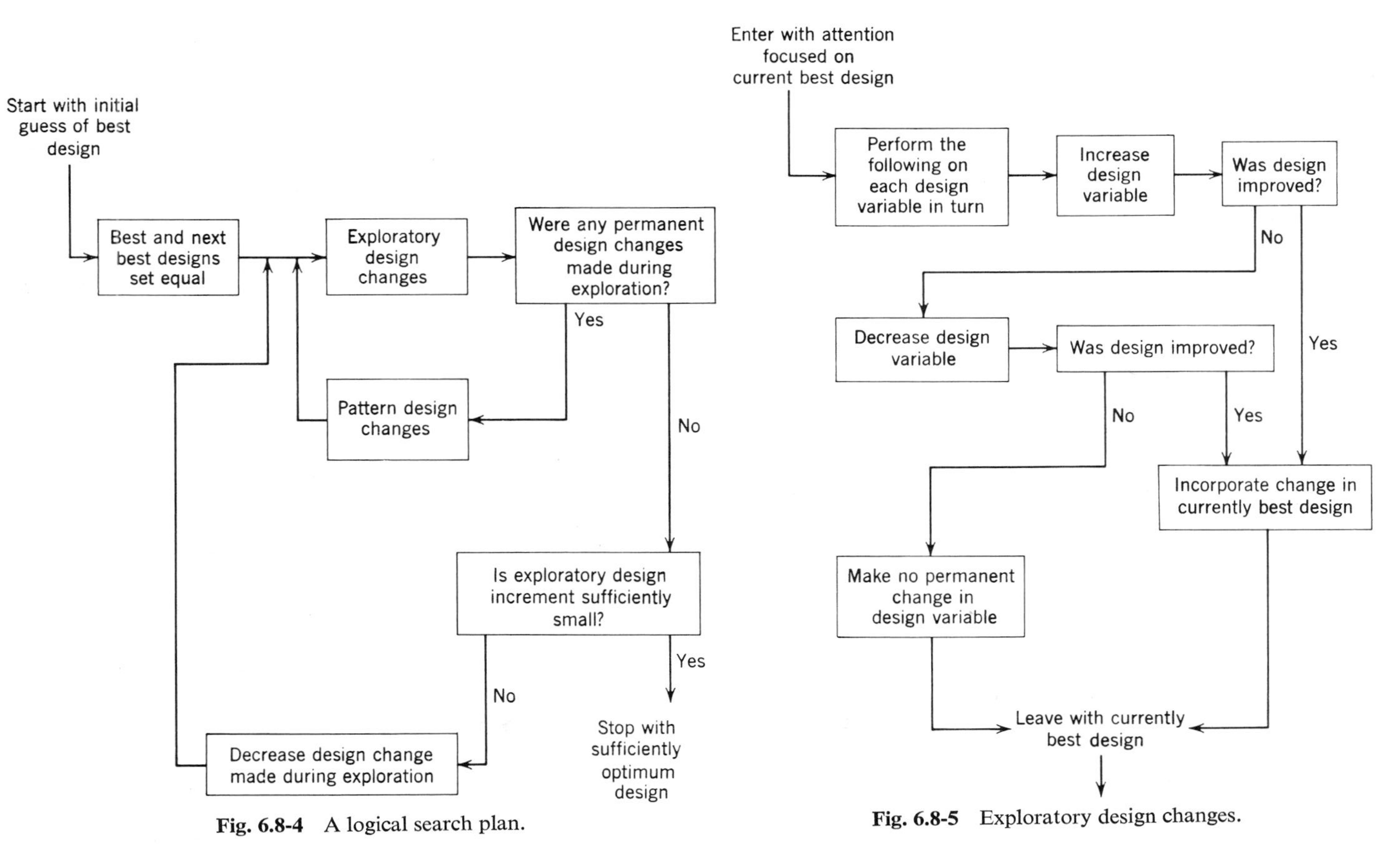

Fig. 6.8-4 A logical search plan.

Fig. 6.8-5 Exploratory design changes.

It appears that search strategies based on this Hooke-Jeeves logical plan can efficiently locate and follow ridges of the objective function, and that this is the reason for their efficiency in process design problems in which ridges appear to be common.

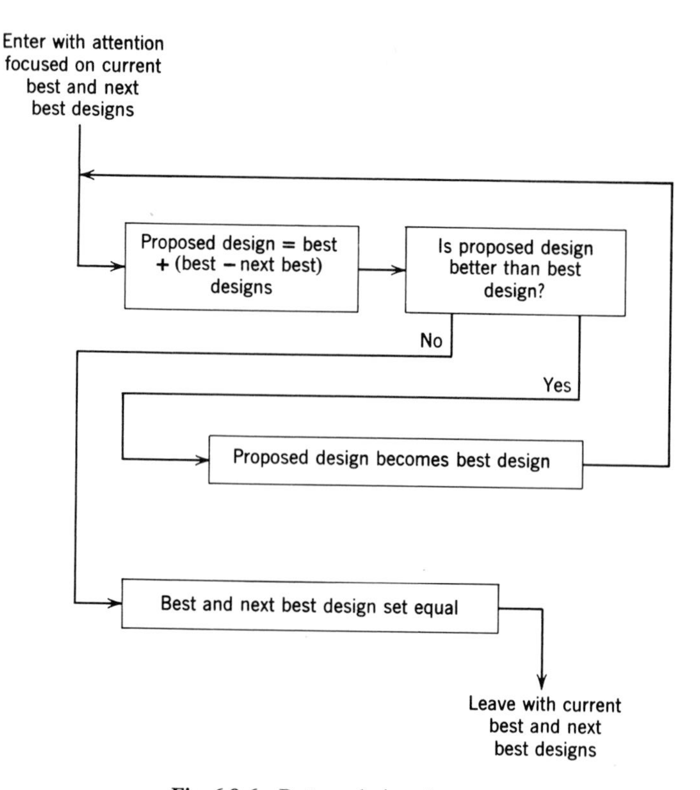

Fig. 6.8-6 Pattern design changes.

6.9 THE OPTIMIZATION OF A REFRIGERATION SYSTEM

In this section we apply the method of logical search to the design of the three stage refrigeration system shown in Fig. 6.9-1, a problem discussed by Fan and Wang in another context.[10] A hot stream is to be cooled in three stages by refrigeration. Each stage consists of a heat exchanger with hot stream on one side of the heat-exchange surface and a boiling refrigerant on the other, cooler side. The temperatures at which the refrigerants boil are known at each

[10] L. T. Fan and C. S. Wang, *The Discrete Maximum Principle*, Wiley, New York, 1964.

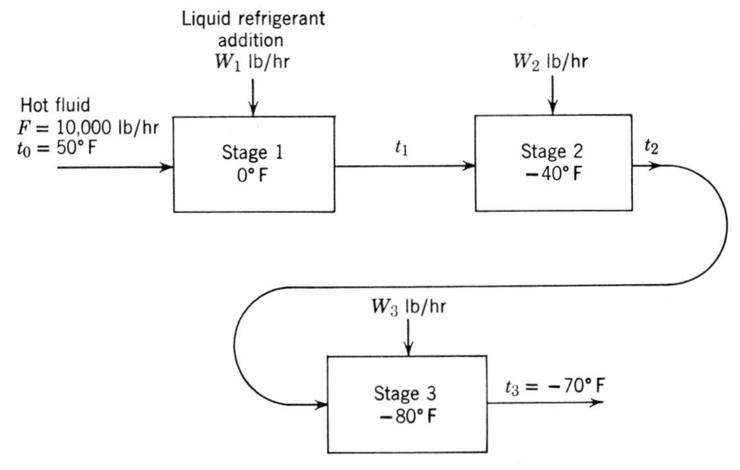

Fig. 6.9-1 Three stage refrigeration system.

stage, and hence the rate of heat transfer is determined by the area of the heat-exchange surface, at a given flow rate and inlet temperature of the hot fluid. In this problem it is necessary to determine the surface areas of the three exchangers most economically to cool 10,000 pounds per hour of the hot fluid from 50°F to −70°F, under the conditions outlined in Table 6.9-1.

Table 6.9-1 Data for system

Symbol	Item	Stage 1	Stage 2	Stage 3
T	Refrigerant temperature, °F	0	−40	−80
λ	Latent heat of refrigerant, Btu/lb	100	100	100
U	Overall heat-transfer coefficient, Btu/hr-ft²-°F	200	200	200
a	Capital cost parameter, \$/hr(ft²)$^{1/2}$	0.05	0.05	0.15
b	Operating cost parameter, \$/lb	2×10^{-4}	3×10^{-4}	4×10^{-4}
	$C_p = 1.0$ Btu/lb°F	$F = 10{,}000$ lb/hr		

The heat transfer at stage i is described by the following equations.

$$Q_i = U_i A_i (\Delta t_i)_{\ln} \qquad \text{heat transfer rate} \tag{1}$$

$$(\Delta t_i)_{\ln} = \frac{t_{i-1} - t_i}{\ln[(t_{i-1} - T_i)/(t_i - T_i)]} \qquad \text{log mean temperature difference} \tag{2}$$

$$Q_i = \lambda_i W_i \qquad \text{energy balance over refrigerant} \tag{3}$$

$$Q_i = FC_p(t_{i-1} - t_i) \qquad \text{energy balance over hot stream} \tag{4}$$

Where Q_i = the amount of heat transferred to the refrigerant at stage i, Btu/hr;

A_i = the area of heat exchange surface at stage i; and the other variables are defined and specified in Fig. 6.9-1 and Table 6.9-1.

The economic objective is to minimize the total cost, which for stage i takes the form

$$C_i = a_i(A_i)^{1/2} + b_i W_i$$

The first term is the capital cost which is correlated to the area of the heat-exchange surface, and the second term is the operating cost of supplying refrigerant. Numerical values of the cost parameters a and b are given in Table 6.9-1.

First we organize the calculations. The flow diagram Fig. 6.9-1 reveals a partial precedence-ordering among the three stages, and we now focus attention on the design equations for a given stage. The structural array of the design equations for stage $i = 1, 2$ is shown in Table 6.9-2.

Table 6.9-2 Structure of Design Equations for Stage $i = 1, 2$

	Variable										
	Q_i	U_i	A_i	$(\Delta t)_{ln}$	t_{i-1}	t_i	T_i	λ	F	C_p	W_i
Equation (1)	X	X	X	X							
(2)				X	X	X	X				
(3)	X							X			X
(4)	X				X	X			X	X	
		$\overset{\uparrow}{\circ}$			$\overset{\top}{\circ}$		$\overset{\uparrow}{\circ}$	$\overset{\uparrow}{\circ}$	$\overset{\uparrow}{\circ}$	$\overset{\uparrow}{\circ}$	

Variables specified by the statement of the problem $\overset{\uparrow}{\circ}$.
Variables specified by the partial precedence-order of Fig. 6.9-1 $\overset{\top}{\circ}$.

Of the eleven variables which enter into the four design equations for either exchanger 1 or 2, the values of five variables (U_i, T_i, λ, F, C_p) are specified by the statement of the problem and the value of one more variable (t_{i-1}) is specified by the previous solution of the upstream stage. Thus, stages 1 and 2 possess $F = 11 - 4 - 5 - 1 = 1$ degree of freedom each. We next apply the design variable selection algorithm of Section 3.6 to the structural array in Table 6.9-3.

Table 6.9-3 Selection of Design Variables for Stages 1 or 2

	Q_i	A_i	$(\Delta t)_{\ln}$	t_i	W_i
Equation (1)	X	X	X		
(2)			X	X	
(3)	X				X
(4)	X			X	

Notice in Table 6.9-3 there are several ways of applying the design variable selection algorithm. These are summarized in Table 6.9-4, and we wish to

Table 6.9-4 Several Orders of Equation Deletion in Table 6.9-3

	Case		
	A	*B*	*C*
Order of deletion			
1	A_i, 1	W_i, 3	W_i, 3
2	$(\Delta t)_{\ln}$, 2	A_i, 1	A_i, 1
3	t_i, 4	$(\Delta t)_{\ln}$, 2	$(\Delta t)_{\ln}$, 2
4	Q_i, 3	Q_i, 4	t_i, 4
Design variable	W_i	t_i	Q_i

select amongst the several feasible design variables. Which of the three candidates for the design variable in Table 6.9-4 is desired as the design variable for reasons other than precedence-ordering?

Case A: W_i as design variable
We have no physically intuitive feel for proper values of refrigerant addition rates. All that can be said is that $0 \leq W_i$.

Case B: t_i as design variable
The temperature leaving a given stage must be greater than the refrigerant temperature,

$$t_0 = 50°\text{F} \geq t_1 \geq 0°\text{F}$$

$$t_1 \geq t_2 \geq -40°\text{F}$$

Case C: Q_i as design variable

Bounds can be established on the maximum heat removed at any stage by making an overall heat balance

$$0 \leq Q_1 + Q_2 \leq FC_p[50 - (-70)] = 1{,}200{,}000 \text{ Btu/hr}$$

However, it is possible inadvertently to select Q_1 or Q_2 which will require outlet temperatures of the hot stream below the refrigerant temperature.

Thus of the three structurally equivalent design variable choices W_i, t_i or Q_i, t_i seems to be best since the bounds on feasible variations are most clearly defined. Figure 6.9-2 shows the precedence order.

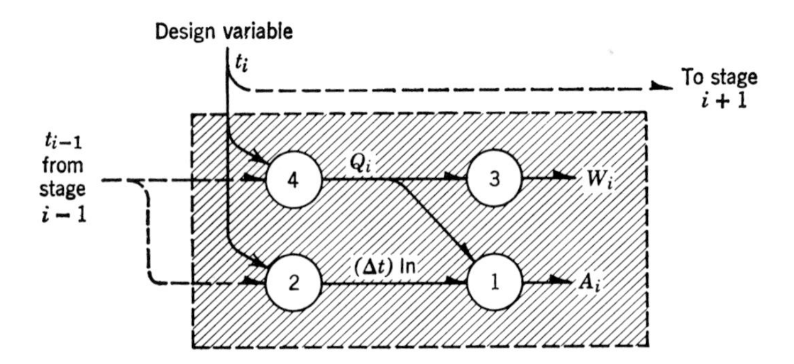

Fig. 6.9-2 Precedence-order within stages 1 or 2.

Stage 3 is a special case since the effluent temperature is specified as $t_3 = -70°$, consuming the last degree of freedom at stage 3. The structural array which must be precedence-ordered is shown in Table 6.9-5, the application of the ordering algorithm of Section 3.6 is summarized in Table 6.9-6 and the precedence-order is shown in Fig. 6.9-3.

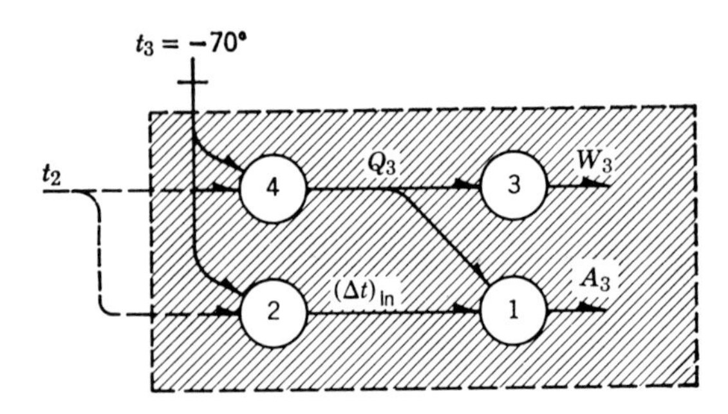

Fig. 6.9-3 Precedence-order of equations for stage 3.

Table 6.9-5 Structural Array for Stage 3

Equation	Q_3	A_3	$(\Delta t)_{\ln}$	W_3
(1)	X	X	X	
(2)			X	
(3)	X			X
(4)	X			

Table 6.9-6 Precedence of Ordering Table 6.9-5

Order of Deletion	Deletion
1	3, W_3
2	1, A_3
3	4, Q_3
4	2, $(\Delta t)_{\ln}$

The precedence-order for the complete systems calculation is shown in Fig. 6.9-4. There are two design variables to be adjusted, t_1 and t_2; the com-

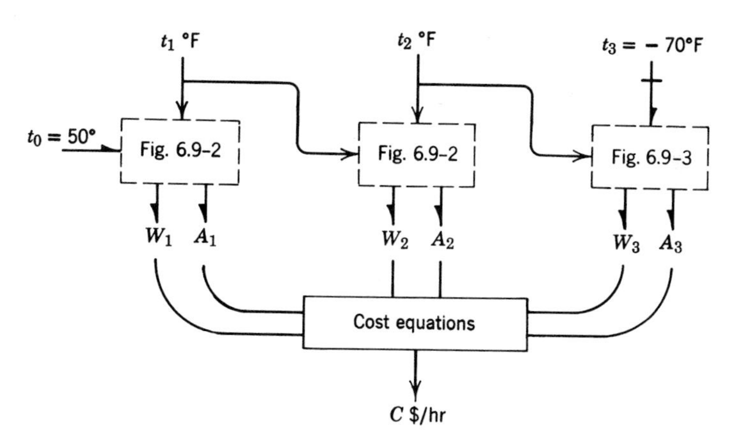

Fig. 6.9-4 Complete precedence order for system calculations.

putation order outlined in Fig. 6.9-4 enables the determination of A_1, W_1, A_2, W_2, A_3, and W_3, handling only one equation at a time. This illustrates the principles of problem set-up on a rather simple set of equations. This kind of preliminary set-up usually saves significant amounts of computation time.

Having established the intermediate temperatures as reasonable design variables, the logical search for the optimum design conditions begins with the selection of a base design and a search increment for each variable. Arbitrarily we might select the midpoint between the constraints as defining the base design and 20 per cent of the range of variation as the initial search increment.

Variable	Range of variation	Base design	Search increment
t_1°F	50 to 0	25	10
t_2	50 to −40	5	18

Table 6.9-7 summarizes the beginnings of a logical search for the optimum design conditions.

Table 6.9-7 Summary of Search for Optimum Conditions

t_1, °F	t_2, °F	C, $/hr	
25	5	5.13	Base case
Begin exploration with $\Delta t_1 = 10$ and $\Delta t_2 = 18$			
35	5	5.18	Failure
15	5	5.06	Success
15	23	—	Violates constraint $t_1 > t_2$
15	−13	5.00	Success
Attempt a pattern move			
5	−31	4.98	Success
Attempt a second pattern move			
−15	−67	—	Violates constraints $t_1 > 0$ $t_2 > -40$
Return to exploration			
15	−31	4.97	Success
15	−13	5.00	Failure
15	−49	—	Violates constraint $t_1 > -40$
Attempt a pattern move			
25	−31	4.98	Failure
Return to exploration with reduced search increment $\Delta t_1 = 5$ and $\Delta t_2 = 9$			
20	−31	4.97	
10	−31	4.98	
15	−22	4.96	

6.10 CONCLUDING REMARKS

Direct search is conducted assuming that the only source of information about the system is through direct computation. This results in efficient search only in the case of a few design variables. The number of function evaluations which attend the search for the optimum may increase rapidly as the number of design variables increases. The point is often reached (Fig. 6.10-1) where

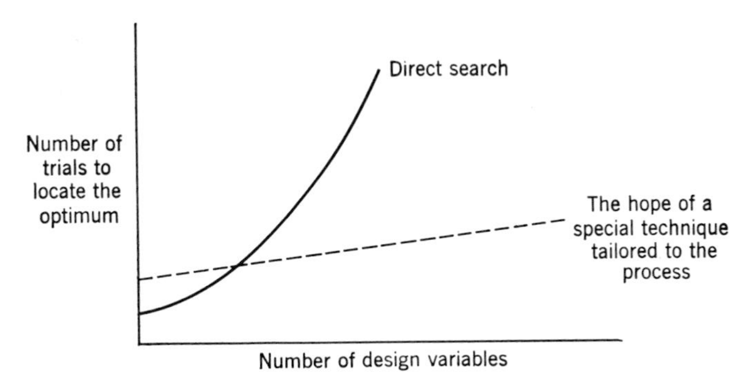

Fig. 6.10-1

the search gets out of hand, and where we must be content with a partial optimization unless more efficient methods of optimization can be brought to bear.

The rate of increase of computation labor with an increase in the number of design variables depends strongly on the search plan employed as well as the nature of the problems being solved. A rough observation has been made on the increase in search difficulty with dimension.

$$W_{DS} = a(b)^N \tag{6.10-1}$$

W_{DS} is the number of trials required to isolate the optimum; a and b are parameters which might take on values of around 2 to 10; and N is the number of variables being searched. Thus, if a single variable search takes 10 trials, we *might* expect a similar three variable search to take something on the order of $W \cong 3(3)^3 \approx 80$ trials, if $a = b = 3$. Such loose computational arguments play an important role in the evaluation of optimization programs.

In the references in this chapter, we mention a number of other search methods which fall in the class of "blind search." In the several chapters which follow we show how the labor of optimization can be reduced greatly by taking advantage of special properties of certain optimization problems.

References

These two books are the primary sources of information on the use of calculus and direct search respectively in optimization.

H. Hancock, *Theory of Maxima and Minima*, Dover Pub., New York, 1960.
D. J. Wilde, *Optimum Seeking Methods*, Prentice-Hall, Englewood Cliffs, N.J., 1964.

An excellent introduction to many methods of optimization is presented by Boaz.

A. H. Boaz, "What is Optimization All About," *Chem. Eng.*, December 10, 1962.
———, "How to Use Lagrange Multipliers," *Chem. Eng.*, January 7, 1963.
———, "How Search Methods Locate Optimum in Univariate Problems," *Chem. Eng.*, February 4, 1963.
———, "Optimizing Multivariable Functions," *Chem. Eng.*, March 4, 1963.
———, "Optimization via Linear and Dynamic Programming," *Chem. Eng.*, April, 1963.

Other search methods include:

Gradient Search: H. E. Zellnik, N. E. Sondak, and R. S. Davis, *Chem. Eng. Progr.*, **58**, 3 (1962).
Gradient Free Search: D. J. Wilde, *A.I.Ch.E. J.* **9**, 2 (1963).
Powell's Method: M. J. D. Powell, *Computer Journal*, **5**, 2 (1962).
Poorman's Optimizer: R. A. Mugele, Proc. Western Joint Computer Conference, 1962.
Rotating Coordinates: H. H. Rosenbrock, *Computer Journal*, **3**, 3 (1960).
Gradient Projection: J. B. Rosen, *SIAM J.*, **8** (1960); *SIAM J.*, **9**, 1962.
Conjugate Gradient: M. R. Hestenes and E. Stiefel, *J. Res. Nat. Bur. Std.*. **48**, (1952).

For some applications in process engineering see:

D. M. Himmelblau, "Process Optimization by Search Techniques," *Ind. Eng. Chem., Fundamentals*, No. 4, **2** (1963).
R. A. Koble and H. W. Goard, "Optimizing a Filter Cake Washing Circuit," *Chem. Eng. Progr.*, No. 12, **58** (1962).
L. J. Hvisdos, "Optimizing Oxygen Recovery in Low Temperature Distillation," *Chem. Eng. Progr.*, No. 11, **60** (1964).
D. H. Moorhead and D. M. Himmelblau, "Optimization of Operating Conditions in a Packed Liquid–Liquid Extraction Tower," *Ind. Eng. Chem., Fundamentals*, No. 1, **1** (1962).
L. A. Reed and W. F. Stevens, "Optimal Design of a Continuous Stirred-Tank Reactor by a Gradient Method," *Can. J. Chem. Eng.*, August 1963.

PROBLEMS

6.A. One of the nation's largest chemical manufacturers stores polyethylene pellets in 1,500 cu ft cylindrical vessels with 60° conical bottoms. The actual size is some 9 ft in diameter and 28 ft high.

Assuming that the per area cost C_T of the flat top and conical bottom is n times that of the cylindrical sides C_S, derive an expression for the minimum cost dimensions for such vessels. In practice, n is on the order of two: how does the theory compare to the practice?

Answer:

$$r^3 = \frac{\text{Volume}}{5.8n}$$

6.B. A storage tank for 50 per cent NaOH solution of specific gravity = 1.525 at 20°C is to be designed. The bottom rests on a concrete pad, is $\frac{1}{4}$ in. plate, and will cost (complete with the pad) \$8.00 per square foot. The top is a $\frac{1}{8}$ in. supported roof, costing \$3.00 per square foot. The thickness, t in., of the cylindrical sides will depend upon the hydrostatic pressure, with a minimum of $\frac{3}{16}$ in. as follows:

$$t \geq \tfrac{3}{16} \text{ in.}$$

$$t = \frac{PD}{2S} = \frac{PD}{24{,}000}$$

for

$S = 12{,}000$ psi, allowable stress
$P =$ pressure, psi
$D =$ tank diameter, inches

The sides cost $20t$ dollars per square foot.

Using calculus, determine the geometry and cost of the optimum tank to store:

(a) 7,000 gallons of NaOH
(b) 700,000 gallons of NaOH.

6.C. What is wrong in the following analysis which hopefully might lead to the lowest cost pressure vessel?

We wish to store W_G pounds of methane gas in a pressure vessel of inside diameter D, length L, and thickness t. The standard relationship for the required thickness of a thinwalled vessel to contain material at a pressure P is

$$t = \frac{PD}{2S}$$

where S is the allowable tensile strength of the material of construction. Assume that methane follows the perfect gas law at the temperature of storage.

$$\frac{P}{\rho_G} = RT$$

where ρ_G is the density of the gas. Thus, the constraint on vessel dimensions and pressure take the form

$$W_G = V\rho_G = \left(\frac{\pi D^2}{4} L\right)\left(\frac{P}{RT}\right)$$

Now suppose that the cost of the vessel is directly proportional to the weight of material W_r used in its construction and that we wish to minimize that weight

$$\min\left[W_v = \rho_v\left(\pi DL + \frac{2\pi D^2}{4}\right)t\right]$$

where ρ_v is the density of the material of construction. The length L can be eliminated from the weight equation using the constraint that W_G pounds of methane must be stored in the vessel

$$W_v = \rho_v\left[(\pi D)\left(\frac{RTW_G}{P\pi D^2/4}\right) + \frac{2\pi D^2}{4}\right]\left(\frac{PD}{2S}\right)$$

This is a two variable minimization problem involving D and P, which is solved by setting the appropriate derivatives to zero.

$$\frac{\partial W_v}{\partial D} = \rho_v \frac{P3\pi D^2}{S4} = 0$$

$$\frac{\partial W_v}{\partial P} = \rho_v \frac{\pi D^3}{4S} = 0$$

This gives $D = 0$! How can this be?

6.D. In storing liquefied gases, vapor losses can be critical. Determine an expression for the proper thickness of insulation for a spherical storage tank to contain a liquefied gas with a boiling point below ambient temperature. A balance must be made between the cost of insulation and the cost of vapor losses.

Data

Boiling point of the liquid	t_b	°F
Ambient temperature	t_a	°F
Amortized cost of insulation	c_i	\$/ft³-yr
Cost of vapor loss	c_v	\$/lb
Volume of liquid stored	V	ft³
Heat of vaporization of liquid	λ	Btu/ft³
Thermal conductivity of insulation	k	Btu/ft³-°F

Heat-transfer resistance of tank is negligible

6.E. Ten thousand pounds per hour of an industrial waste contains 0.002 pounds of benzoic acid per pound of water. It is proposed to recover the acid by washing with benzene, as in problem 5.H. The benzoic acid is worth \$0.40 per pound when in the benzene phase, and the benzene costs \$0.01 per pound.

The benzene is soluble in the water phase in the amount of 0.07 pounds of benzene per pound of water. The industrial waste is of no value either before or after the benzene has been removed.

The benzoic acid distributes itself between the two phases according to the distribution $y = 4x$, where y and x are defined in Fig. 6.E-1.

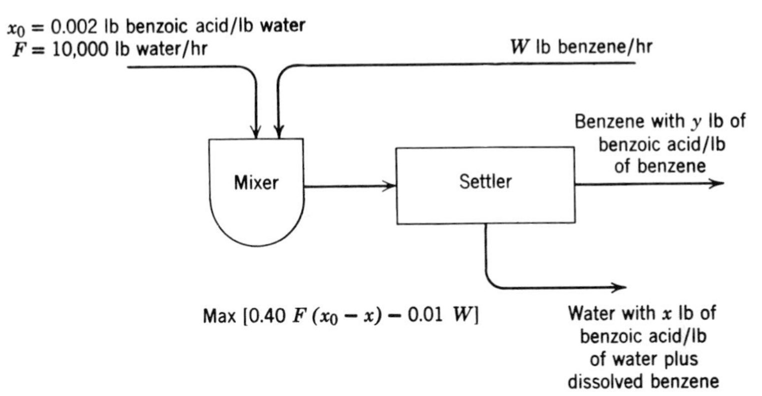

Fig. 6.E-1

Determine by calculus the benzene addition rate which maximizes the profitability of the process.

6.F. It has been proposed to remove a valuable solute from its waste carrier solvent by a series of batch equilibrium extractions with a wash solvent. S pounds of the carrier solvent containing X_0 pounds of solute per pound of solvent are to be placed in a vessel to which W_1 pounds of an immiscible wash solvent is to be added. After the two solvents have been well mixed and the solute distributed between the phases (according to the *linear* equilibrium expression $Y = \alpha X$, where Y is the pounds of solute per pound of wash solvent in equilibrium with the carrier solvent with concentration X pounds of solute per pound of solvent), the two phases are allowed to separate and the wash solvent phase is decanted off as the valuable product.

(a) Derive the following expression for X_i, the concentration of solute in the carrier solvent S after a series of washings with $W_1, W_2, W_3, \ldots W_i$ pounds of wash solvent

$$X_i = X_0 \left(\frac{1}{1 + \alpha W_1/S}\right) \left(\frac{1}{1 + \alpha W_2/S}\right) \cdots \left(\frac{1}{1 + \alpha W_i/S}\right)$$

(b) Suppose we wish to maximize the value of the extracted material minus a charge for the wash solvent used and a labor charge for each washing.

$$\max\,[c_1(X_0 - X_N) - c_2(W_1 + W_2 + \cdots W_x) - c_3 N]$$

The left-hand term in the maximand is the value of the solute extracted, the next term is the cost of N different amounts of solvent, and the right-hand term is labor cost for the N washings.

Demonstrate that equal amounts of wash water should be used.

$$W_1 = W_2 = \cdots = W_N$$

Hint. Differentiate the objective function first with respect to wash W_K and then with respect to wash W_L, and set the results to zero. Show that this leads to $W_K = W_L$, and that K and L can take on any value 1, 2, 3, ... N. Thus, equal wash additions are recommended.

(c) Taking advantage of the result in Part (b), derive an expression for W^*, the proper amount of wash to use in each washing for a given number of washings N.

(d) Determine N^*, the proper number of washings.

6.G. Waste heat in the flue gas from a blast furnace is to be used to evaporate water from a dilute brine solution (see Fig. 6.G-1). The flue gas heats a bed

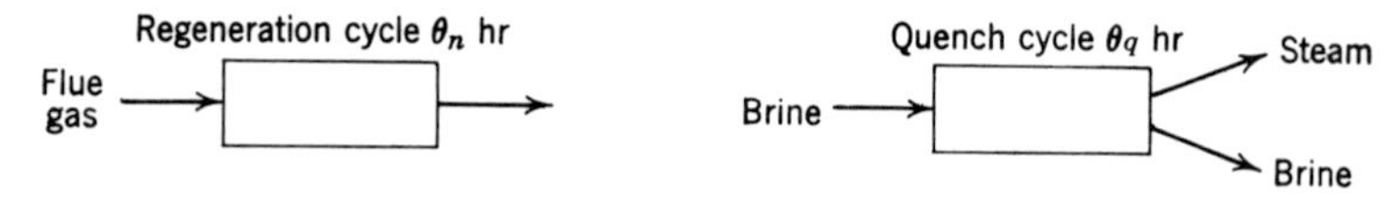

Fig. 6.G-1

of pebbles. The brine is then introduced and the accumulated heat in the pebbles boils off the water. We wish to determine the temperature to which the pebbles must be heated in order that a maximum rate of water removal will be achieved.

Data

The time to heat the bed by the flue gas is approximated by

$$\theta_h = A\left(\frac{T_h - T_0}{T_f - T_h}\right)$$

where

$T_f =$ the flue gas temperature
$T_0 =$ the temperature of the bed after quenching
$T_h =$ the temperature to which the bed is heated

The time of the quench cycle is observed to be relatively independent of the bed temperature.

Answer:

$$T_h = \frac{T_f + T_0\sqrt{A/\theta_q}}{1 + \sqrt{A/\theta_q}}$$

6.H. The productivity of a catalyst in a reactor decreases with time due to carbon fouling. It is necessary to recharge the reactor with fresh catalyst after a given period of time. Determine the optimal catalyst cycle time θ^* from the following data.

Decay of the catalyst productivity P (\$ product/hr), over θ hours $= P_0\, e^{-a\theta}$.
Cost of recharging the catalyst $C =$ \$/charge.
The production time lost during recharging is negligible.

Answer:

$$\frac{C}{P_0} = -\theta^* e^{-a\theta^*} + \frac{1}{a}(1 - e^{-a\theta^*})$$

6.I. Sulfur is removed from gas oils in a refinery by *hydrodesulfurization* in which the sulfur is converted to hydrogen sulfide and then removed (see Fig. 6.I-1).

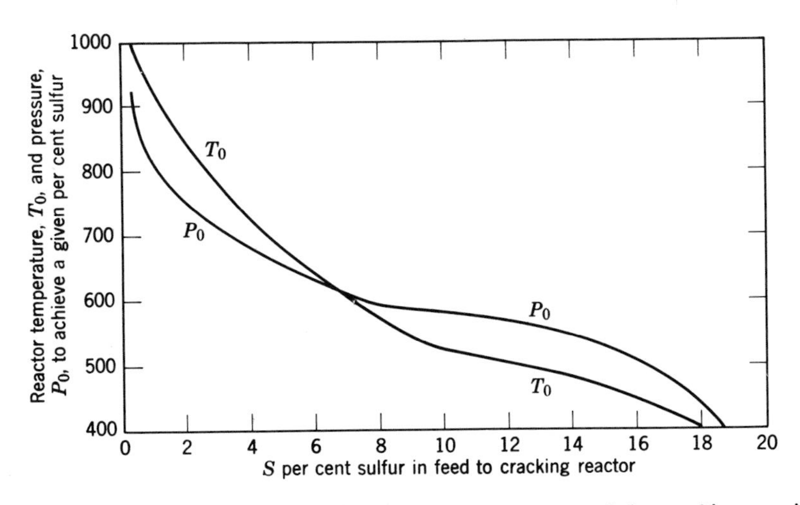

Fig. 6.I-1 Desulfurization reactor temperature and pressures needed to achieve a given per cent sulfur in feed to cracker.

The sulfur acts as a permanent poison to the catalyst used in cracking the gas oils, and therefore the economic justification for the desulfurization unit resides in increasing the life of the cracking catalyst. However, to remove all of the sulfur requires high operating temperatures and pressures in the desulfurization reactor and more extensive hydrogen sulfide removal systems. The following empirical data are available. From these data, search for the minimum cost sulfur content in the feed to the catalytic cracking reactor.

Desulfurization reactor operating costs

$$C_d = 4.0 \times 10^{-2} T_0 + 10^{-2} P_0 \qquad \text{¢ per barrel of gas oil}$$

Hydrogen sulfide removal costs

$$C_n = 0.5S \qquad \text{¢ per barrel of gas oil}$$

Catalyst replacement cost for catalytic cracker

$$C_c = 3.0S \qquad \text{¢ per barrel of gas oil}$$

6.J. A hot oil side stream is presently being drawn off a distillation column at 450°F at a rate of 300,000 pounds per hour. The stream is cooled with water and sent to storage. An operator suggests that a heat exchanger be installed and the hot oil be used to preheat crude oil which enters a process at a rate of 600,000 pounds per hour. Should this be done? What area exchanger should be installed?

Data

Value of heat in process: $0.20/$10^6$ Btu
Heat capacity of fluids: 0.50 Btu/lb °F
Heat-transfer coefficient in exchanger: 60 Btu/hr-°F-ft²
Heat-transfer coefficient is defined by log mean temperature difference

$$\text{Delivered cost of exchanger } I_F = \$35{,}000\left(\frac{A\ \text{ft}^2}{4{,}000}\right)^{0.8}$$

Minimum acceptable return on invested capital: 0.20 $/$-yr

(a) Compute the area of the heat exchanger needed to cool the hot oil to 350°F, the value of the heat recovered in the crude oil, and the investment required in the heat exchanger. Assume that the installed cost of the heat exchanger is 2.5 times the delivered cost.

(b) Search over the proper design variable to find the most economical design.

6.K. Saline water is to be concentrated from 3.5 to 7.0 per cent at a rate of 100,000 pounds per hour in a two stage evaporation system (see Fig. 6.K-1).

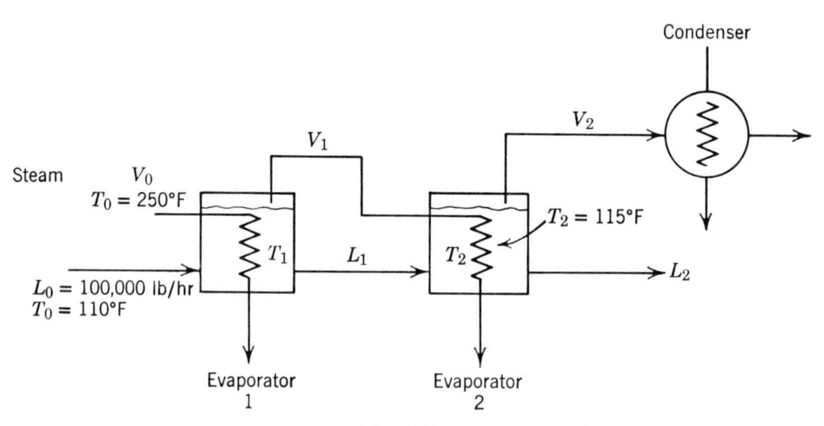

Fig. 6.K-1

Steam is available at 250°F and the seawater enters at 110°F. To avoid the formation of salt precipitates which would foul the heat-transfer surfaces, the temperature in the last stage is to be maintained at 115°F. This is accomplished by providing sufficient heat-exchange area in the vapor condenser. The heat losses and boiling point rise are assumed to be negligible, and the latent heat of vaporization, the specific heat, and the overall heat-transfer coefficents are assumed constant. Determine the heat-transfer surface areas for the two evaporators which minimize the consumption of steam. Design equations for stage i, $i = 1, 2$ are

$$V_{i-1}\lambda + C_p L_i (T_{i-1} - T_i) = V_i \lambda \qquad \text{energy balance}$$

$$L_{i-1} = L_i + V_i \qquad \text{material balance}$$

$$A_i = \frac{V_{i-1}\lambda_{i-1}}{U(T_{i-1} - T_i)} \qquad \text{heat-transfer area}$$

$$C_p = 1.0 \text{ Btu/lb-°F}$$

$$\lambda = 1{,}000 \text{ Btu/lb}$$

$$U = 100 \text{ Btu/ft}^2\text{-°F}$$

6.L. Repeat Problem 6.K for the counter-current evaporator system. Which of the two alternatives, cocurrent or countercurrent, is optimal according to the minimum steam consumption criterion? (See Fig. 6.L-1).

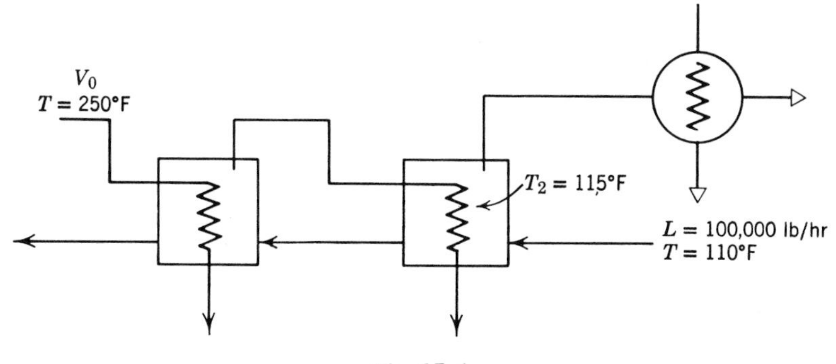

Fig. 6.L-1

6.M. Can you demonstrate without any calculations that a three stage evaporator system will at least be as economical of steam as the two stage system of the same structure? Prove this in the special case of Problem 6.K, by optimizing a three stage cocurrent process.

6.N. Optimize a three stage countercurrent evaporator system patterned after the system in Problem 6.L. Which of the four systems proposed for saline water concentration is most economical in the consumption of steam?

6.O. The minimization of steam consumption is not a valid design criterion, because this criterion fails to account for the costs of cooling water and for the investment in process equipment. We might attempt to repair this deficiency by estimating the cost of the evaporators, the condenser, and cooling water for the minimum steam consumption designs, and then computing a venture profit. The design which gives the maximum venture profit would then be selected as optimal. Demonstrate why this reasoning is erroneous.

7

LINEAR PROGRAMMING

The term *linear programming* describes a large and important class of optimization problems which involve linear objective functions subject to linear constraints. These problems arise frequently in process engineering, for example, in the scheduling of oil refinery operations. The direct search methods of linear programming are so efficient and useful that problems involving as many as 25,000 variables are solved in the normal course of industrial activity.

Typical Problems

The engineers of an oil company wish to blend a number of components: alkalytes, isopentane, straight run gasoline, and so forth. From these they wish to obtain a number of marketable products: aviation gasoline, automotive gasoline, jet fuel, and the like. Given the costs and physical property specifications of the number of components and products, how should the blend be performed to be most economical?

7.1 INTRODUCTION

The general linear programming problem can be stated as: *given a set of m linear equations and/or inequalities involving n variables, find the nonnegative values of these variables which satisfy the equations and inequalities and also maximize or minimize a linear objective function.*

In this chapter we shall see how this kind of problem arises in engineering and how it can be solved efficiently. First, a special class of the linear programming problems called the *transportation problem* is introduced and a method of solution called the *stepping stone* method is developed. Based on the experience drawn from this method of solution the *simplex method* is introduced for solving a more general form of the general linear programming problem.

These methods of direct search are so efficient that the bottleneck in the solution of linear optimization problems does not reside in the methods of solution, but rather in the gathering and keeping up to date of the economic and physical data needed to define the problem, and in the transformation of the industrial optimization problem into the linear mathematical form. Were this only true in all facets of process engineering, the engineer could focus attention on the technical and economic problems of process designs without the constant concern of forming mathematical problems which are too difficult to solve. Unfortunately, the mathematics has reached this ideal level of usefulness perhaps only in the area of linear programming.

7.2 THE TRANSPORTATION PROBLEMS

We shall now introduce the transportation problem first posed by Hitchcock[1] in 1941, and the stepping-stone method of solution introduced by Charnes and Cooper[2] in 1954. Possible applications of the transportation problem methods to the utilization of process equipment will suggest themselves to the engineer. However, the principles are easiest to visualize in the transport context.

Suppose we are confronted with the problem of dispatching ten barges from three points of origin A, B, and C to four points of destination I, II, III, and IV. The number of barges available and required at each point is given in the following table, along with the cost of transport.

			Destination			
			I	II	III	IV
			1	1	4	4
	A	5	$ 700	1,800	2,100	1,400
Origin	B	3	2,400	900	800	1,100
	C	2	1,600	2,300	1,500	1,000

For example, there are two barges at point C, four barges required at point IV, and a cost of transport between these two points of $1,000 per barge. We wish to find the allocation of barges that satisfies the requirements at a minimum total cost.

The transportation problem is a special case of the linear programming problem. Let x_{ij} be the number of barges to be dispatched from point of origin i to point of destination, j, c_{ij} the transportation cost per barge over that route,

[1] F. L. Hitchcock, "The Distribution of a Product from Several Sources to Numerous Localities," *J. Math. Phys.*, **20**, 224–230 (1941).

[2] A. Charnes and W. W. Cooper, "The Stepping-Stone Method of Explaining Linear Programming Calculations in Transportation Problems," *Management Sci.*, **1**, 1 (1954).

a_i the barges available at i, and b_j the barges needed at j. This transportation problem can then be written as

$$\operatorname*{Min}_{(x_{ij})}\left[\sum_i \sum_j c_{ij} x_{ij}\right] \tag{7.2-1}$$

subject to

$$x_{ij} \geq 0, \qquad \text{for all } i \text{ and } j$$

$$\sum_i x_{ij} = b_j, \qquad \text{for all } j$$

$$\sum_j x_{ij} = a_i, \qquad \text{for all } i$$

a linear optimization problem subject to linear constraints.

The first step in our optimization plan might be to determine a reasonably low cost allocation by dispatching the lowest cost barges first. For example, the cost of transport between points A and I is the lowest, \$700 per barge; we can dispatch one barge yielding the reduced problem.

		II	III	IV
		1	4	4
A	4	1,800	2,100	1,400
B	3	900	800	1,100
C	2	2,300	1,500	1,000

The lowest allocation now involves dispatching three barges from B to III at a cost of \$800 per barge, yielding the further reduced problem.

		II	III	IV
		1	1	4
A	4	1,800	2,100	1,400
C	2	2,300	1,500	1,000

Following through with this plan of dispatching the lowest cost barges first, yields the following feasible[3] allocation of barges. The total cost of this allocation is \$11,800. Is this optimal?

		I	II	III	IV
		1	1	4	4
A	5	1	1	1	2
B	3	0	0	3	0
C	2	0	0	0	2

We now present a test for optimality for any *feasible solution* satisfying the following two conditions.

[3] A feasible solution is one in which all the barges are allocated to meet the constraints but not necessarily at a minimum total cost.

No more than $n + m - 1$ routes are being used, where n is the number of points of origin and m is the number of points of destination.

It is impossible to increase or decrease any individual allocations without either changing the routes used or violating column or row restrictions. This is the concept of *independence* of the feasible solution.

Notice that the feasible allocation of barges above satisfies these conditions. The general method of obtaining the initial feasible allocation of barges by allocating *all* the barges possible over the lowest cost route first will yield an independent feasible solution and it can be shown that the optimal allocation of barges must also be an independent feasible solution. Thus, these two restrictions on the test of optimality do not limit its usefulness at all.

In the test for optimality, fictitious costs of transport $c_i + d_j$ are defined for each route employed in a feasible solution. First we define these costs and then show how they are used in the search for an optimal allocation.

The fictitious costs c_i are appended to each row in the allocation array, and the d_j to each column. The numerical values are then adjusted so that

$$c_i + d_j = c_{ij}$$

for each route *used in the feasible solution*, where c_{ij} is the actual cost of transport. This is done by defining some one c_i as zero and building up the remaining fictitious costs from there, as illustrated below for the previous feasible solution.

	I	II	III	IV	c_i
A	700	1,800	2,100	1,400	0
B	●	●	800	●	−1,300
C	●	●	●	1,000	−400
d_j	700	1,800	2,100	1,400	

(● Unused route)

Notice that for each route used in the previous allocation, the sum of the two components of the fictitious cost equals the actual cost.[4]

We have shown how these costs c_i and d_j are determined, and we now show how they can be used to test for optimality.

Consider any change in a feasible solution which involves a new route, for example, the dispatch of a barge from B to I. This must be accompanied by a corresponding change in the number of barges dispatched over some of the existing routes, to maintain feasibility. For example, we might move a barge

[4] Should the case arise where less than $n + m - 1$ routes are used in the independent feasible solution, we merely introduce a route over which nothing is transported in the proper place in the table so that fictitious costs can be calculated.

from cell A, I to cell B, I and a barge from cell B, III to cell A, III. The change in total cost of transport is computed to be

$$\Delta c_t = c_{B,\mathrm{I}} - c_{A,\mathrm{I}} + c_{A,\mathrm{III}} - c_{B,\mathrm{III}}$$

Using the definition of fictitious costs

$$\begin{aligned} \Delta c_t &= c_{B,\mathrm{I}} - (c_A + d_{\mathrm{I}}) + (c_A + d_{\mathrm{III}}) - (c_B + d_{\mathrm{III}}) \\ &= c_{B,\mathrm{I}} - (c_B + d_{\mathrm{I}}) \end{aligned} \tag{7.2-2}$$

Therefore, in Eq. 7.2-2, if the sum of the fictitious costs, $c_B + d_{\mathrm{I}}$, is greater than the actual cost, $c_{B,\mathrm{I}}$, this change in allocation will reduce the total cost.

Thus, a simple test for optimality has been devised.

Compute fictitious costs c_i and d_j so that $c_i + d_j = c_{ij}$ for each route used in a feasible allocation. If $c_i + d_j > c_{ij}$ for any unused route, this feasible solution is not an optimal solution, and can be improved.

We now test the initial feasible solution for optimality.

	I	II	III	IV	c_i
A	×	×	×	×	0
B	2,400	900	×	1,100	−1,300
C	1,600	2,300	1,500	×	−400
d_j	700	1,800	2,100	1,400	

(× Route used)

One unused route exists for which $c_i + d_j > c_{ij}$, route C, III. Our initial feasible solution is *not* optimal.

Having detected nonoptimality in the initial feasible solution we now determine how the nonoptimal solution can be modified to improve the allocation. We shall use a stepping stone device for this modification. First we ask if it is possible to define a *loop* by stepping from the cell where the optimality test failed to only used cells and finally back to the original cell. No diagonal steps are allowed. It will always be possible to find one or more such loops for an independent feasible solution. An improved allocation will be found by moving barges around a loop into the unused cell at which the test for optimality failed.

For example, the loop shown below:

	I	II	III	IV
A	×	×	×	×
B			×	
C			□	×

would suggest the new allocation.

		I	II	III	IV
		1	1	4	4
A	5	1	1	0	3
B	3	0	0	3	0
C	2	0	0	1	1

The total cost of this second feasible solution is $11,600. The second allocation is less costly than the first.

There are special techniques for determining exactly how this transfer should be made around a loop, but these are not needed in the small problems discussed here. However, these techniques save manipulations in larger problems: the reader is referred to the discussion by Hadley.[5]

A test for the optimality of this improved solution follows:

Construction of fictitious costs

					c_i
	700	1,800	●	1,400	0
	●	●	800	●	−1,100
	●	●	1,500	1,000	−400
d_j	700	1,800	1,900	1,400	

Test for optimality

	×	×	2,100	×	0
	2,400	900	×	1,100	−1,100
	1,600	2,300	×	×	−400
d_j	700	1,800	1,900	1,400	

In this new feasible allocation there exists no unused route for which the fictitious cost is greater than the actual cost. A new route cannot be introduced to lower the cost for this feasible solution. This allocation is optimal.

Notice the ease and rapidity of solution. An optimization problem with four degrees of freedom has been solved in two steps. This astounding efficiency derives from the linearity of the problem.

Next we focus attention on a more general linear programming problem, involving inequality constraints. A general strategy of solution is called the *simplex method.* To develop the simplex method rigorously, we need a background in the theory of convex point sets, which is beyond this text. Hence, our presentation will be suggestive. Keep in mind that the simplex technique differs only in detail, and not in principle, from the simple approach of this section. To use the simplex method effectively requires no sophisticated mathematics for the method has been reduced to rote.

[5] G. Hadley, *Linear Programming*, Addison-Wesley, Reading, Mass., 1962, Chap. 9.

7.3 THE SPECIAL PROPERTY OF EXTREME POINTS

In this section we demonstrate that the solution of a linear optimization problem is to be found only at one of several distinguished locations, called *extreme points*. The extreme points are defined by the constraints on the problem and are relatively few in number. The linear programming algorithms such as the simplex method derive their efficiency by limiting search to extreme points.

Suppose we wish to blend three metals, A, B, and C, to form ten tons of an alloy. The alloy must satisfy certain specifications. Namely, the alloy must contain at least 25 per cent lead, not more than 50 per cent tin, and at least 20 per cent zinc. The compositions and costs of the three metals are shown in Table 7.3-1.

Table 7.3-1

Metal	A	B	C
Component			
Lead	0.1	0.1	0.4
Tin	0.1	0.3	0.6
Zinc	0.8	0.6	0.0
$/ton	1,400	2,000	3,000

What blend of these metals will produce an alloy which satisfies the specifications at minimum cost?

This minimum cost blending problem can be cast as a linear programming problem. Let x_A and x_B be the tons of metal A and B used in the blend. By difference, the tons of metal C is $10 - (x_A + x_B)$. The cost of the alloy, to be minimized, is then

$$\begin{aligned} \text{Cost} &= 1{,}400x_A + 2{,}000x_B + 3{,}000[10 - (x_A + x_B)] \\ &= 30{,}000 - 1{,}600x_A - 1{,}000x_B \end{aligned} \tag{7.3-1}$$

The constraints are:

Lead: $0.1x_A + 0.1x_B + [10 - (x_A + x_B)]0.4 \geq 10(0.25)$

or

$$0.3x_A + 0.3x_B \leq 1.5 \tag{7.3-2}$$

Tin: $0.1x_A + 0.3x_B + [10 - (x_A + x_B)]0.6 \leq 10(0.50)$

$$0.5x_A + 0.3x_B \geq 1.0 \tag{7.3-3}$$

Zinc: $$0.8x_A + 0.6x_B \geq 10(0.2) = 2.0 \tag{7.3-4}$$

Amounts of A, B, and C which form an alloy that meets the three specifications are said to form a *feasible solution* to the blending problem. The region of feasible solutions is shown in Fig. 7.3-1. Notice the five points at the extremes of the region of feasible solutions. These are the *extreme points*, one of which corresponds to the minimum cost blend.

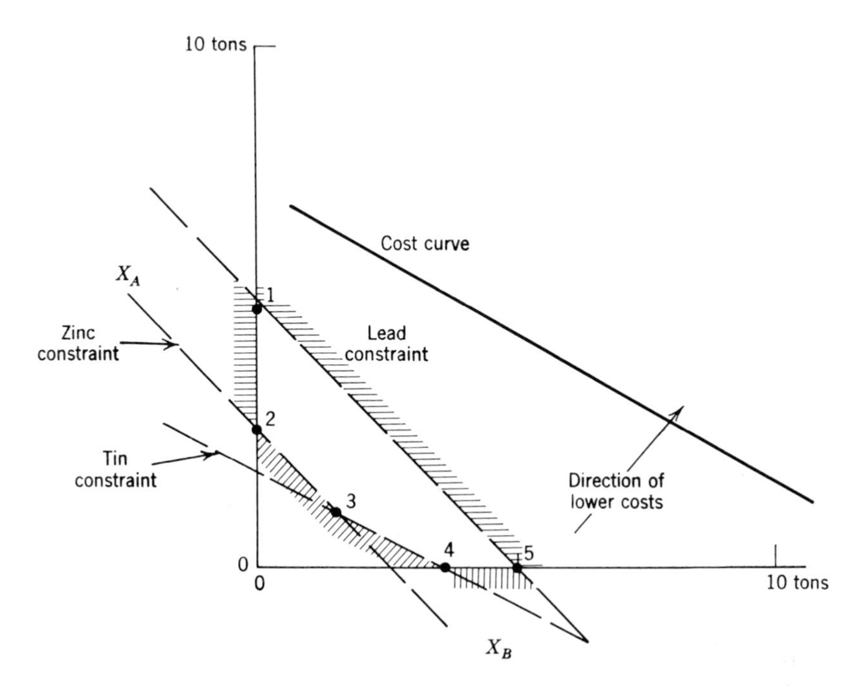

Fig. 7.3-1 The five extreme points.

The cost function

$$\text{Cost} = 30{,}000 - 1{,}600x_A - 1{,}000x_B$$

is a linear curve with slope $S = -10/16$. The *minimum cost* blend is achieved when $x_A = 10$ tons, $x_B = 0$, and $x_C = 0$; but, this solution is *not feasible* since the constraints are violated. We see that the minimum cost feasible blend is reached at extreme point 1, with $x_A = 5$ tons, $x_B = 0$, and $x_C = 5$ tons.

In general, for any cost of the three metals, we need search over only these five extreme points. Thus, if a method is available for detecting the extreme points, the search for the solution of a linear programming problem is greatly simplified.

7.4 THE SIMPLEX ALGORITHM

The geometric approach of the previous section is useful only pedagogically to show that the solution must fall on an extreme point. A nongeometric method is required for problems involving more than a couple of variables. The simplex technique offers such an approach.

The simplex technique developed by Dantzig[6] in 1947 involves manipulations not altogether different from those developed in Section 7.2 for the solution of the transportation problem. This technique has been programmed for nearly every digital computer and dominates the linear programming scene. We present this method by means of a simple example.

The form of the simplex method to be presented now is designed to solve linear programs with the inequalities shown in Eq. 7.4-1.

$$\max\,(c_1x_1 + c_2x_2 \dots c_nx_n)$$

subject to

$$\begin{aligned} &x_1 \geq 0;\ x_2 \geq 0; \dots x_n \geq 0 \\ &a_{11}x_1 + a_{12}x_2 + \dots a_{1n}x_n \leq b_1 \\ &a_{21}x_1 + a_{22}x_2 + \dots a_{2n}x_n \leq b_2 \\ &\vdots \qquad\quad \vdots \\ &a_{m1}x_1 + a_{m2}x_2 \quad \dots a_{mn}x_n \leq b_m \end{aligned} \tag{7.4-1}$$

where $b_1, b_2, b_3 \dots b_m$ are not less than zero. Notice that Eq. 7.4-1 is but one form of the general linear programming problem stated in Section 7.1.

For the purposes of illustration we shall solve the following trivial problem.

$$\max\,(6x_1 + 4x_2)$$

subject to

$$\begin{aligned} x_1, x_2 &\geq 0 \\ 2x_1 + 4x_2 &\leq 14 \\ 6x_1 + 3x_2 &\leq 11 \end{aligned} \tag{7.4-2}$$

The six steps of the simplex algorithm are:

Step 1. Convert the inequalities to equalities by the introduction of the appropriate nonnegative slack variables and place the entire problem into the tabular form of Table 7.4-1.

[6] G. Dantzig, *Activity Analysis of Production and Allocation*, T. C. Koopman, ed., Wiley, New York, 1951, Chap. XXI.

Table 7.4-1

c_i	x_i	x_1	x_2	x_n	x_{n+1}	x_{n+2}	x_{n+m}	B
0	P_{n+1}	a_{11}	a_{12}	a_{1n}	1	0	0	b_1
0	P_{n+2}	a_{21}	a_{22}	a_{2n}	0	1	0	b_2
0	P_{n+m}	a_{m1}	a_{m2}	a_{mn}	0	0	1	b_m
	c_j	c_1	c_2	c_n	0	0	0	
Solution		0	0	0	b_1	b_2	b_m	
Δ_j		Δ_1	Δ_2	Δ_n	0	0	0	

Example:
Eqs. 7.4-2 are transformed to equalities by the introduction of two slack variables x_3 and x_4

$$2x_1 + 4x_2 + x_3 = 14$$

$$6x_1 + 3x_2 + x_4 = 11$$

The tabular form is shown in Table 7.4-2.

Table 7.4-2

Basic set c_i	x_i	x_1	x_2	x_3	x_4	B	b_i/a_{ie}
0	x_3	2	4	1	0	14	14/2
0	x_4	6	3	0	1	11	11/6 ← Leave
	c_j	6	4	0	0		
Solution		0	0	14	11		
Δ_j		6	4	0	0		
		↑ Enter					

Step 2. Read off a *feasible* solution and insert it as the second row below the table. An obvious feasible solution involves only those variables whose columns consist of one (1) and the rest (0), the solution being the value of b at the row where the single entry appears. The variables which form this feasible solution are called the *basic set* and are noted in the two columns on the left of the table.

Example:
The feasible solution $x_1 = 0$, $x_2 = 0$, $x_3 = 14$, and $x_4 = 11$ in Table 7.4-2 corresponds to the origin in Fig. 7.4-1.

Step 3. The feasible solution is now tested for optimality. Each variable which is not part of the feasible solution above is *evaluated* by computing

$$\Delta_j = c_j - \sum_i a_{ij} c_i$$

If one or more of the Δ_j are positive, the feasible solution is non-optimal. If $\Delta_j \leq 0$ for all j, the solution has been found, and the simplex method terminates.

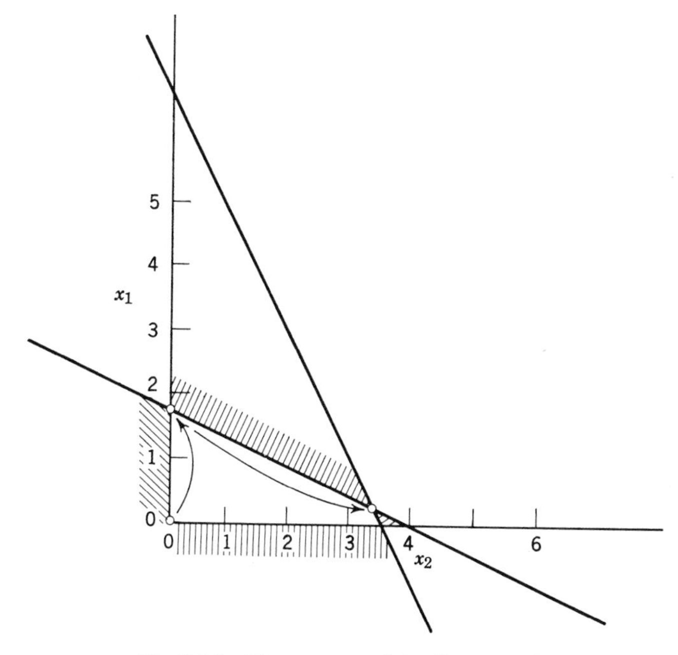

Fig. 7.4-1 The progress of the direct search.

Example:
In Table 7.4-2, Δ_1 is computed by taking the sum of the products of the elements in the first and third columns of the entire table and subtracting that sum from the cost appended below the third column

$$
\begin{aligned}
0 \times 2 &= 0 \\
0 \times 6 &= 0 \\
\hline
\text{Sum} \quad &= 0 \\
\Delta = 6 - 0 &= 6
\end{aligned}
$$

Δ_2 is calculated in a similar manner. There are two Δ_j greater than zero: $\Delta_1 = 6$, $\Delta_2 = 4$. The solution is nonoptimal.

Step 4. The variable with the maximum Δ_j is selected to enter the solution in the place of some variable already in the solution. The variable to leave the solution is found by computing b_i/a_{ie} where e is the entering variable. The variable with the smallest b_i/a_{ie} will leave.

Example:
Variable x_1 will enter the solution and x_4 will leave.

Step 5. The table is now modified so that the entering variables will be distinguished by having a 1 in the row of the departing variable and zeros elsewhere. This is done by row manipulations in the table, multiplying rows by constants, and subtracting from other rows until the job is done.

Example:
The second row of coefficients is divided by six to get the one at the (2, 1) position, and twice this row is subtracted from row 2 to remove the 2 in the (1, 1) position. Table 7.4-3 results.

Table 7.4-3

c_i	x_i	x_1	x_2	x_3	x_4	B	b_i/a_{ie}
0	x_3	0	3	1	−2/6	31/3	31/9 ← Leave
6	x_1	1	3/6	0	1/6	11/6	11/3
	c_j	6	4	0	0		
Solution		11/6	0	31/3	0		
Δ_j		0	1	0	−1		
			↑ Enter				

Step 6. Return to Step 3 and continue until the solution tests for optimality.

Example:
In Table 7.4-3 a Δ_j is found greater than zero, hence, the solution is nonoptimal, and variable x_2 must enter the solution. Step 4 indicates that variable x_3 is to leave, and Step 5 leads to Table 7.4-4: which tests for optimality with $x_1 = 1/9$ and $x_2 = 31/9$.

Table 7.4-4

c_i	x_i	x_1	x_2	x_3	x_4	B
4	x_2	0	1	1/3	−1/9	31/9
6	x_1	1	0	−1/6	−2/9	1/9
	c_i	6	4	0	0	
Solution		1/9	31/9	0	0	
Δ_j		0	0	−1/3	−8/9	

It is interesting to trace the progress of the simplex method on Fig. 7.4-1. Notice that the successive feasible solutions are the extreme points on the region of feasible solutions and that not all extreme points were examined.

It can be shown that the simplex method generally converges in between m and 2m trials, where m is the number of inequalities. In the example where m = 2, convergence should then occur between 2 and 4 trials; it took three trials.

The six simple rules can be easily programmed, as they have been for nearly all computers. It is not unusual to find simplex problems involving several hundred constraints and involving even more variables being solved during the optimization of industrial processes.

7.5 PETROLEUM REFINERY SCHEDULING

The process engineer has only a passing interest in the mathematical aspects of linear programming for they are down pat for the most part. The real problem in the application of linear programming is that of problem formulation rather than problem solution. We shall now examine this difficulty by considering an optimization problem in refinery scheduling.

Any attempt to define realistically the operation of a large oil refinery can easily lead to 100 or even 500 constraining equations; a problem of such magnitude that we would only attempt to formulate it (much less attempt the solution) if considerable benefits were expected. In such an industrial problem, however, the benefits resulting from more efficient operation may amount to hundreds of thousands of dollars per year. We now present an extremely simplified version of a refinery scheduling problem to expose the principles of analysis.

The sources from which this example is drawn are the text of Hadley,[7] and the monographs of Manne[8] and Symonds.[9] Figure 7.5-1 illustrates a simplified oil refinery, the operation of which will be examined.

The atmospheric distillation of crude oil yields a variety of petroleum fractions such as naphthas, gas oils, and bottoms. These petroleum fractions may be blended directly to form the products which range from aviation gasoline to fuel oils, or a fraction may be processed further before blending. The gas oils may be sent to a cracking unit to break the long chain hydrocarbons, and the

[7] G. Hadley, *op. cit.*

[8] A. Manne, *Scheduling of Petroleum Refinery Operations*, Harvard Economic Studies, Vol. 48: Harvard University Press, 1956.

[9] G. H. Symonds, *Linear Programming; The Solution of Refinery Problems*, Esso Standard Oil Company, New York, 1955.

naphthas may be reformed to improve the combustion properties. In a real refinery problem there will be hundreds of streams corresponding to the few illustrated in Fig. 7.5-1.

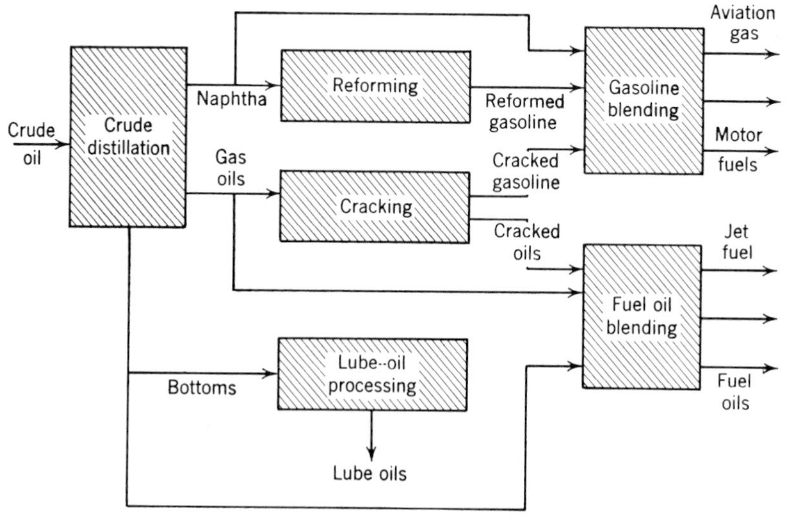

Fig. 7.5-1 An oil refinery block diagram.

An actual refinery may use a dozen or so different crude oils, but we shall assume that only two are available to the refinery and that they differ in chemical composition. Let the availability of these two crudes be θ_1 and θ_2 barrels per day, and let x_1 and x_2 be the yet to be determined amounts to be used by the refinery. Thus, the following constraints arise:

$$0 \le x_1 \le \theta_1$$
$$0 \le x_2 \le \theta_2$$

If the crude distillation tower has a limited capacity of A barrels per day, the following constraint must be imposed.

$$x_1 + x_2 \le A$$

The naphtha stream produced by the crude distillation contains light, medium, and heavy naphthas. Suppose that one barrel of crude i produces a_{1i} barrels of the light naphtha, a_{2i} barrels of the medium naphtha, and a_{3i} barrels of the heavy naphtha. Thus if y_1, y_2, and y_3 barrels per day of the respective naphtha fractions are produced, the following three equations arise

$$y_1 = a_{11} x_1 + a_{12} x_2$$
$$y_2 = a_{21} x_1 + a_{22} x_2$$
$$y_3 = a_{31} x_1 + a_{32} x_2$$

Each naphtha stream is split into two parts: one for gasoline blending and one for reforming. Let y_j^b be the barrels per day of naphtha j sent to blending and y_j^r be the barrels per day sent to reforming. The following equations arise

$$y_1 = y_1^b + y_1^r$$
$$y_2 = y_2^b + y_2^r$$
$$y_3 = y_3^b + y_3^r$$

The crude distillation also produces oils; a light oil (1) and a heavy oil (2). One barrel of crude i produces b_{1i} of light oil and b_{2i} of heavy oil, and if u_j^b and u_j^c are the barrels per day of oil j which are sent to fuel-oil blending and cracking, respectively, then

$$u_1^b + u_1^c = b_{11}x_1 + b_{21}x_2$$
$$u_2^b + u_2^c = b_{21}x_1 + b_{22}x_2$$

If the bottom stream from the crude-distillation is of a single composition and if one barrel of crude i produces d_i barrels of bottoms and if v^p barrels are routed to lube-oil processing and v^b barrels are routed directly to fuel-oil blending, then the material balance on the bottom stream is

$$v^p + v^b = d_1x_1 + d_2x_2$$

Now we look at the reformer. If B is the maximum number of barrels per day that can be processed by the reformer, then

$$y_1^r + y_2^r + y_3^r \leq B$$

If one barrel of naphtha j can be reformed into f_1 barrels of reformed gasoline, and if w is the barrels per day of reformed gasoline produced by the reformer, the following balance arises

$$w = f_1 y_1^r + f_2 y_2^r + f_3 y_3^r$$

Now we examine the cracking operation. If one barrel of oil j can be cracked into g_i barrels of cracked gasoline and h_j barrels of cracked oils, and if r barrels per day of cracked gasoline and s barrels per day of cracked oils are produced, the following equations arise

$$r = g_1u_1^c + g_2u_2^c$$
$$s = h_1u_1^c + h_2u_2^c$$

Further, if D barrels per day is the capacity of the cracking unit, then,

$$u_1^c + u_2^c \leq D$$

For the lube-oil processing unit one barrel of bottoms yields k barrels of lube oil, and the capacity of the lube-oil unit is E barrels per day. Hence, if q is the production of lube-oil in barrels per day

$$q = kv^p$$
$$v^p \leq E$$

Suppose further that the refinery must supply at least S barrels per day of lube-oil to its subsidiaries.

$$q \geq S$$

We now move to the blending operation and examine the quality restrictions on the final products. Suppose that for each gasoline there is but a single restriction; the aviation, premium, and regular gasolines must have at least the octane numbers p_1, p_2, and p_3. Let m_j^n be the octane numbers of the three naphthas from the crude-distillation, m_r be the octane number of the reformed gasoline, and m_c be the octane number of the cracked gasoline. If these octane numbers blend linearly,[10] and if w^a, w^p, w^r, r^a, r^p, and r^r are the barrels per day of the reformed gasoline and the cracked gasoline blended into aviation, premiums, and regular gasolines, the following equations arise

$$w = w^a + w^p + w^r$$
$$r = r^a + r^p + r^r$$

The total aviation gasoline produced is $\hat{a}$ barrels per day, premium gasoline $\hat{p}$ barrels per day, and regular gasoline $\hat{r}$ barrels per day, where

$$\hat{a} = w^a + r^a + y_1^{ba} + y_2^{ba} + y_3^{ba}$$
$$\hat{p} = w^p + r^p + y_1^{bp} + y_2^{bp} + y_3^{bp}$$
$$\hat{r} = w^r + r^r + y_1^{br} + y_2^{br} + y_3^{br}$$

where y_j^{bk} is the barrels per day of naphtha j which is to be blended into gasoline, where $k = a, p, r$ (aviation, premium, and regular gasoline). To match the total naphthas available for blending

$$y_1^b = y_1^{ba} + y_1^{bp} + y_1^{br}$$
$$y_2^b = y_2^{ba} + y_2^{bp} + y_2^{br}$$
$$y_3^b = y_3^{ba} + y_3^{bp} + y_3^{br}$$

The octane constraints may be written thus

$$\hat{a}p_a \leq w_a m_r + r_a m_c + m_1^n y_1^{ba} + m_2^n y_2^{ba} + m_3^n y_3^{ba}$$
$$\hat{p}p_p \leq w_p m_r + r_p m_c + m_1^n y_1^{bp} + m_2^n y_2^{bp} + m_3^n y_3^{bp}$$
$$\hat{r}p_r \leq w_r m_r + r_r m_c + m_1^n y_1^{br} + m_2^n y_2^{br} + m_3^n y_3^{br}$$

[10] This will not generally be true over a wide range of compositions, for the blending may be nonlinear. However an iterative procedure may be used employing a "blending" octane number which is defined to blend linearly near the optimum.

In addition to the octane constraints on the gasolines, there may be constraints on the amounts produced; for example, only a limited amount of aviation gasoline might find a market. Also, there will be limits on vapor pressure, boiling range, aromatic content, and sulfur content in some or all of the products. Tetraethyl lead might have to be blended in to improve the anti knock qualities of motor fuels.

We shall not detail the formulation of the additional constraints alluded to above nor shall we analyse the fuel oil blending problem. Let it suffice to be said that the scheduling problem of a large refinery can be cast into the form of a large set of linear equations and linear inequalities. These equations involve the variety of variables which describe the flow of petroleum fractions through the refinery.

The economic factors are then brought in to form the detailed optimization problem. Data will be required on the sales price for all the refinery products, the cost of the crude oils, and the costs of processing a barrel of material in each of the several units. From these data a linear profit objective function is formed involving nearly all of the variables.

It becomes obvious that the formulation of an industrially significant linear programming problem requires considerable energy and technical know-how on the part of the engineer and requires access to a vast amount of economic and technical data. These are the obstacles which must be overcome. It is not at all unusual for a number of parameters (such as a_{32}, the barrels of heavy naphtha that will be produced from a barrel of crude oil) to be unknown, and part of the linear programming problem then becomes that of gathering performance data on the refining equipment.

However, all of this effort appears to be worthwhile, since nearly all major refineries are now scheduled by the solution of linear programs. This example in petroleum refinery optimization has been chosen to illustrate the principles of problem formulation. The general method of problem formulation should now be apparent to the engineer.

7.6 CONCLUDING REMARKS

Upon concluding this brief introduction to linear programming we must make an observation that is not obvious from our method of describing the simplex method of direct search. The simplex method is derived from the modern theory of linear algebra, and without this theoretical base it is doubtful if an efficient method of solving linear programming problems would have been developed. In the simplex method we witness the tremendous impact that rigorous mathematical theory can have on such eminently practical problems as operating oil refineries. There is nothing as practical as a good theory, and it behooves the engineer to maintain contact with both camps—the theory and the practice—for a wedding of the two often has astounding results.

Clearly, our discussion of linear programming has been only suggestive, and the vast literature of linear programming deserves study. We offer an introduction to that literature through the following brief list of references.

References

There are several excellent texts on linear programming.

G. Hadley, *Linear Programming*, Addison-Wesley, Reading, Mass., 1962.
G. Dantzig, *Linear Programming and Extensions*, Princeton University Press, Princeton, N.J., 1963.
S. I. Gass, *Linear Programming*, McGraw-Hill, New York, 1958.
S. Vajda, *Readings in Linear Programming*, Wiley, New York, 1958.

PROBLEMS

7.A. Solve the following transportation problem by the stepping-stone method after using the minimum cost route first method of obtaining an initial feasible solution. A degeneracy occurs in which the number of routes used becomes less than $m + n - 1$. This offers no difficulty if the footnote on page 191 is studied.

		2	2	4	3	1	2
	7	100	60	170	70	110	70
Origin	6	30	50	60	200	40	30
	1	95	200	10	40	600	20

		2	2	4	3	1	2
Answer:	7		2		3		2
	6	2		3		1	
	1			1			

7.B. The method of obtaining an initial feasible solution to the transportation problem by using the minimum cost route first might be called the *matrix minima* method. Investigate the properties of the following methods for generating an initial solution to the transportation problem 7.A.

(a) *Column minima.* Starting at the left-hand column of the cost matrix select the minimum cost route. Allocate over that route as many items as possible. If the demand is satisfied at that column move to the next column. If the supply is depleted over the column minimum cost route, select the next lowest route in that column and allocate items over that route and so forth until demand is satisfied at column 1. Sweep through all columns in the same manner.

Answer: *initial feasible solution*

	2	2	4	3	1	2
7			1	3	1	2
6	2	2	2			
1			1			

(b) *Row minima.* The row minima method assigns items in a way similar to the column minima method, the only difference being that the role of the row and column are interchanged in the description of column minimum method.

Answer: *initial feasible solution*

	2	2	4	3	1	2
7		2		3		2
6	2		3		1	
1			1			

Notice that this method accidentally solved the transportation problem

(c) *Vogel's method.* The following method was suggested by W. R. Vogel.[11] For each row i find the lowest cost c_{ie} and the next lowest cost c_{in}, and compute $\Delta_i = c_{in} - c_{ie}$. Proceed in the same way for the columns,

[11] N. V. Reinfeld and W. R. Vogel, *Mathematical Programming*, Prentice-Hall, Englewood Cliffs, N.J., 1958.

computing $\Delta_j = c_{aj} - c_{ne}$ for each column j. Choose the largest of these differences first; this focuses attention on either a row or column. Now allocate first over the route in that row or column that has the minimum cost. This is repeated until the initial feasible solution is obtained.

Answer: *initial feasible solution*

	2	2	4	3	1	2
7		2		3		2
6	2		3		1	
1			1			

7.C. Solve the following transportation problem using the following three methods for generating an initial feasible solution.
(a) Matrix minima method.
(b) Column minima method.
(c) Row minima method.
(d) Vogel's method.

Destination

		50	20	70
Origin	100	$2,000	7,000	4,000
	40	3,000	14,000	1,000

Answer: minimum cost solution of allocation $400,000

7.D. Solve the following transportation problem using the method you deem best for generating the initial feasible solution.

Destination

		6	5	1	8
	3	$70	30	40	60
Origin	7	60	20	200	10
	10	10	70	160	70

7.E. This transportation problem is distinguished by the fact that there are more items at the points of origin than there are items demanded at the points of destination. Some items will therefore not be transported. The simple methods of solution cannot be used directly on the problem in its present form.

		Destination		
		5	3	2
	6	\$100	20	80
Origin	2	70	40	10
	7	100	70	30

Show how the problem can be altered so that the simple stepping stone method of solution can be applied. Introduce a fictitious destination which demands the surplus items and assign zero transportation costs to that destination. When this altered problem is solved, those items that were transported to the fictitious destination are those items that should be considered as surplus in the original problem.

7.F. Solve the following linear programming problem by the simplex method and trace the progress of iteration on a graph of the constraints

$$\max\,(6x_1 + 10x_2)$$

subject to

$$x_1, x_2 \geq 0$$
$$x_1 + x_2 \leq 1$$
$$4x_1 + x_2 \leq 2$$

7.G. The following problem is of the form solvable by the simplex method, but degenerates as the algorithm fails to converge.

$$\max\,(3x_1 + 7x_2 + 1x_3 + 13x_4 + 2x_5 + 11x_6 + 4x_7)$$

subject to

$$x_1, x_2, x_3, x_4, x_5, x_6, x_7 \geq 0$$
$$13x_1 + 2x_2 + 7x_3 + 11x_4 + 2x_5 + x_6 - x_7 \leq 10$$
$$x_3 \leq 1$$

7.H. A refinery has four different crude oils A, B, C, and D which are to be refined to form four different products G, H, L, and F. The crude oils are available in limited amounts and limits are placed on the maximum amount of each product that can be sold. The limits are

Crude oil availability

$$0 \leq A \leq 100 \text{ Mbpw}$$
$$0 \leq B \leq 100$$
$$0 \leq C \leq 200$$
$$0 \leq D \leq 100$$

Product market

$$0 \leq G \leq 170$$
$$0 \leq H \leq 85$$
$$0 \leq L \leq 20$$
$$0 \leq F \leq 85$$

Also available are data on the conversion of a given crude oil into the many products, the operating cost for the refinery, and the value of the final products.

	Barrels of product produced per barrel of crude oil				
	A	*B*	*C*	*D*	Value of product $/barrel
Products *G*	0.6	0.5	0.4	0.3	3.00
H	0.2	0.2	0.1	0.3	2.00
L			0.2		4.00
F	0.1	0.2	0.2	0.3	1.00
Operating cost $/barrel of crude	1.00	2.00	0.50	1.50	

For example, 0.2 barrels of *F* can be produced from one barrel of crude *B* at a cost of $2.00 per barrel of *B* processed. The value of *F* is $1.00 a barrel. The crude oils are available at $0.50 a barrel, except for crude *C* which costs $1.00 per barrel.

Schedule the operation of this refinery for maximum profit.

7.I. The following crude oil distribution problem was suggested by Symonds.[12]

Ten crude oils are available in stated quantities from 10,000 to 30,000 barrels per day each, with a total availability of 200,000 barrels per day. Three refineries *X*, *Y*, and *Z* have required capacities which total 180,000

[12] G. H. Symonds, *Linear Programming: The Solution of Refinery Problems*, Esso Standard Oil Co., New York, 1955.

barrels per day. Of the available crude, 20,000 barrels per day will not be processed. One refinery has two means of processing crude oil X_1 and X_2 at different efficiencies. The net profit or loss for each barrel of crude processed in each refinery is given. Allocate the crude oils to the refineries for maximum profit.

					Crude						Required
Refinery	*a*	*b*	*c*	*d*	*e*	*f*	*g*	*h*	*i*	*j*	Mbpd
X_1	−6¢/barrel	3	17	10	63	34	15	22	−2	15	30
X_2	−11	−7	−16	9	49	16	4	10	−8	8	40
Y	−7	3	16	13	60	25	12	19	4	13	50
Z	−1	0	13	3	48	15	7	17	9	3	60
Available Mbpd	30	30	20	20	10	20	20	10	30	10	

Answer: maximum profit $25,400 per day

7.J. A one hundred pound mix of a certain type of iron is to be formed, containing a minimum of 2 per cent carbon, and 2 per cent manganese. Four materials are available for blending. Determine the least cost blend.

	Material			
	A	*B*	*C*	*D*
Per cent carbon	2	3	4	5
Per cent manganese	3	2	1	0
Per cent iron	95	95	95	95
Cost, $/100 lb	10	15	8	7

8

THE SUBOPTIMIZATION OF SYSTEMS WITH ACYCLIC STRUCTURE

Dynamic programming is a strategy whereby large optimization problems which exhibit no recycle of information can be decomposed into a sequence of simpler suboptimization problems. Under certain conditions, the suboptimization of small parts of the problem leads to order-of-magnitude reductions in the difficulty of optimization. Moreover, this concept of suboptimization is the basis of the more advanced methods of optimization to be presented in Chapters 9 and 10.

Typical Problem

How might the following process system be optimized by confining attention to one design variable at a time?

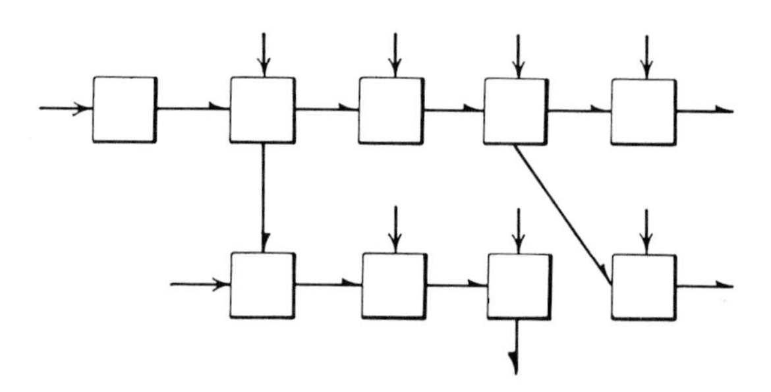

8.1 THE PRINCIPLE OF OPTIMALITY

In Chapter 6 we saw that *direct search* for the optimum design can become quite cumbersome and laborious when a moderate number of variables are to be manipulated. This lack of efficiency derives from the inability of direct search to take advantage of any special properties of the particular optimization problem. In Chapter 7 we observed the great efficiency *linear programming* derived from the special property of linear optimization problems: that the optimum must fall on an extreme point defined by the constraints. In this chapter we examine *dynamic programming*, an optimization plan based on a special structural property of systems with no recycle.

A system is said to have a *serial* or *acyclic* structure when its information flow structure is head-to-tail with no recycle. In such systems, a change in the design of a given component can only influence components downstream. This is the special property of acyclic systems from which dynamic programming derives its efficiency.

In the early 1950's Richard Bellman stated the *Principle of Optimality*.

An acyclic system is optimized when its downstream components are suboptimized with respect to the feed they receive from upstream.

In its general form, this principle and the terms upstream and downstream refer to the flow of design information as discussed in Chapter 3, and this may differ from the direction of material flow in the process flow diagram.

That is, we must do the best with what we have from any point in the system on to the end. This principle is illustrated in Fig. 8.1-1, where larger and larger numbers of components are considered as the *downstream components*. In each case the group of downstream components must be designed for maximum combined objective function, i.e., suboptimized with respect to the feed they receive from upstream.

It is important to understand that suboptimization only applies to the *end group of components* in a system. For example, suppose we attempted to tear the interior boiling water heat exchanger from the system shown in Fig. 8.1-2 to suboptimize its design based on its own direct contribution to the system profitability. The design criterion might be tentatively assigned as:

$$\underset{\text{(area)}}{\text{Maximize}}\ (\text{value of steam} - \text{operating and investment costs})$$

The difficulty is obvious. There is no simple way of including the effects of a change in the design of this interior component on the operation of the downstream components it influences. In this example, the heat exchanger strongly influences the operation of the still. Local economic losses might be tolerated at the exchanger for more than compensating gains in the still downstream. Only the end component can be suboptimized, for *only* there is there nothing downstream to be disrupted.

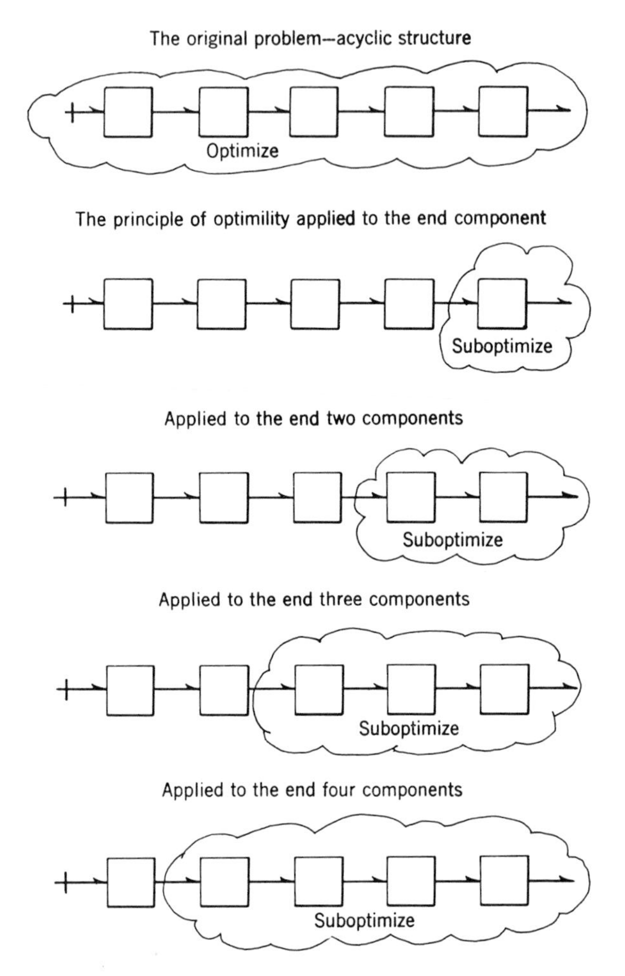

Fig. 8.1-1 The principle of optimality: A guide to suboptimization.

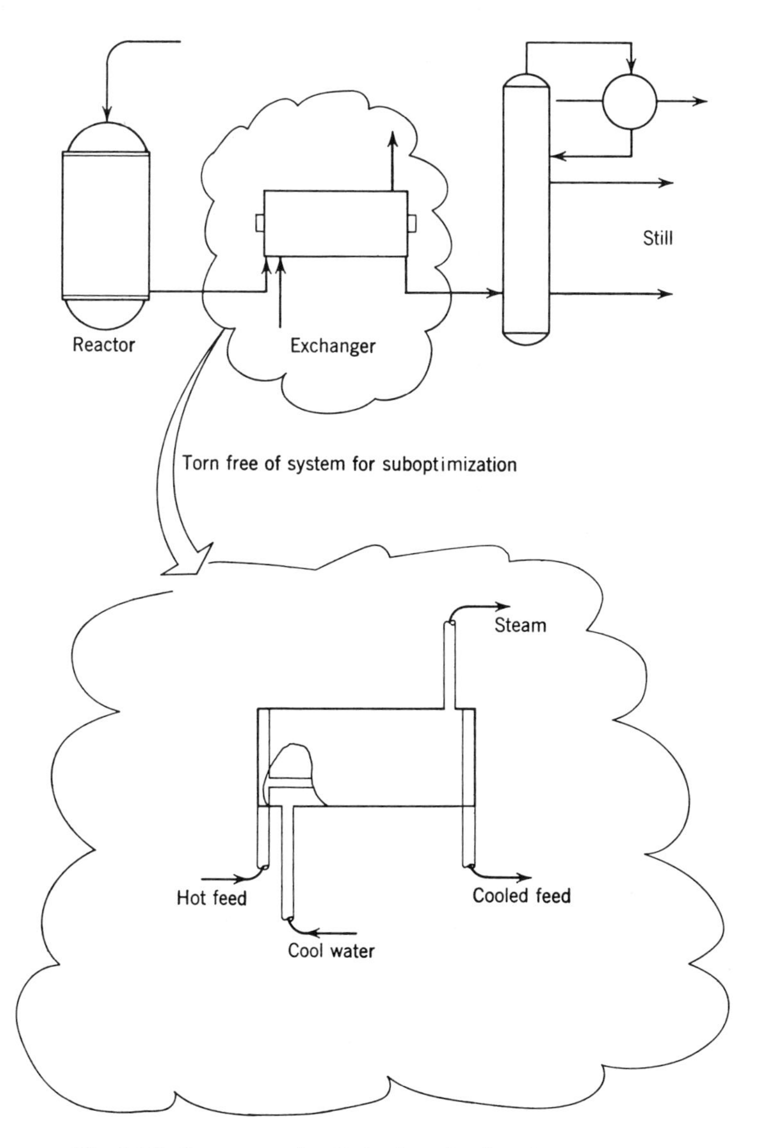

Fig. 8.1-2 Improper suboptimization of an interior component.

8.2 DYNAMIC PROGRAMMING—A SEQUENCE OF SUBOPTIMIZATIONS

We now show how the principle of optimality leads to the *sequential suboptimization* of larger and larger groups of end components, a strategy called dynamic programming. Figure 8.2-1 defines the notation; notice how the

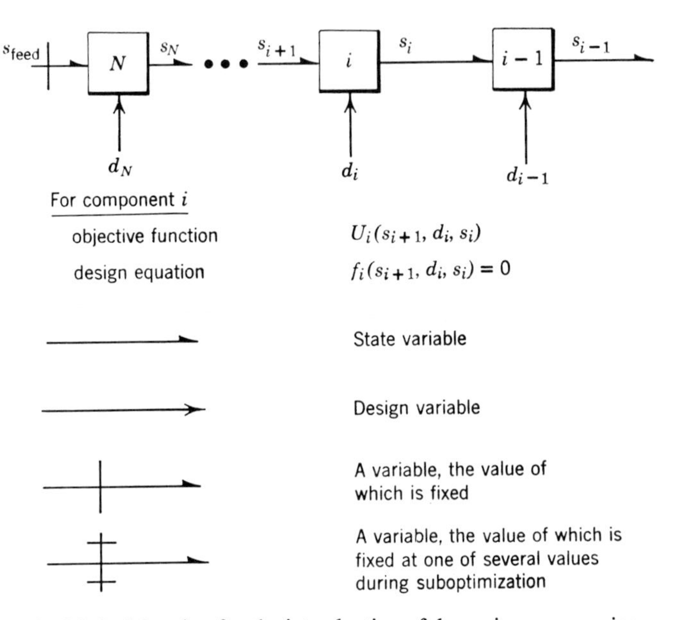

Fig. 8.2-1 Notation for the introduction of dynamic programming.

components are numbered in reverse order to simplify notational problems in dealing with the end components first.

Dynamic programming begins with the suboptimization of the end component numbered 1. This involves the solution of the problem

$$\max_{\{d_1\}} [U_1(s_2, d_1)] = \max{}_1 (s_2)$$

The best value of the design variable, denoted as $d_1{}^*$, and the value of the maximum itself, denoted as $\max_1$, depend on the condition of the feed that this last component receives from upstream (i.e., on s_2). This dependence can be denoted symbolically by $d_1{}^* (s_2)$ and $\max_1 (s_2)$. Since the particular value of s_2 which will obtain after the upstream components have been optimized cannot be known at this stage, this initial suboptimization problem is solved for a range of possible values for s_2 and the results tabulated as in Fig. 8.2-2.

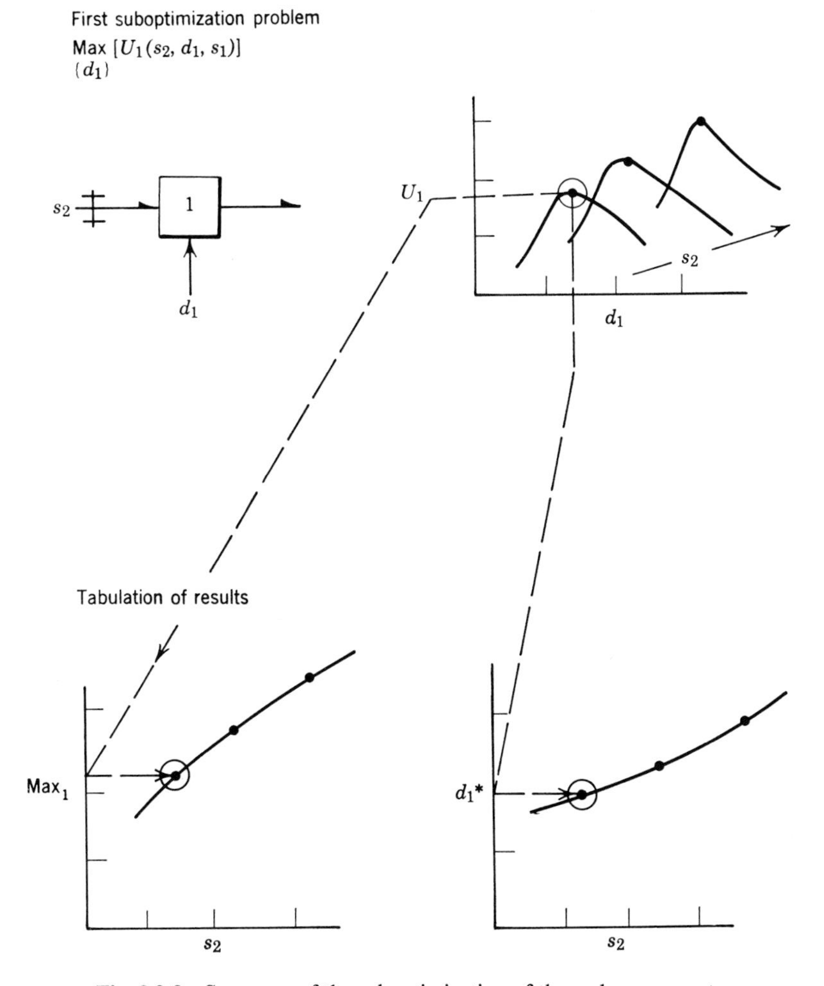

Fig. 8.2-2 Summary of the suboptimization of the end component.

These tables contain a complete summary of the suboptimizations of stage 1 and there is no further need to study that stage in detail.

The next step in the suboptimization strategy is to move up the serial chain to include the *last two* components. The principle of optimality states that the design variables d_1 and d_2 must be adjusted so that these two stages *together* are suboptimized with respect to the feed they receive from upstream, as shown in Fig. 8.2-3. That is

$$\max_{\{d_2, d_1\}} [U_2(s_3, d_2) + U_1(s_2, d_1)] = \max_2(s_3)$$

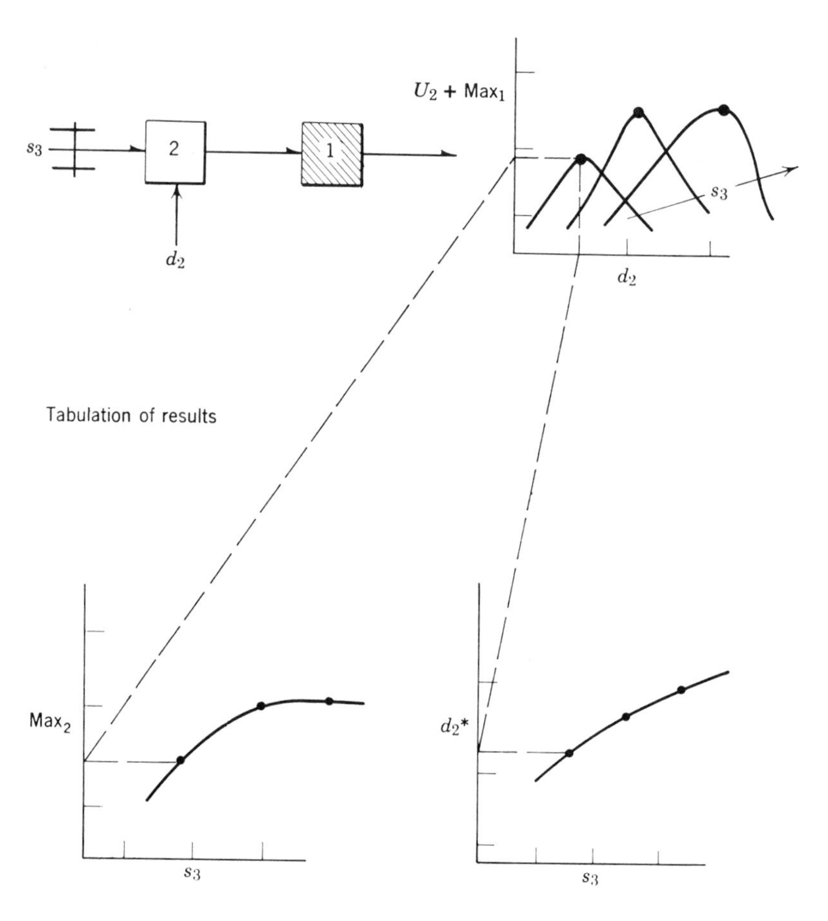

Fig. 8.2-3 Summary of the suboptimization of the end two components.

But, we know all about the optimum design of component 1 and that information is encoded in the graph of $\max_1(s_2)$. This then can be substituted for U_1 to give the following form for the suboptimization of the last two components.

$$\max_{\{d_2\}} [U_2(s_3, d_2) + \max_1(s_2)]$$

Notice that this simple maneuver has reduced the number of variables that need be considered during maximization from *two* (d_1 and d_2) to *one* (d_2). Reference to component 1 is now through the table $\max_1(s_2)$ rather than through a maximization problem.

Now again, we do not know the value of the state variable s_3 which passes information down from upstream, since the upstream components are yet to be optimized. So a range of values of s_3 is considered, as illustrated in Fig. 8.2-3 which (along with Fig. 8.2-2) summarizes the suboptimization of these two components.

Suppose now that this suboptimization sequence has been carried on to include $i - 1$ of the end components, and the next step is to suboptimize the i end components.

$$\max_{\{d_i d_{i-1} \ldots d_1\}} (U_i + U_{i-1} + \cdots U_1) \equiv \max_i$$

Since by this time the $i - 1$ end components would have already been suboptimized, the function

$$\max_{\{d_{i-1} \ldots d_1\}} (U_{i-1} \cdots U_1) = \max_{i-1}$$

would be available, and, hence, could be used to reduce the dimensionality of the i component suboptimization to

$$\max_{\{d_i\}} [U_i + \max_{i-1}(s_i)]$$

where the design equations for stage $i, f_i(s_{i+1}, s_i, d_i) = 0$, provide the link between d_i and s_i.

Again, notice that an i dimensional optimization problem has been reduced to a one dimensional optimization problem by means of a summary of the solutions to the downstream suboptimization problem. This sequential suboptimization strategy of dynamic programming is shown in Fig. 8.2-4. Notice that the design variables are analyzed one at a time rather than simultaneously.

Once stage N is reached the optimization is complete and we need only retrace the steps through the tables to gather up the complete set of d^*'s for the system. Rather than belabor the details at this point we present an example problem which should be considered a medium for transferring the subtle points of dynamic programming.

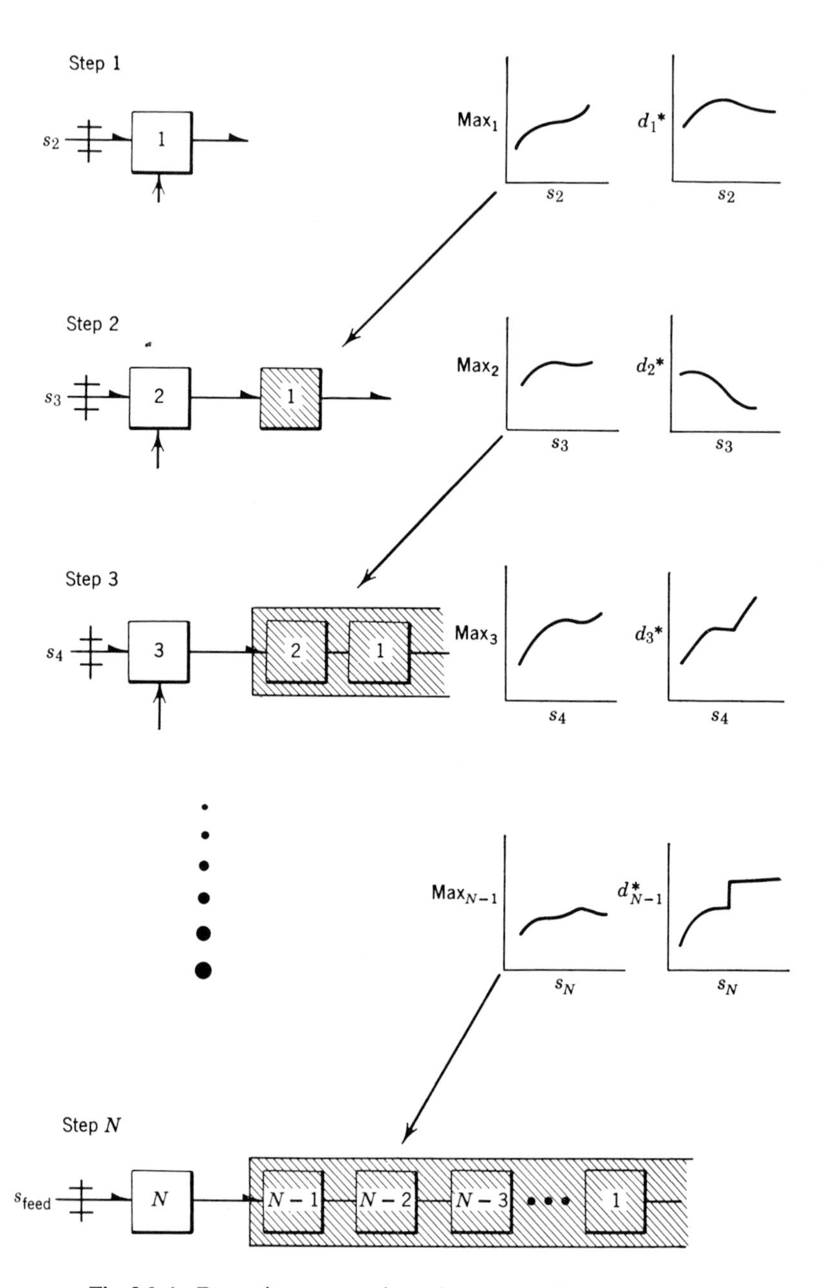

Fig. 8.2-4 Dynamic programming—A sequence of suboptimizations.

8.3 SOLVENT ALLOCATION TO A CROSSCURRENT EXTRACTOR[1]

We illustrate dynamic programming by considering the problem of allocating an expensive solvent to the three stage crosscurrent extractor system shown in Fig. 8.3-1. A carrier solvent contains a valuable solute in a concentration $x_F = 0.20$ pounds solute per pound solvent, and this solution is to be processed at a rate $Q = 1{,}000$ pounds of carrier solvent per hour. The processing consists of extraction by an immiscible wash solvent in a multistage crosscurrent extraction system. A certain amount of wash solvent and carrier solvent are mixed in an extractor and allowed to separate, the solute having reached an equilibrium distribution between the two phases. The carrier solvent phase is sent to a downstream extractor for further processing, and the wash solvent phase containing part of the solute is considered as the valuable product from that given extractor.

We wish to determine the allocation of wash solvent which maximizes the following economic objective function

$$\max_{\{W_1, W_2, W_3\}} \left[Q(x_F - x_0) - \lambda \sum_{i=1}^{3} W_i \right] \tag{8.3-1}$$

The term $Q(x_F - x_0)$ is the pounds per hour of solute removed from the carrier solvent, $\sum W_i$ is the total amount of wash solvent used per hour to achieve that solute removal rate, and λ is the cost of a unit amount of wash solvent divided by the value of a unit amount of solute extracted ($\lambda = 0.05$ in this example).

The first step in the analysis of any optimization problem is the establishment of the design relationships which describe the performance of each stage in the system.

Material balances over extractor i

$$\left.\begin{array}{l} \text{Carrier solvent:} \\ Q_{\text{in}} = Q_{\text{out}} = Q \\ \text{Wash solvent:} \\ W_{\text{in}} = W_{\text{out}} = W \end{array}\right\} \begin{array}{l}\text{a result of the assumed} \\ \text{immiscibility of solvents}\end{array}$$

Solute:

$$Qx_{i+1} = Qx_i + Wy_i \tag{8.3-2}$$

[1] R. Aris, D. F. Rudd, and N. R. Amundson, "On Optimum Cross-Current Extraction," *Chem. Eng. Sci.*, **12**, 88 (1960).

A typical extractor—stage 2

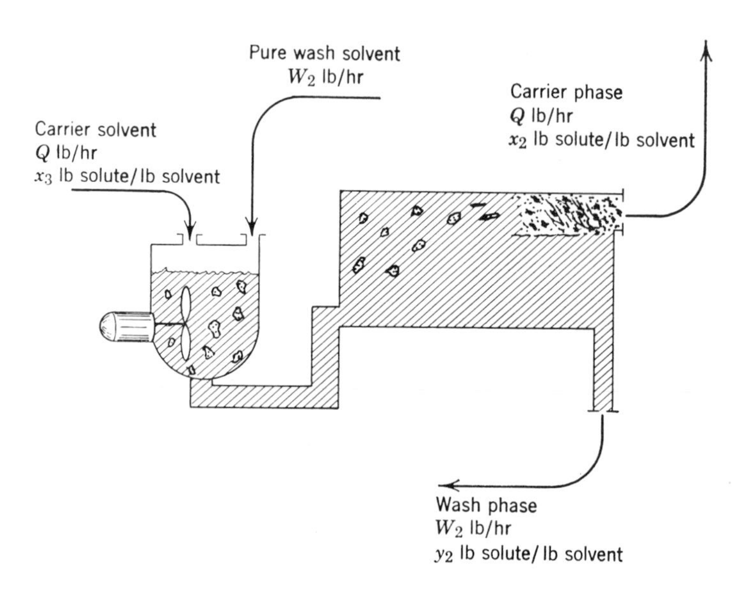

System structure

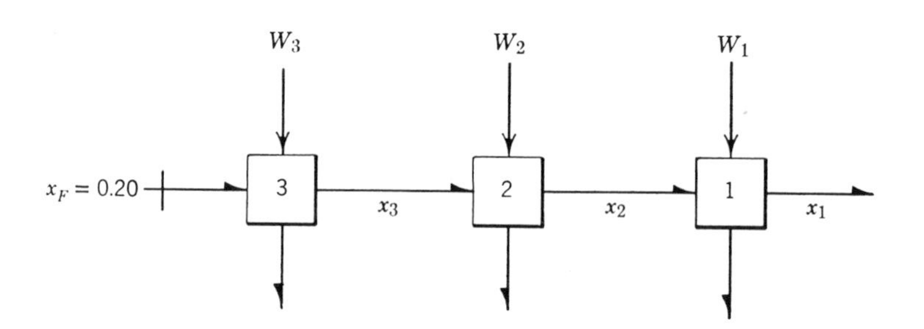

Fig. 8.3-1 A crosscurrent extraction system.

Equilibrium relationship between leaving phases

$$\phi(x_i, y_i) = 0 \qquad \text{(See Fig. 8.3-2)} \tag{8.3-3}$$

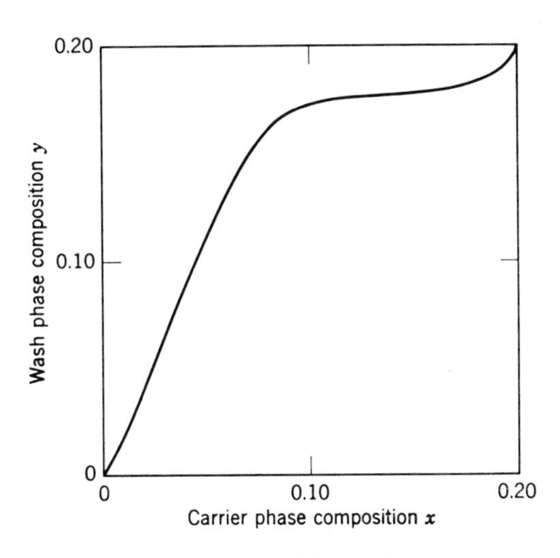

Fig. 8.3-2 Equilibrium data.

Thus for each extractor we have two independent equations (the solute balance and the equilibrium relation) and five variables (Q, W, x_{i+1}, x_i, y_i). The local degrees of freedom are $5 - 2 = 3$. When the extractors are connected to form a system, Q will be specified by the required processing rate, $Q = 1{,}000$ pounds of carrier solvent per hour, and the concentration of the solute in the incoming carrier phase x_{i+1} will be determined from upstream. Thus, two degrees of freedom are consumed, and only one economic design variable remains for each stage. We might elect W_i as the design variable. The information flow structure in Fig. 8.3-1 results, an acyclic structure.

Step 1. Suboptimize Extractor 1

Dynamic programming begins by considering the end stage in the structure, extractor 1. The principle of optimality states that:

> *For any composition of the solute x_2 in the carrier phase received from extractor 2, the wash solvent must be added to extractor 1 at a rate $W_1{}^*$ which maximizes the economic objective function for extractor 1.*

The suboptimization problem is

$$\max_1 = \max_{\{W_1\}} [Q(x_2 - x_1) - \lambda W_1] \tag{8.3-4}$$

where

$$Qx_2 = Qx_1 + W_1 y_1$$

$$\phi(x_1, y_1) = 0 \qquad \text{(See Fig. 8.3-2)}$$

and

$$Q = 1{,}000 \text{ lb/hr.}$$
$$\lambda = 0.05$$
$$x_2 = \text{value specified during dynamic programming}$$

Notice that this single variable search problem requires the solution of two simultaneous equations to obtain x_1 and y_1. However, if we elect to search over x_1 instead of W_1 the problem reduces to

$$\max_1 = \max_{\{x_1\}} [Q(x_2 - x_1) - \lambda W_1] \tag{8.3-5}$$

where

$$\phi(x_1, y_1) = 0$$

can be solved for y_1 directly. This then allows the solution of

$$W_1 = \frac{Q(x_2 - x_1)}{y_1}$$

A new selection of design variable has simplified the computation. Moreover, the new design variable is bounded, $0 < x_1 < x_2$, a happy situation in direct search. This is an application of the principles of design variable selection mentioned first in Chapter 3.

Table 8.3-1 summarizes a typical direct search for the optimal solvent allocation; for $x_2 = 0.10$. Since the proper value of x_2 is not known yet, (the upstream allocations have not been made) the suboptimization must be performed for a number of x_2 values, as summarized in Table 8.3-2.

Table 8.3-1 Typical Extractor 1 Suboptimization Problem
$x_2 = 0.10$

x_1	W_1	$Q(x_2 - x_1) - 0.05W_1$	Remarks
0.00	∞	$-\infty$	Lower bound
0.10	0	0	Upper bound
0.03	1,000	20	
0.06	270	27	Eliminate 0.00–0.03
0.05	400	29	Eliminate 0.06–0.10
⋮	⋮	⋮	⋮
0.045	500	30	To be entered in Table 8.3-2

Table 8.3-2 The Optimal Solvent Allocations for Extractor 1

x_2	$x_1{}^*$	$W_1{}^*$	Max_1
0.03	0.024	83	1
0.04	0.027	171	3
0.05	0.030	254	6
0.06	0.036	274	10
0.07	0.038	342	14
0.08	0.040	394	19
0.09	0.043	437	24
0.10	0.045	500	30
0.11	0.045	584	35
0.12	0.046	641	40
0.13	0.049	682	46
0.14	0.051	724	52
0.15	0.053	759	58
0.16	0.055	788	64
0.17	0.057	841	70
0.18	0.059	878	77
0.19	0.059	929	83
0.20	0.060	1,000	90

Step 2. Suboptimize the Extractor 2-1 Subsystem

The principle of optimality states that the end component must be optimized for any feed it receives from upstream. In Step 2 of dynamic programming we view the last two extractors as the end component. The proper objective function for the extractor 2-1 subsystem is

$$\max_2 = \max_{\{W_1, W_2\}} [Q(x_3 - x_2) + Q(x_2 - x_1) - \lambda(W_1 + W_2)] \tag{8.3-6}$$

Since Table 8.3-2 summarizes all we need to know about the optimal allocation of solvent to extractor 1, the two variable direct search problem above reduces to the single variable search problem below.

$$\max_2 = \max_{\{W_2\}} [Q(x_3 - x_2) - \lambda W_2 + \max_1 (x_2)] \tag{8.3-7}$$

where

$$Q = 1{,}000 \text{ lb/hr}$$
$$\lambda = 0.05$$
$$\phi(x_2, y_2) = 0$$
$$Qx_3 = Qx_2 + W_2 y_2$$

and x_3 is to be assigned a specific value during suboptimization.

Table 8.3-3 summarizes a typical direct search for the optimal value of W_2. Again we change the design variable from W_2 to x_2 to simplify the calculations.

Table 8.3-3 Typical Extractor 2 Suboptimization Problem
$x_3 = 0.20$

x_2	W_2	Max_1 (From Table 8.3-2)	$Q(x_3 - x_2) - \lambda W_2 + \text{max}_1$
0	∞	0	$-\infty$
0.20	0	90	90
0.08	730	19	101
0.12	450	40	98
0.10	590	30	100
⋮	⋮		⋮
0.076	775	—	102

Table 8.3-4 Optimal Solvent Allocations to the Two Extractor System

x_3	$x_2{}^*$	$W_2{}^*$	max_2
0.03	0.026	56	1
0.04	0.031	104	4
0.05	0.037	139	8
0.06	0.041	181	13
0.07	0.044	227	18
0.08	0.047	278	24
0.09	0.053	284	29
0.10	0.057	311	36
0.11	0.059	357	42
0.12	0.060	427	48
0.13	0.064	447	55
0.14	0.067	486	62
0.15	0.067	553	68
0.16	0.071	573	75
0.17	0.075	588	82
0.18	0.075	650	89
0.19	0.075	714	95
0.20	0.076	775	102

Table 8.3-4 is a summary of the results of a number of such single variable search problems, each for different values of the incoming feed. Now, for any feed concentration x_3 we can obtain the optimal allocation of solvent to extractor 2 and extractor 1. However, we do not yet know the proper value of x_3; that is obtained by considering the allocation of solvent to extractor 3.

Step 3. Optimize the Three Extractor System

The point has been reached where we can consider the system optimization problem.

$$\max_{\{W_1 W_2 W_3\}} [Q(x_F - x_1) - \lambda(W_1 + W_2 + W_3)] \quad (8.3\text{-}8)$$

which reduces to the following, taking advantage of the information contained in Table 8.3-4.

$$\max_{\{W_3\}} [Q(x_F - x_3) - \lambda W_3 + \max{}_2(x_3)] \quad (8.3\text{-}9)$$

$$\phi(x_3, y_3) = 0$$
$$Qx_F = Qx_3 + W_3 y_3$$
$$Q = 1{,}000, \qquad \lambda = 0.05, \qquad \text{and } x_F = 0.20$$

The direct search solution of this single variable optimization problem is summarized in Table 8.3-5.

Table 8.3-5 Direct Search for the Optimal Solvent Allocation to Extractor 3

x_3	W_3	$\max_3$	$Q(x_F - x_3) - \lambda W_3 + \max_2$
0	∞	0	$-\infty$
0.20	0	102	102
0.08	730	24	108
0.12	450	48	105
⋮	⋮		⋮
0.09	650		108

Step 4. Recovery of Optimal Plan

Thus, the allocation of 650 pounds per hour of solvent to extractor 3 is part of the optimal allocation plan. This yields a feed of concentration $x_3 = 0.09$ for the two stage extraction system following extractor 3. Table 8.3-4 indicates that with that feed concentration an allocation of $W_2{}^* =$ 284 pounds per hour to extractor 2 is recommended, yielding a feed of concentration $x_1 = 0.053$ for extractor 1. Table 8.3-2 indicates that, at that feed concentration, $W_1{}^* \cong 260$ pounds per hour.

Thus, in summary, the optimal solvent allocation plan is

$$W_1 = 260 \text{ lb/hr}$$
$$W_2 = 284 \text{ lb/hr}$$
$$W_3 = 650 \text{ lb/hr}$$

Notice how we have avoided the need for searching over several variables simultaneously. The original three variable optimization problem has been transformed into a sequence of one variable problems. Since single variable search problems are particularly easy, this transformation seems desirable. However, observe how we were forced to solve suboptimization problems for feed concentrations which did not eventually enter into the optimal plan. We had to *imbed* the original problem into a larger class of problems involving a variety of intermediate compositions. Dynamic programming is a recommended optimization strategy when a proper balance can be reached between the reduction of dimensionality on the one hand and the imbedding on the other hand. This topic is discussed towards the end of this chapter.

Finally, we must remark that while the cross current extraction problem may be an excellent medium for illustrating dynamic programming, *optimum crosscurrent extraction* may not be synonymous with *optimum extraction*. In confining attention to the crosscurrent concept of wash solvent flow, we may well have excluded superior concepts. For example, consider the *countercurrent* extraction system shown in Fig. 8.3-3; the

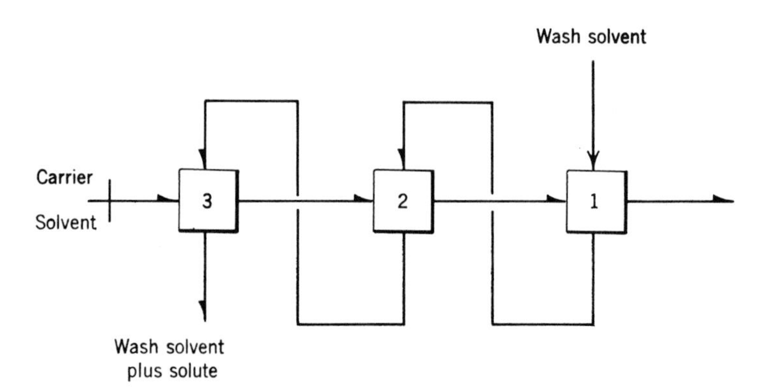

Fig. 8.3-3 A countercurrent extraction system.

wash solvent leaving extractor 1 is passed counter to the carrier solvent flow to provide the wash for extractor 3. This gives a single variable optimization problem, the free variable being the wash solvent rate. The optimization problem is

$$\max_{\{W\}} [Q(x_F - x_1) - \lambda W] \tag{8.3-10}$$

subject to the material balances about each stage and the equilibrium relation between phases. Frequently countercurrent extraction is superior to crosscurrent extraction. This illustrates the point made earlier in the text, that there are two planes of optimization; the *mathematical* and the *innovative*. Dynamic programming is an example of mathematical optimization, and the consideration of the different concept (countercurrent extraction) is an example of innovative optimization. Both must be in action during process engineering.

8.4 THE OPTIMIZATION OF A MORE COMPLEX SYSTEM[2]

In this section we further illustrate dynamic programming by optimizing the chemical processing system shown in Fig. 8.4-1. This particular problem has been solved in *detail* by Mitten and Nemhauser and we now show how we might *rough-out* an approximation to the optimal design. A comparison of this approximate solution to the detailed solution of Mitten and Nemhauser gives insight into the sensitivity of dynamic programming to the fineness of the state variable grid.

The processing system is to receive 50,000 lb per year of raw materials which are blended, heated, and exposed to a catalyst, thereby generating a mixture of valuable product and unreacted raw material from which at least 15,000 lb per year of pure product must be separated. The processing alternatives involve

1. The type of blender (mixer A, B, or C).
2. The degree of mixing (1.0, 0.8, 0.6, 0.5).
3. The temperature to which the blend is heated (650, 700, 750, 800°F).
4. The type of primary reactor (reactor I_A, I_B, or I_C).
5. The catalyst contained therein (catalyst 1 or 2).
6. The type of clean-up reactor (reactor II_A, II_B, or none).
7. The type of separator (one large separator or two small separators).

There exist a total of (3)(4)(4)(3)(2)(3)(2) = 1,728 different ways of implementing this processing plan, and with dynamic programming we shall find the optimal plan by investigating the equivalent of less than 15 processing plans. Now we present a bit more detail about each component in this system.

Mixer. Three types of mixers, A, B, and C, are available. Table 8.4-1 contains the initial investment for each mixer, and the yearly operating costs required to achieve a given degree of mixing. One is free to select the type of mixer and the degree of mixing to be achieved by that mixer.

[2] L. G. Mitten and G. H. Nemhauser, "Multistage Optimization," *Chem. Eng. Progr.*, January 1963.

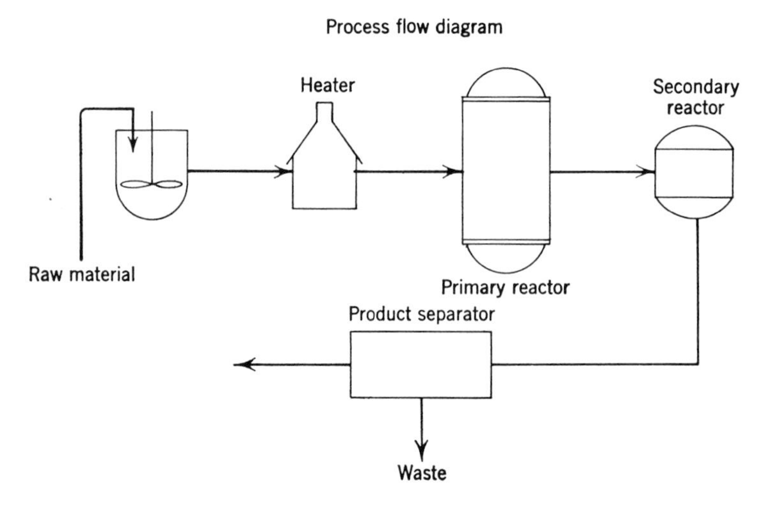

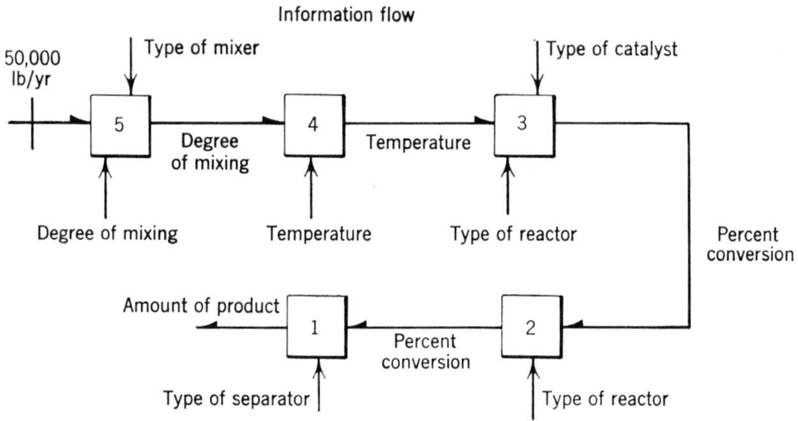

Fig. 8.4-1 Processing system.

Table 8.4-1 Data on the Mixing Operation

Mixer	Initial Investment ($)	Operating Costs ($/yr)			
		Degree of Mixing			
		1.0	0.8	0.6	0.5
A	10,000	12,000	6,000	3,000	2,000
B	15,000	8,000	4,000	2,500	1,500
C	25,000	5,000	3,000	2,000	1,000

Heater. The ease by which the blended material can be heated depends upon the degree of mixing for a given type of heater. A standard heater can be used to raise the temperature to 700°F or less; however, a specially designed and more costly heater would be required should the engineer elect higher outlet temperatures for the blend leaving the heater. Table 8.4-2 contains the necessary data on heater performance and costs.

Table 8.4-2 Data on Heating Operation[a]

Degree of Mixing	Operating Costs ($/yr)			
	Temperature			
	650	700	750	800
1.0	500	1,000	6,000	10,000
0.8	1,000	1,500	8,000	12,000
0.6	1,500	2,500	10,000	16,000
0.5	2,000	3,000	12,000	20,000

[a] Initial Investment ($). *Standard heater.* 5,000 for temperatures at or below 700°F. *Special heater.* 20,000 for temperatures exceeding 700°F.

Primary reactor. The blended and heated raw materials are brought into contact with a catalyst to effect partial conversion to the desired product. The per cent conversion achieved depends upon the type of reactor and the type of catalyst, as well as the temperature at which the raw materials enter the reactor. The degree of mixing is not a significant variable at these high temperatures. Table 8.4-3 describes the various alternative modes of reaction, and Table 8.4-4 contains the pertinent economic data.

Table 8.4-3 Per Cent Conversion in Primary Reactor

Temperature	650		700		750		800	
Catalyst	1	2	1	2	1	2	1	2
Reactor I_A	30	25	40	30	50	45	60	50
I_B	25	20	30	25	45	40	50	45
I_C	20	15	25	20	40	30	45	40

Table 8.4-4 Primary Reactor Costs

	Initial Cost ($)	Operating Cost ($/yr)
Reactor		
I_A	40,000	4,000
I_B	20,000	2,000
I_C	5,000	1,000
Catalyst		
1	—	10,000
2	—	4,000

Clean-up reactor. Should the conversion in the primary reactor be insufficient, one of two clean-up reactors may be used to effect further reaction. Table 8.4-5 contains data on the degree of conversion of material leaving the several clean-up reactors as a function of the degree of conversion of the feed from the primary reactor. Table 8.4-6 contains the economic data.

Table 8.4-5 Conversion after the Clean-Up Reactor

Conversion in Primary Reactor	15	20	25	30	40	45	50	60
Clean-up reactor								
II_A	30	40	50	60	80	85	90	95
II_B	45	60	75	85	90	95	95	95
None	15	20	25	30	40	45	50	60

Table 8.4-6 Cost of Clean-Up Reactors

Clean-up Reactor	Investment ($)	Operating Cost ($/yr)
II_A	60,000	10,000
II_B	80,000	20,000
None	0	0

Separators. Finally, the pure product must be separated from the unreacted raw materials and waste. We have the option of employing one large separator or two smaller separators. The separators are capable of recovering all of the pure product, in each case. Table 8.4-7 summarizes the cost of separation.

Table 8.4-7 Separation Costs

Per Cent Conversion	One Large Separator		Two Small Separators	
	Initial Cost ($)	Operating Cost ($/yr)	Initial Cost ($)	Operating Cost ($/yr)
30	12,000	2,500	15,000	3,000
40	12,000	3,000	15,000	3,000
45	12,000	4,000	15,000	3,000
50	15,000	4,000	18,000	3,000
60	15,000	5,000	18,000	4,000
75	20,000	6,000	24,000	4,000
80	20,000	6,500	24,000	4,000
85	20,000	7,000	24,000	4,000
90	20,000	7,500	24,000	5,000
95	20,000	8,000	24,000	5,000

Market conditions. The market is rather flexible and the selling price of the pure product is expected to vary with the amount produced by this system. To satisfy internal demands the system must produce at least 15,000 lb per year. Table 8.4-8 summarizes the market conditions.

Table 8.4-8 Expected Selling Price

Production (Pounds per Year)	Selling Price ($/lb)
47,500	3.2
45,000	3.3
42,500	3.4
40,000	3.6
37,500	3.8
30,000	4.6
25,000	5.0
22,500	5.2
20,000	5.3
15,000	5.5

Economic analysis. This system should be designed for maximum venture profit.

$$V = (S - C) - (S - C - dI)t - eI - i_m I$$

Table 8.4-9 Summary of an Haphazard Design

Component	Input	Design	Investment	Operating Cost	Output
Raw materials	50,000 lb per year	—	—	\$50,000 per year	—
Mixer	50,000 lb per year of of raw material	Mixer *A* 0.5 degree of mixing	\$10,000	\$ 2,000 per year	50,000 lb per year with 0.5 degree of mixing
Heater	50,000 lb per year with 0.5 degree of mixing	Temperature 750°F	\$20,000	\$12,000 per year	50,000 lb per year at 750°F
Primary reactor	50,000 lb per year at 750°F	Reactor I_A Catalyst 2	\$40,000	\$ 8,000 per year	50,000 lb per year; 45 per cent converted to product
Clean-up reactor	50,000 lb per year; 45 per cent converted	Reactor II_B	\$80,000	\$20,000 per year	50,000 lb per year; 95 per cent converted
Separator	50,000 lb per year; 95 per cent converted	One large separator	\$20,000	\$ 8,000 per year	47,500 lb per year of pure product
Market	47,500 lb per year of pure product	at \$3.20 per pound, a sales income of \$152,000 per year.			

Total investment $I = \$170{,}000$.
Total operation cost $C = \$100{,}000$ per year.
Total sales income $S = \$152{,}000$ per year.
Venture profit $V = 0.4(S - C - 0.2I)$
$= \$7{,}200$
Thus, by investing in this design we stand to make \$7,200 per year more than by investing the \$170,000 in other company activities.

This reduces to the following, in the economic environment into which this particular system must integrate,

$$V = 0.4(S - C - 0.2I)$$

where S is the sales income, (\$/yr), C the manufacturing costs (\$/yr) and I the total investment (\$).

The problem is now well defined and we are faced with the problem of selecting among the numerous alternative designs. First consider the results of an haphazard selection of design variables summarized in Table 8.4-9. There are some 1,700 similar design alternatives and a trial-and-error selection strategy would be far too inefficient. We must refine our selection strategy to take advantage of the serial structure of this system. Dynamic programming is now brought to bear using a coarse state variable grid during the suboptimization. This effects an approximate or "roughed-out" solution to the design, and a comparison to the more nearly exact solution of Mitten and Nemhauser is made later.

Step 1. Design the Separator

The Principle of Optimality indicates that the last stage, the separator, must be optimal with respect to the feed it receives from upstream. That is, the separator which maximizes $V_S = 0.4(S - C_S - 0.2I_S)$ for a given per cent conversion in the feed from the clean-up reactor is optimal. A quick check of Table 8.4-7 indicates that two separators minimize $C_S + 0.2I_S$ for feed conversion equal to or greater than 45 per cent, and one large reactor minimizes that term otherwise. This policy, then, constitutes the optimal design for a given feed.

Since the feed conversion is unknown several values must be assumed to span a reasonable range of conversions to summarize the design of this last unit. We select 30, 60, and 90 per cent as reasonable values. Table 8.4-10 summarizes the optimization calculations and Fig. 8.4-2 the "roughed-out" design.

Table 8.4-10 Summary of Calculations Leading to the Design of Separator

s_2 Per cent Conversion after Clean-Up	d^* Optimal Design	S Value of Product, \$/yr	I_S Investment in Separator, \$	C_S Operating Cost, \$/yr	Max$_1(s)$ $0.4(S - C - 0.2I_S)$, \$/yr
0.3	One separator	82,500	12,000	2,500	31,040
0.6	Two separators	130,400	18,000	4,000	52,160
0.9	Two separators	148,500	24,000	5,000	55,480

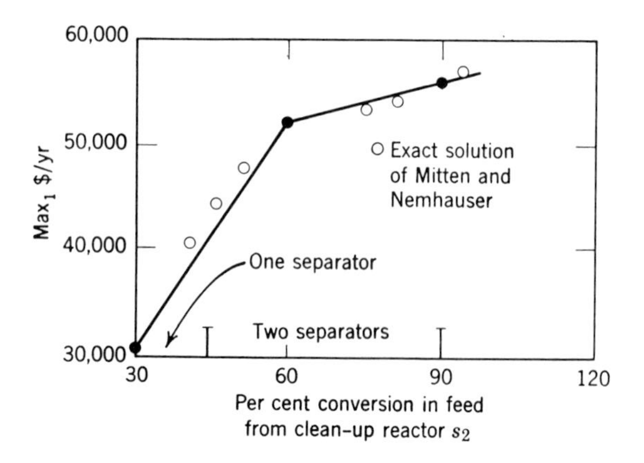

Fig. 8.4-2 The "roughed-out" design of the separator.

Step 2. Design Clean-Up Reactor

The clean-up reactor must be selected to maximize the venture profit from the separator–clean-up reactor subsystem. $V = \max_1 - 0.4\,(C_c + 0.2I_c)$. Figure 8.4-2 is the source of information on the optimal design of the separator. The calculations which lead to the clean-up reactor design are contained in Table 8.4-11, and the optimal design is summarized in Fig. 8.4-3. Conversions 15, 30, and 60 in the feed from the primary reactor are used to span the input state variable for this subsystem.

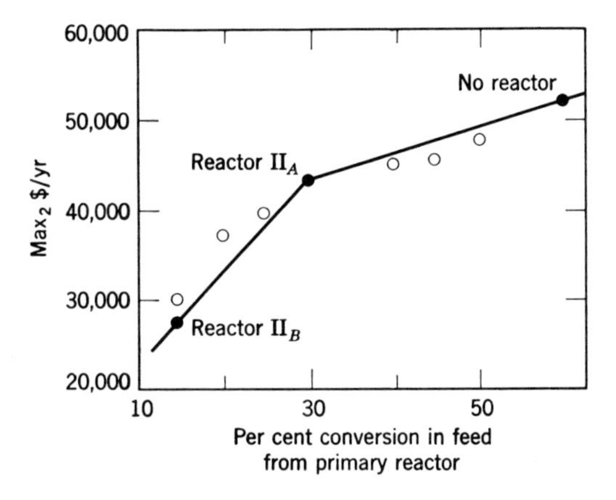

Fig. 8.4-3 The "roughed-out" design of the clean-up reactor.

Table 8.4-11 Summary of Calculations Leading to the Design of the Clean-Up Reactor

Per Cent Conversion after Primary Reactor s_3	Clean-Up Reactor Type	Per Cent Conversion after Reactor s_2	Max_1	Investment I_c, \$	Operating Cost C_c, \$/yr	$\text{Max}_1 - 0.4(C_c + 0.2I_c)$, \$/yr
15	II_A	30	31,040	60,000	10,000	22,240
	II_B	45	41,500	80,000	20,000	27,100[a]
	None	15 ← Not acceptable since less than 15,000 lb per year of product will be produced				
30	II_A	60	52,160	60,000	10,000	43,360[a]
	II_B	85	54,800	80,000	20,000	40,400
	None	30	31,040	—	—	3,040
60	II_A	95	56,500	60,000	10,000	47,700
	II_B	95	56,500	80,000	20,000	No need to calculate
	None	60	52,160	—	—	52,160[a]

[a] Denotes Max_2.

Step 3. Design Primary Reactor

The primary reactor must be designed to maximize the venture profit for the primary reactor–clean-up reactor–separator subsystem.

$$V = \max_2 - 0.4(C_p + 0.2I_p)$$

Figure 8.4-3 is the source of information on the optimal design of the clean-up reactor–separator subsystem. The calculations which lead to the primary reactor design are contained in Table 8.4-12 and the optimal design is summarized in Fig. 8.4-4. Temperatures of 650 and 750°F for the feed from the heater are used to span the input state variable to this subsystem.

Table 8.4-12 Summary of the Calculations which Lead to the Optimal Design of the Primary Reactor

s_4, Temperature after Heating	d_3 Design of Primary Reactor and Catalyst	s_3 Conversion	Max_2	I_p, \$	C_p, \$/yr	$\text{Max}_2 - 0.4(C_p + 0.2I_p)$, \$/yr
650	I_A-1	30	43,360	40,000	14,000	35,560[a]
	I_B-1	25	37,500	20,000	12,000	31,100
	I_C-1	20	32,300	5,000	11,000	27,500
	I_A-2	25	37,500	40,000	8,000	31,100
	I_B-2	20	32,300	20,000	6,000	28,300
	I_C-2	15	27,100	5,000	5,000	24,700
750	I_A-1	50	49,420	40,000	14,000	40,600
	I_B-1	45	48,000	20,000	12,000	41,600
	I_C-1	40	46,500	5,000	11,000	41,700
	I_A-2	45	48,000	40,000	8,000	41,600
	I_B-2	40	46,500	20,000	6,000	42,500[a]
	I_C-2	30	43,360	5,000	5,000	40,960

[a] Denotes Max_3.

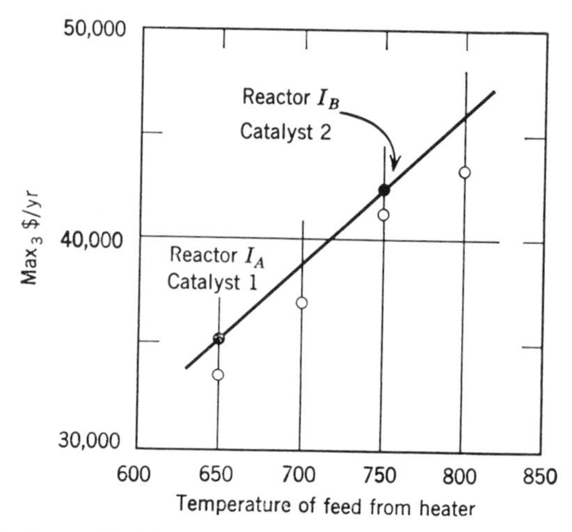

Fig. 8.4-4 The "roughed-out" design of the primary reactor.

Step 4. Design Heater

The heater must be designed to maximize the venture profit for the heater–primary reactor–clean-up reactor–separator subsystem.

$$V = \max_3 - 0.4(C_H + 0.2I_H)$$

Figure 8.4-4 is the source of information on the design of the primary-reactor–clean-up reactor–separator subsystem. The calculations which lead to the heater design are contained in Table 8.4-13, and the optimal design is summarized in Fig. 8.4-5. The degree of mixing of the feed from the blender is taken as 0.5 and 0.8 to span the input state variable.

Table 8.4-13 Summary of the Calculations which Lead to the Design of the Heater

s_5 Degree of Mixing of Feed from Blender	d_4 and s_4 Temperature to which the Blend Is To Be Heated	Max$_3$	I_H, $	C_H, $/yr	Max$_3 - 0.4(C_H + 0.2I_H)$, $/yr
0.5	650	35,560	5,000	2,000	34,360
	700	38,800	5,000	3,000	37,200[a]
	750	42,500	20,000	12,000	36,100
	800	46,000	20,000	20,000	36,400
0.8	650	35,560	5,000	1,000	34,760
	700	38,800	5,000	1,500	37,300
	750	42,500	20,000	8,000	37,700
	800	46,000	20,000	12,000	39,600[a]

[a] Denotes Max$_4$.

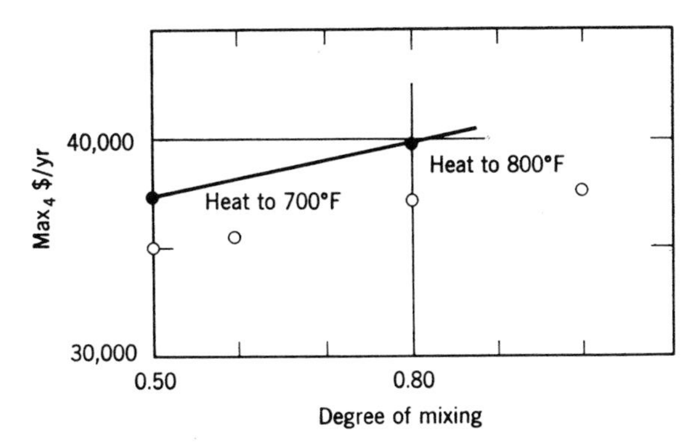

Fig. 8.4-5 The "roughed-out" design of the heater.

Step 5. Design the Mixer

The last step in this optimization strategy is the design of the mixer to maximize the venture profit for the entire system.

$$V = \max_4 - 0.4(C_B + 0.2I_B + 50{,}000)$$

where the $50,000 per year is the raw material cost. Figure 8.4-5 contains the summary of the design of all downstream components. The feed to the blender is known, and we must select among the three blenders and the degree of blending to specify the design. The calculations leading to the blender design are summarized in Table 8.4-14.

Table 8.4-14 Summary of the Calculations Leading to the Design of the Blender

Mixer	Degree of Mixing	Max_4	I_B, $	C_B, $/yr	$\text{Max}_4 - 0.4(C_B + 0.2I_B + 50{,}000)$, $/yr
A	1.0	41,200	10,000	12,000	15,600
	0.8	39,600	10,000	6,000	16,200
	0.6	38,000	10,000	3,000	16,000
	0.5	37,200	10,000	2,000	15,600
B	1.0	41,200	15,000	8,000	16,800
	0.8	39,600	15,000	4,000	16,800
	0.6	38,000	15,000	2,500	15,800
	0.5	37,200	15,000	1,500	15,400
C	1.0	41,200	25,000	5,000	17,200[a]
	0.8	39,600	25,000	3,000	16,400
	0.6	38,000	25,000	2,000	16,200
	0.5	37,000	25,000	1,000	14,800

[a] Denotes Max_5, the roughed-out maximum profit for the system.

Step 6. Recovery of Optimal Plan

The final step in this approximate optimization strategy is to recover the optimal plan from the "roughed-out" design summaries for each subsystem, as follows:

(a) Table 8.4-14 indicates that *Mixer C* operating with a degree of mixing of 1.0 is the "best" mixer design.

(b) Entering Fig. 8.4-5 with a degree of mixing of 1.0 indicates that the best heater design involves heating to what appears to be 800°F.

(c) Entering Fig. 8.4-4 at 800°F indicates that the closest design of the primary reactor is *Reactor I_B—Catalyst 2.* Now it may well pay to rough-out further detail on Fig. 8.4-4 near 800°F since extrapolating this design might be rather dangerous. But, nevertheless, let us take the above design.

(d) Table 8.4-3 indicates that the effluent from a reactor I_B with Catalyst 2, for which the feed is at 800°F, will be 45 per cent converted. Entering Fig. 8.4-3 at that point indicates that the optimal clean-up reactor design could be either *Reactor II_A* or *no reactor.* A quick calculation at 45 per cent conversion in Table 8.4-11 indicates that *no reactor* is recommended.

(e) The effluent from the primary reactor then passes directly to the separator with a 45 per cent conversion. Figure 8.4-2 indicates that *two separators* are recommended.

Table 8.4-15 A Comparison of the Roughed-Out Design and the Optimal Design[a]

Component Feed	Input	Design	Investment ($)	Operating Cost ($/yr)
Mixer	—	Mixer *C* 1.0 Degree of mixing (Mixer *B* 0.8 Degree of mixing)	25,000 (15,000)	5,000 (4,000)
Heater	1.0 (0.8)	800°F (800°F)	20,000 (20,000)	10,000 (12,000)
Primary reactor	800°F (800°F)	Reactor I_B (?) Catalyst 2 (Reactor I_A Catalyst 1)	20,000 (40,000)	6,000 (14,000)
Clean-up reactor	45% (60%)	No reactor (No reactor)	— —	— —
Separator	45% (60%)	2 Small separators (2 Small separators)	15,000 (18,000)	3,000 (4,000)
Product value		22,500 lb/yr worth $116,500 per year (30,000 lb/yr worth $138,000 per year)		

[a] Parentheses indicate the *optimal* design, by Mitten and Nemhauser.

Total Investment:	I = $80,000 (93,000)	Sales Income:	S = $116,500 per year (138,000)
Total Operating Cost:	C = $74,000 per year (84,000)	Venture Profit:	$V = 0.4(S - C - 0.2I)$ = 10,600 (14,100)

A complete summary of this "roughed-out" design and the optimal design obtained by Mitten and Nemhauser is contained in Table 8.4-15, with the optimal design denoted by brackets. Notice how the original haphazard design in Table 8.4-9 yields a *venture profit* of $7,200 per year on an investment of $170,000, the roughed-out design yields a *venture profit* of $10,600 per year on an investment of $80,000, and the optimal design yields $14,100 per year on an investment of $93,000. Recall that the venture profit is the return on invested capital *above* the return obtained had the capital been invested in other activities of the company. The *extra* rate of return then amounts to 4.2 per cent, 13 and 15 per cent, respectively; the "roughed-out" design is a sufficient approximation to the optimal design.

That the roughed-out design differs from the optimal design is an indication of the effects of the state variable approximation. This is also shown by the difference between the actual venture profit of $10,600 per year for the roughed-out design and $\max_5$ of $17,200 per year. The last difference is the result of the interpolation errors (which accumulate and appear as a difference between the "max" curves for the approximate and exact solutions) and the inability to distinguish between designs during the recovery of the optimal plan [for example, Step 6(c)].

In any case, a complete and exhaustive search of all of the design alternatives would have required the analysis of over 1,700 distinct designs, whereas the exact dynamic programming solution of Mitten and Nemhauser required an analysis of an equivalent of about 15 designs, and our roughed-out design required even less work.

8.5 COMPARISON OF DIRECT SEARCH AND DYNAMIC PROGRAMMING

When does it pay to take advantage of the serial structure of an optimization problem and use dynamic programming? This is a logical question to ask, since dynamic programming requires that we span a range of the system state variables and solve a large number of suboptimization problems which do not eventually become part of the optimal plan, an apparent waste of effort. On the other hand, dynamic programming reduces the number of variables over which we must search at any given time, an apparent advantage. Under what conditions do the advantages outweigh the disadvantages?

To give a partial answer to this question, we consider the case when there are N stages in the system (with D design variables per stage and S state variables connecting the stages) and when optimizations must be performed by direct search. The direct search of the DN design variables could then be classified as a DN dimensional search problem.

Dynamic programming reduces the search problem to D dimensions. However, at each stage a total of $(C)^s$ state variable combinations must be considered, where C is the number of points that adequately span the range of each of the S state variables. Thus, each of the first $N - 1$ steps in dynamic programming involves $(C)^s$ D dimensional searches. The input state variables are known for the last stage to be optimized; hence, the final step requires one D dimensional search.

In summary, should we elect to consider this problem as structured, the location of the optimum would involve

$$1 + (N - 1)(C)^s \quad D \text{ dimensional searches; } \textit{Dynamic Programming}$$

But if the problem is considered as structureless, the location of the optimum would involve

$$1 \quad ND \text{ dimensional search; } \textit{Direct Search}$$

The selection between these two alternatives requires an estimate of how rapidly the labor of direct search might increase with the number of variables being searched, and this increase depends upon the nature of the surfaces being searched and upon the particular direct search strategy employed.

Suppose, however, that the number of function evaluations W_{DS} required to locate the optimum by direct search increases with the number of variables ND according to Eqn. 6.10-1

$$W_{DS} = 3(5)^{ND}$$

and suppose that a grid of 10 variables spans the state variables quite adequately,

$$C = 10$$

The number of function evaluations using dynamic programming is then

$$W_{DP} = [1 + (N - 1)(10)^s]3(5)^D$$

The ratio W_{DS}/W_{DP} is a measure of the relative advantages of one approach over the other, and this ratio is computed in Table 8.5-1 for several typical values of the parameters:

N the number of components
D the number of design variables per component
S the number of state variables between two components

The observation can be made that dynamic programming is strongly favored in systems with a large number of components and *few state variables* between components. In general, dynamic programming should not be applied to problems which involve only a few components or which require several state variables to connect the components, even when the structure is acyclic. However, it must be kept in mind that this general statement is only valid when all optimizations are to be performed by direct search. Even a two stage system can be attacked efficiently by dynamic programming if the suboptimization problem is easily solved, as is the case in several problems at the end of this chapter.

Table 8.5-1 Comparison of Dynamic Programming and Direct Search[a]

Work by Direct Search/Work by Dynamic Programming

Number of state variables between components (S) \ Number of Components (N)	2	5	10
1	0.5	15	22,000
3	0.0005	0.15	220
5	0.000005	0.0015	2

$D = 1$, only one design variable per stage

[a] Hatched areas indicate dynamic programming favored.

8.6 CONCLUDING REMARKS

We have presented the simple theory of dynamic programming with none of the refinements which may reduce further the labor of optimizing processes which exhibit acyclic structure. The important lessons to be learned from a study of this chapter are as follows.

1. Dynamic programming is a strategy of optimization, rather than a rote such as direct search or linear programming. In fact, the methods of direct search are used to solve the suboptimization problems which dynamic programming supplies; dynamic programming is more a way of thinking than of computing.

2. The price we pay for being able to suboptimize and consider only parts of the whole problem at a given time, is the need for imbedding the problem at hand into the larger class of problems involving a range of feed conditions for each suboptimization problem. Under certain conditions this price may be too high, and thus an estimate of the labor of optimization should be made, along the lines of Section 8.5, before computation begins.

3. The experienced dynamic programmer uses the principle of optimality as a base of operations and attempts to improve on the simple theory of dynamic programming by making computational approximations such as using a coarse grid in the suboptimization to rough out a solution to be further refined, as in Section 8.4.

Finally we must remark that other special purpose methods are building for the optimization of processes with acyclic structure, such as Pontryagin's maximum principle. While there is no doubt that these methods are of great use to the process engineer, they do not constitute part of what might be called the elementary theory of optimization, and hence, in this text, we shall not discuss these methods. Rather we refer the reader to other published sources of information.

References

For historical purposes we must consider the basic reference on dynamic programming:

R. Bellman, *Dynamic Programming*, Princeton University Press, Princeton, N.J., 1957.

Other texts are now available at an introductory level, such as:

G. H. Nemhauser, *Introduction to Dynamic Programming*, Wiley, New York, 1966.
R. Aris, *Discrete Dynamic Programming*, Blaisdell, New York, 1961.
S. M. Roberts, *Dynamic Programming in Chemical Engineering and Process Control*, Academic Press, New York, 1964.
R. A. Howard, *Dynamic Programming and Markov Processes*, Wiley, New York, 1960.

A study of the following text will yield ideas for improving the simple theory:

R. Bellman and S. E. Dreyfus, *Applied Dynamic Programming*, Princeton University Press, Princeton, N.J., 1962.

This text is an advanced study of the application of dynamic programming to reactor design:

R. Aris, *The Optimal Design of Chemical Reactors*, Academic Press, New York, 1961.

Pontryagin's maximum principle is an alternate way of approaching acyclic optimization problems:

Pontryagin, Boltyamskii, Gamkrelidze, and Mishchenko, *The Mathematical Theory of Optimal Processes*, Interscience, New York, 1962

For a variety of worked examples of the maximum principle see:

L. T. Fan and C. S. Wang, *The Discrete Maximum Principle*, Wiley, New York, 1964.
L. T. Fan, *The Continuous Maximum Principle*, Wiley, New York, 1966.

PROBLEMS

8.A. A simple exercise in suboptimization is afforded by the following routing problem. A load of material is to be shipped from point A to point B along the minimum cost path through the network shown in Fig. 8.A-1.

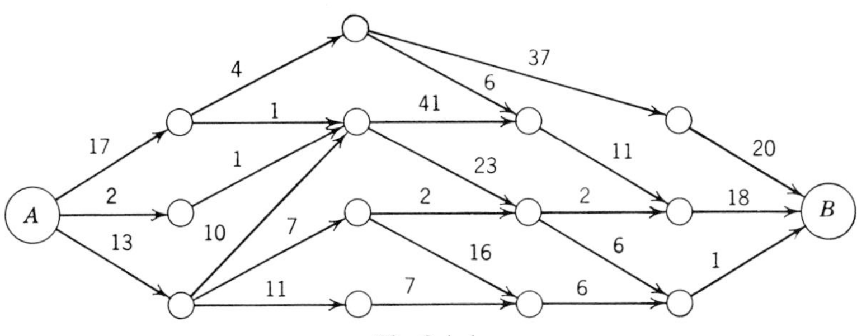

Fig. 8.A-1

The cost of transportation from junction to junction is affixed to the arrows. For example, the cost of the upper route is $17 + 4 + 37 + 20 = 78$. Find the lowest cost route by suboptimizing from right to left through the network, at each phase computing the minimum cost of going from a given junction to point B. Only a fraction of the routes need be considered in this way.

8.B. G. Nemhauser[3] suggests the following problem to be solved by dynamic programming. Four sizes of coal are to be separated in a three stage screening operation (see Fig. 8.B-1). The feed containing all sizes of coal is to be placed on the largest size screen from which size 4 is removed and sizes 3, 2 and 1 pass on to the next screen which removes size 3, and then on to the final screen which removes size 2. However, at each screen a certain amount of undersize is removed with the larger product. For example, the nominal size 4 product contains coal of sizes 3, 2, and 1 as carry-over. The amount of this carry-over can be altered by changing the operation of the screen. The amount of undersize that is passed on to the next screen is given by the following relation:

$$\text{Amount of undersized passed on to next screen} = \alpha_i \text{ (amount of undersize entering } i\text{th screen)}$$

where

$$0 < \alpha < 1 \text{ depends on the operation of the screen}$$

The cost of separation on the ith screen is given as:

$$\text{Separation cost} = \text{(amount entering screen) } k_i \ln \left(\frac{1}{1 - \alpha_i}\right)$$

[3] G. Nemhauser, *Introduction to Dynamic Programming*, Wiley, New York, 1966.

where

$$k_4 = 1.5 \quad \$/\text{ton}$$
$$k_3 = 1.0$$
$$k_2 = 0.5$$

If the feed coal has a composition of:

1 ton of coal of size 4
3 ton of coal of size 3
1 ton of coal of size 2
4 ton of coal of size 1

determine the operating level for each screen (i.e., α_4, α_3, and α_2) which maximizes the profit of the system.

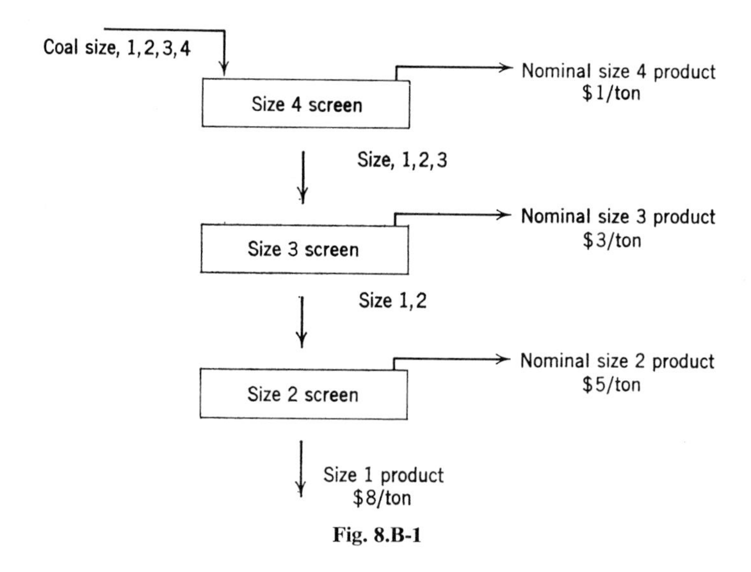

Fig. 8.B-1

8.C. A two stage process is envisioned in which 1,000 pounds per hour of raw material A is to be converted into a product P in solvent S. The first stage is a chemical reactor in which A is to be converted into P. The fractional conversion c of A to P has been correlated to the amount of heat added to the reactor, q Btu per pound of A fed to the reactor. The second stage is an extractor in which P is extracted from the A phase into a solvent phase S, A being insoluble in S.

An economic study of this process results in a design objective function which consists of terms involving the value of the product, the cost of the solvent and the cost of the heat addition. This information is available in Fig. 8.C-1.

(a) How many degrees of freedom exist in this design? Demonstrate that the optimization problem exhibits acyclic structure.

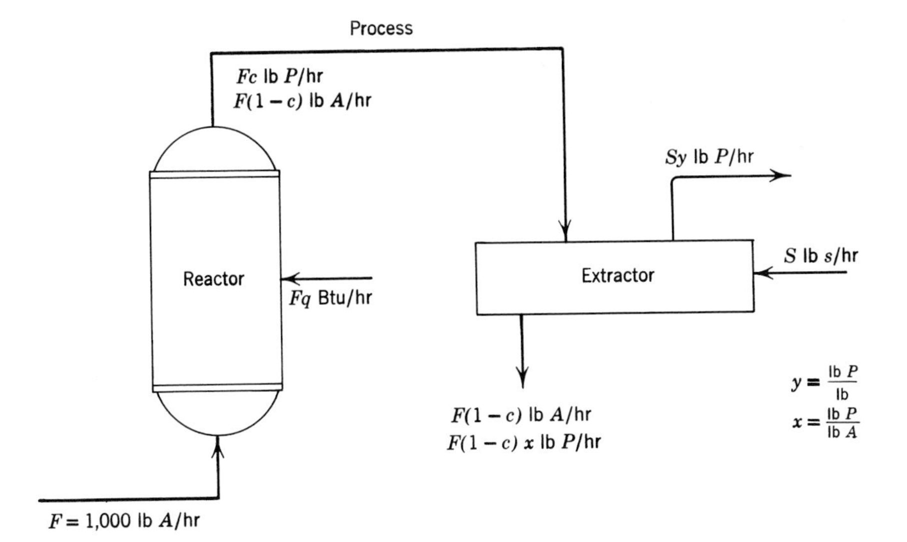

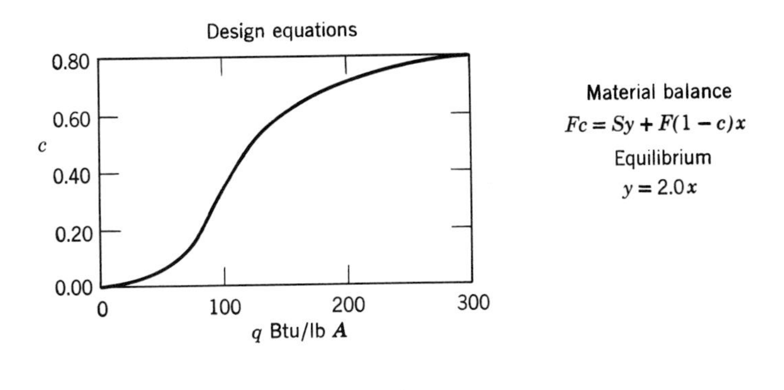

Design criterion

Max $(0.10Sy - 0.05S - 0.25q)$

Fig. 8.C-1

(b) Suboptimize the extractor by determining the solvent addition rate which maximizes the value of the product minus the cost of the solvent, for a given conversion.

$$\text{Answer: } S^* = F\left(\sqrt{c(1-c)} - \frac{1-c}{2}\right)$$

(c) Apply the principle of optimality and the results of the extractor suboptimization to a direct search over the reactor conversion.

8.D. In this problem we can recognize a serial information flow structure and achieve a decomposition of the problem into two parts by use of the *principle of optimality*. One part of the problem takes the form of the *transportation problem* which can be suboptimized. An optimization strategy can then be tailored expressly for this problem, in which the transportation problem forms a subordinate part of a main optimization problem. This illustrates how several methods of optimization can be fitted to the characteristics of a given problem.

Six terminals are to be supplied from two refineries. The refineries are to be scheduled to satisfy the demands at the terminals at minimum manufacturing plus transportation costs. The data are given in Table 8.D-1. For example, terminal II requires 50 M barrels per week and the transportation costs from refinery A to terminal II are 15 cents per barrel. If refinery A is operating at the 500 M barrel per week level, the manufacturing costs are \$1.00 per barrel. How should the two refineries share the manufacturing load?

Table 8.D-1

The Transportation Costs
(¢ per Barrel)

		Terminal I	II	III	IV	V	VI
Demand	$\frac{\text{M barrels}}{\text{week}}$	100	50	200	50	300	10
Refinery	A	10	15	6	14	22	5
	B	11	10	15	5	11	16

The Manufacturing Costs
(\$ per Barrel)

Refinery	
A	$1.00 + \left(\frac{Q_A - 500}{500}\right)^2$
B	$2.00 - \left(\frac{Q_B - 500}{500}\right)$

$Q =$ M barrels per week produced.

(a) Show that the information flow structure can be cast into a serial form with a manufacturing optimization problem feeding information into a transportation problem. Applying the principle of optimality show that the transportation problem can be suboptimized for any given manufacturing schedule.

(b) To illustrate how the transportation problem is to be suboptimized, use the stepping-stone method to obtain the total cost of supplying the terminals in the case where the refineries share the load equally.

(c) Develop in detail an attack on the manufacturing problem which follows this pattern.

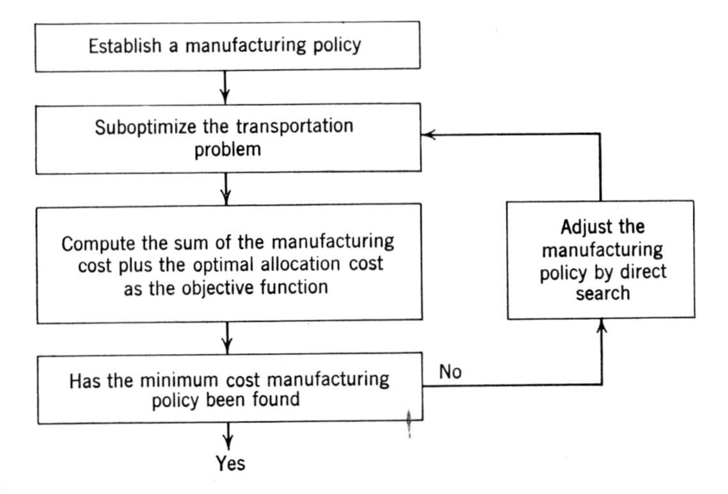

(d) Solve the refinery scheduling problem.

8.E. Frequently computational advantages can be achieved in dynamic programming by first sketching through a rough optimization of a process, to be followed by a more refined optimization. This problem illustrates one approach that can be taken.

(a) Solve the three stage crosscurrent extraction problem (Section 8.3) roughly using the very crude grid span of $x = 0.08, 0.04$ during suboptimization. This will lead to an approximation to the optimal solvent allocation but will involve only five single variable direct searches.

(b) In an attempt to refine the approximate solution above, select grid points above and below the state variable obtained above by 0.01. Repeat the course suboptimization as in Part (a), and compare your answer with the more accurate answer of Section 8.3.

8.F. Converse[4] applied a nonimbedding technique based on the principle of optimality to the solution of the crosscurrent extraction problem. The

[4] A. Converse, *Chem. Eng. Progr., Symp. Ser.* No. 46, **59** (1963).

idea is that a base allocation of solvent be selected and the intermediate concentrations computed for that allocation. Then suboptimization is performed assuming the feed concentrations to be those of the base case. Suboptimization is iterated using the updated estimate of the intermediate concentrations. After reading the article, compare ordinary dynamic programming and Converse's nonimbedding method. What convergence problems might arise?

8.G. The information accumulated during suboptimization can be of great use in solving similar design problems. Consider the five stage system in Fig. 8.4-1, and suppose that a blender is now available with the following cost characteristics.

Investment $	*Operating costs* $/*yr*			
	Degree of mixing			
	1.0	0.8	0.6	0.5
5,000	22,000	2,000	2,000	500

Determine the design of the complete process which best takes advantage of the availability of this mixer. There is no need to solve any suboptimization problems because sufficient information was accumulated during the optimization of the process without this mixer. How extensive an optimization problem would arise if a third separator were available? If a new catalyst were discovered for the primary reactor?

8.H. This problem illustrates how a hybrid optimization plan can be established by means of the principle of optimality to obtain a quick solution to the optimum crosscurrent extraction problem of Section 8.3. Notice in Fig. 8.3-2 that the equilibrium curve for concentrations of $x \leq 0.08$ is linear, and that in Problem 6.F it was shown how the crosscurrent extraction problem could be solved analytically by means of calculus in the special case of linear equilibrium data. This means that suboptimization problems are particularly simple for certain feed concentrations.

(a) Obtain the optimum allocation of solvent to the three stage system of Section 8.3 in the case of $X_F = 0.10$ by conducting a direct search over the allocation of solvent to the first stage *only*, using the principle of optimality to take advantage of the analytical solution for the suboptimization of the two downstream stages.

(b) For what feed compositions would you expect this hybrid optimization plan to deteriorate?

9

MACROSYSTEM OPTIMIZATION STRATEGIES

The optimization methods of calculus, direct search, linear programming, and dynamic programming offer efficient approaches to a wide variety of optimization problems. Unfortunately, the chemical process in its original form often is so complex as to be unassailable by these powerful methods. Complex nonlinearities prevent the use of calculus and linear programming, involved recycle invalidates the dynamic programming approach, and direct search becomes inefficient for problems with a large number of degrees of freedom. In this chapter we present certain macrosystem optimization strategies which are being developed to handle large industrial process design problems. Soon these methods and their offspring will doubtless be established as part of industrial practice.

Typical Problem

Our design group is confronted with the problem of adjusting the design details of a process with the following dominant information flow structure. How might we proceed?

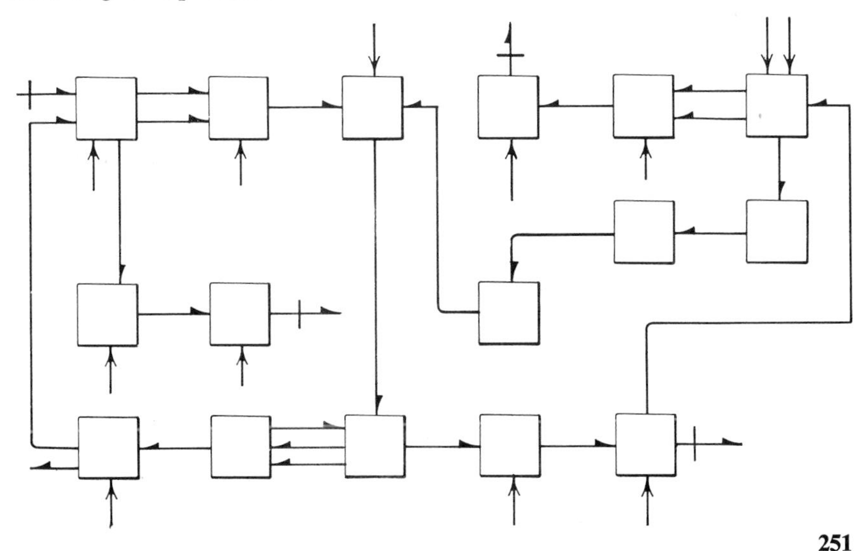

9.1 THE QUEST FOR A MACROSYSTEM STRATEGY

The intuitive skills of the experienced engineer can be extremely effective, whereby the engineer senses a reasonable approach to hopelessly large and complicated problems. Approximations are made, certain details ignored, the main problem is blocked out, subproblems defined, secondary design variables are fixed at seemingly arbitrary values, partial and tentative solutions assumed based on experience, and other poorly defined but deceptively effective techniques are brought to bear. The mathematical community has yet to devise techniques as effective, though specialized. Unfortunately, without some formalization these heuristic tools are apt to remain inaccessible to the mass of problem solvers.

The *principle of optimality appeared* to offer guidance to the optimization by the *mathematical formalization* of common sense. Doubtless, the attention that dynamic programming has received (arising from the hope that dynamic programming and suboptimization might be the forerunner of a larger class of techniques eventually leading to a clearer understanding of the heuristic approach of the experienced problem solver) is more than is justified by its computational advantages. While this hope has not been realized, and while this whole area is in the midst of active research, some real progress has been made. We wish to acquaint the reader with some of this thinking.

First we address the problem of sensitivity, showing how we must test the design variables to block out the dominant design problem, suppressing the minor details. We next investigate more closely the concept of information flow reversal. Then, the strategy of combining components to reduce the labor of optimization is discussed, followed by an introduction to the strategy for cutting recycle loops. Finally, some "rules of thumb" for attacking macrosystems are discussed.

It will be interesting, a few years from now, to review the literature and observe the progress made in this active research area. Doubtless, new methods of analysis will appear even before this book reaches print. Time has the frustrating property of making what is puzzling now seem trivial in the all too near future. This chapter is a progress report on an area of active research.

9.2 SENSITIVITY AND THE DOMINANT PROBLEM

In this section we indicate how we must detect numerically or empirically the dominant design variables and information flow structure to remove the superfluous detail which accompanies every large problem. Once the core problem has been identified and solved, the solution can be "fine-tuned" by the systematic recovery of as much detail as is justified. Often a quick analysis of the main features of a design is all that is really justifiable considering the uncertainties surrounding the data.

In a sensitivity analysis we select some reasonable base design, and vary each design variable and input variable in turn by some fraction of the allowable range of variation. The resulting variations in the objective function and the output variables reveal the importance of each tested variable. The variables to which the system appears to be insensitive are fixed at reasonable values, and attention is focused on the more important design variables. Once the *dominant problem*, involving only these important design variables, has been solved, a retesting of sensitivity is in order to see if some secondary variables take on importance near the optimum. The analysis we describe is very similar to the multidimensional direct search techniques described in Chapter 6. Milton D. Marks in *Sensitivity of a Steam-Methane Reformer*[1] analyzed the sensitivity of the design equations for an industrial steam-methane reformer, identifying a dominant design relation which is sufficiently accurate for preliminary design studies and surprisingly simple. We use this as an example of how superfluous detail should be eliminated in systems studies.

The primary steam-methane reformer generates hydrogen from light hydrocarbons. Although many alkanes, cycloalkanes, alkenes, and aromatics can be used as the source of the hydrogen molecule, natural gas consisting of 93 per cent methane is the feed in the reformer under study.

The methane reforming reactions are

$$\text{Heat} + CH_4 + H_2O \rightleftharpoons CO + 3H_2 \qquad \textit{methane reforming}$$

$$CO + H_2O \rightleftharpoons CO_2 + H_2 + \text{Heat} \qquad \textit{water-gas shift}$$

The equilibrium is influenced by temperature, pressure, and the composition of the initial reactants. Qualitatively, the methane reforming equilibrium is favored by high temperature, low pressure, and an excess of steam; and the water-gas shift reaction equilibrium is favored by low temperature, and an excess of steam. We might suspect that the system should be operated at low pressures. However, it is economical to operate at high pressure with correspondingly higher temperatures and steam to methane ratios.

The equilibrium relationships for these reactions are:

$$K_{p_1} = \frac{(CO)(H_2)^3 p^2}{(CH_4)(H_2O)}$$

$$K_{p_2} = \frac{(CO_2)(H_2)}{(CO)(H_2O)}$$

[1] Milton D. Marks, *Chem. Eng.*, 750 report, April 1966, Univ. of Wisc.

where the equilibrium constants K_{p_1} and K_{p_2} are temperature dependent. When the following variables are defined and substituted into the equilibrium relations, a set of system design relations results:

A = moles of CH_4 input
B = moles of H_2O input
C = moles of CO input
D = moles of H_2 input
E = moles of CO_2 input
S = total moles input
X = moles of water reacting by reforming reaction
Y = moles of water reacting by shift reaction
p = pressure in atmospheres
T_e = equilibrium temperature, °F

$$K_{p_1} = \frac{(C + X - Y)(D + 3X + Y)^3 p^2}{(A - X)(B - X - Y)(S - 2X)^2}$$

$$K_{p_2} = \frac{(E + Y)(D + 3X + Y)}{(C + X - Y)(B - X - Y)}$$

The temperature dependence of the equilibrium constants is approximated by

$$K_p = a[\ln (T_e)] + b[\ln (T_e)]^2 + c$$

where a, b, and c are constants adjusted to fit experimental reaction data. It is assumed that the system approaches equilibrium conditions in the reactor at a temperature lower by 50°F than the actual reactor temperature T

$$T_e = T - 50°\text{F}$$

Thus, a complex set of design equations is available enabling an estimate of hydrogen production, given feed compositions, operating temperature, and pressure. The feed may contain various amounts of steam, methane, carbon monoxide, carbon dioxide, hydrogen, nitrogen, and oxygen by recycle from the system that the steam-methane reformer services.

By a sensitivity analysis of this processing component, Marks demonstrated that these complex design relations are, in fact, dominated by a much simpler design relationship in the case of pure methane-steam feed, and he showed that this simpler relationship is approximately valid for other feed compositions.

The per cent hydrogen in the product stream was computed for a total of 96 different values of R_{sm} (the steam to methane ratio), p (the operating pressure), and T_e (the equilibrium temperature) in the range

Steam/methane $R_{sm} = 3\text{–}6$
Pressure $p = 300\text{–}600$ psia
Temperature $T_e = 1{,}050\text{–}1{,}550°\text{F}$

A statistical analysis of these data led to the following simplified design relationship

$$(\text{Per cent H}_2) = \frac{(T_e)^{2.66} \cdot \exp(-0.045R_{sm} + 0.00158p - 10.17)}{p} \tag{9.2-1}$$

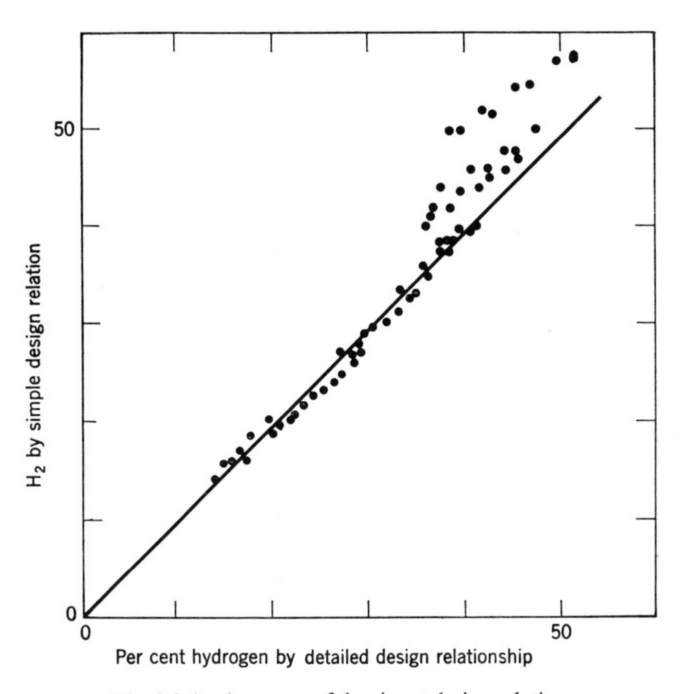

Fig. 9.2-1 Accuracy of dominant design relation.

In Fig. 9.2-1 the per cent hydrogen computed by the simple relationship is compared to that computed by the detailed design relations, showing that the much simpler form dominates in the region

$$R_{sm} = 3\text{–}6$$
$$p = 350\text{–}600 \text{ psia}$$
$$T_e = 1{,}050\text{–}1{,}400°\text{F}$$

Table 9.2-1 summarizes the effects of other components in the feed, indicating further the range of validity of this approximate design equation.

Table 9.2-1

Base Case	Range of Variation	Sensitivity[a]
Equilibrium temperature, $T_e = 1{,}300°F$	500°F	0.86[b]
Operating pressure, $p = 450$ psia	300 psia	0.17[b]
Steam/methane ratio, $R_{sm} = 4.5$	3	0.12[b]
Mole per cent CO in., 0.0	3	0.04
Mole per cent H_2 in., 0.0	5	0.03
Mole percent CO_2 in., 0.0	3	0.06
Mole per cent N_2 in., 0.0	5	0.03
Mole per cent H_2 produced 28.94		

[a] $\dfrac{\text{Fractional change in } H_2 \text{ production}}{\text{Change in variable as fraction of expected range of variation}}$

[b] Dominant variables

Figure 9.2-2 shows the dominant information flow for the steam-methane reformer, indicating that the initial picture of the component was far too detailed and, indeed, would lead unnecessarily to an impossibly complex information flow diagram for the system, which has a steam-methane reformer as a component.

Such reductions in complexity *must* be performed during the design of complex processes. In practice the reduction is achieved either by statistical studies such as that illustrated here or by means of information drawn from engineers familiar with the operation of the components. Either is a valid approach.[2]

The use of simple, although approximate, design relationships such as Eq. 9.2-1 in the place of more detailed design relationships is frequently desirable even now in the days of high speed computation. A large fraction of the extensive design calculations needed in process design are performed on the computer using subroutines stored on magnetic tape or in the memory of the computer. However, for example, the solution of a detailed distillation problem by the computer might involve say one half hour of computing time, and this is far too costly when the design of the system in which the

[2] K. D. Manchanda and D. R. Woods, "Significant Design Variables in Continuous Gravity Decantation," *Ind. Eng. Chem.* Vol.7, No.2, 1968. See also, M. J. Shaw, "An Approach to Model Simplification in Process Optimization and Control," paper 4.6, A.I.Ch.E.-I.Chem.E. meeting, London, June 1965. T. J. Stanford, "Comparison of Complete and Simplified Models for the Dynamic Optimization of Steam-Raising Plant," paper 4.7, A.I.Ch.E.–I.Chem.E. meeting, London, June 1965.

distillation tower is a component is only in the planning stage and the condition of the feed to the tower has not yet been determined in detail. In such a case, a simple, empirical design equation is perfectly adequate, even though only roughly approximate.

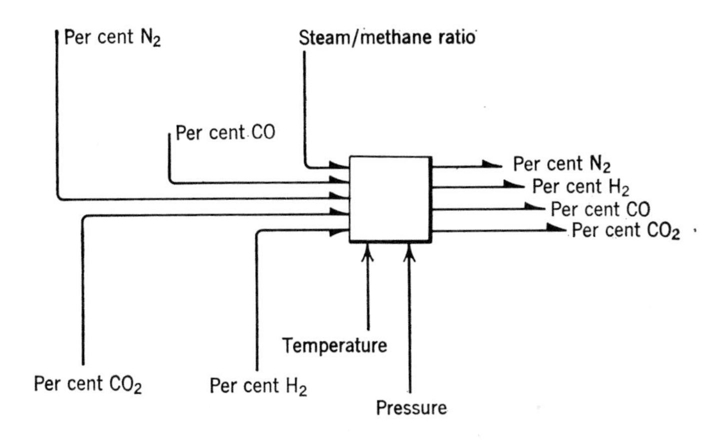

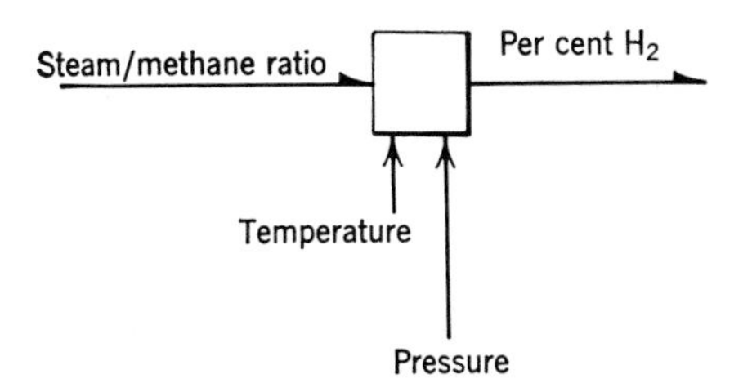

Fig. 9.2-2 The dominant information flow for the steam-methane reformer.

It is becoming more and more common for the computer to supply itself with such approximate relationships by correlating the performance of the more exact design subroutine. For example, R. Cavett of Pure Oil reports success with a design computer program for hydrocarbon flashing systems in which after a few passes through detailed recycle calculations the exact thermodynamic equations are replaced by empirical correlations (which are computed by the correlation of the results of the first few passes) to speed convergence. Once the desired convergence is achieved using the empirical correlation, the exact thermodynamic equations are brought in for fine-tuning the answer.

Further, the use of the dominant design relationship tends to reduce the difficulties of reversing information flow through a processing unit. For example, it is simple to specify the per cent of hydrogen, steam to methane ratio, and operating pressure in Eq. 9.2-1 and compute the operating temperature. This reversal of information flow may well be exceedingly difficult in the case of the more detailed design equations. We shall see what implications ease of reversal of information flow carries.

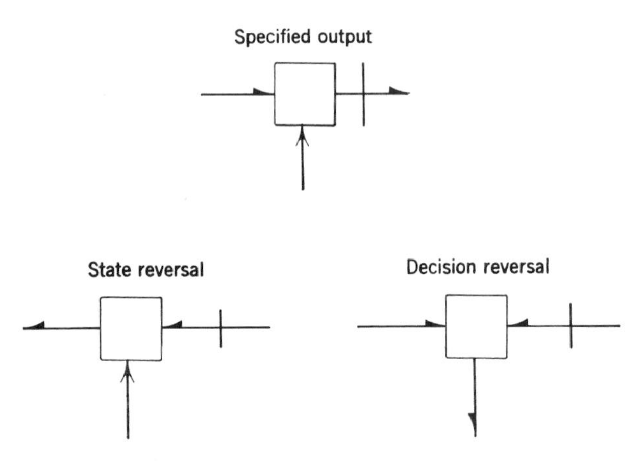

Fig. 9.3-1 State and decision reversal.

9.3 ATTACKING MORE COMPLEX ACYCLIC STRUCTURES

We focused attention on initial value acyclic optimization problems in Chapter 8, and now we expand our vision to include other serial structures.

First a brief discussion of the information flow. The arrows leaving a block in the information flow diagram denote dependent variables, the values of which are determined once values of all of the input arrows are specified. The number of independent variables must remain constant during any manipulations with the structure since this equals the local degrees of freedom. For example, in Fig. 9.3-1 if we specify the value of the output variable, this amounts to changing it from a dependent variable to an independent variable.

This passes information into the component, and the direction of the information flow has been reversed. Now, all variables associated with this component cannot be independent, since the design relationships must be satisfied. We must reverse the direction of either an input state variable or a design variable to preserve the number of output variables. This is called state and decision reversal, respectively. In a number of practical situations such reversals are extremely difficult or even impossible to achieve. In such cases we are stuck with the original structure. Suppose, however, that such reversals are possible.

We are now in a position to attack terminal value serial optimization problems. Consider the acyclic system shown in Fig. 9.3-2 in which the value of the output state variable has been specified. Decision reversal leads to the combination of the last two components into one component and results in the recovery of an initial value problem. On the other hand, had we elected state reversal, the output from the second to the last component would have to become an input, and state or decision reversal would be required at that component. Thus, we see that a number of intermediate structures can be obtained, with a completely reversed initial value serial structure as the limiting case. For example, we might attack the last structure in Fig. 9.3-2 by beginning at stage 4 and suboptimizing in the order 4, 3, 2, 1.

The concept of state or decision reversal has even more far-reaching implications, as we might suspect by observing that a combining branch structure appeared in Fig. 9.3-2 on the way from the terminal value to the initial value problem.

An initial value diverging branch structure is rather easily handled, using ordinary dynamic programming, as shown in Fig. 9.3-3. We begin the suboptimization at the ends of both branches and continue to the junction. The junction suboptimization problem is modified to the extent that both branches are included in the contribution of the downstream components.

$$\max_A = \max_{\{d_A\}} [U_A + \max_B (s'_A) + \max_B (s_A)]$$

Upstream of the junction, dynamic programming is used without modification. Such structure offers no difficulty.

The converging branch structure shown in Fig. 9.3-4 can be transformed into a common serial form by a reassignment of design variables followed by decision and state reversal; it also offers no great difficulty. Notice that these structural manipulations are similar to those introduced in Chapter 3.

In summary, by the use of decision or state reversal it is possible to alter significantly the structure of a systems optimization problem, and since the concept of suboptimization is dependent on certain structural features of a problem, the altering of system structure may offer advantages we can exploit.

Terminal value serial strategy

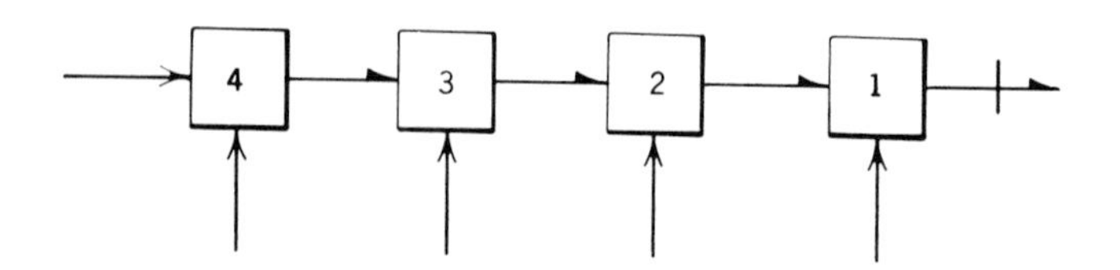

Decision reversal–initial value problem recovered

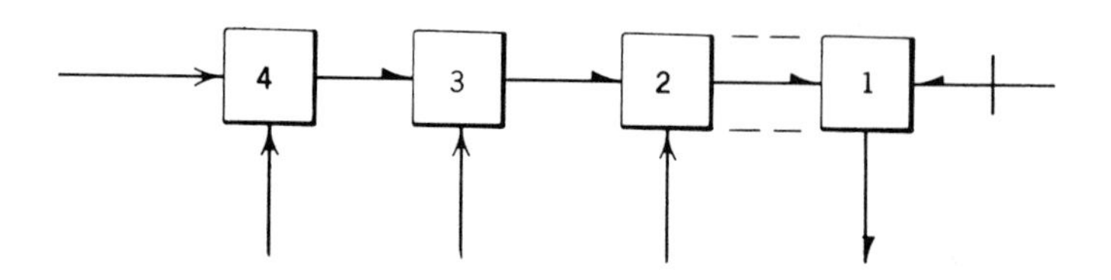

State reversal followed by decision at component 2–combining branch problem appears

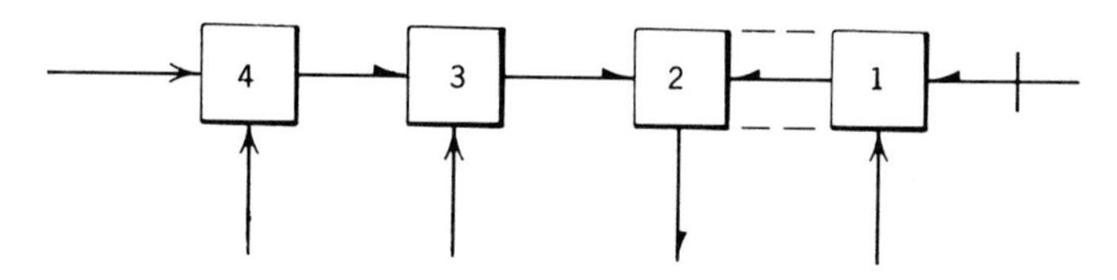

Continued state reversal with a decision reversal at stage 4–initial value problem recovered

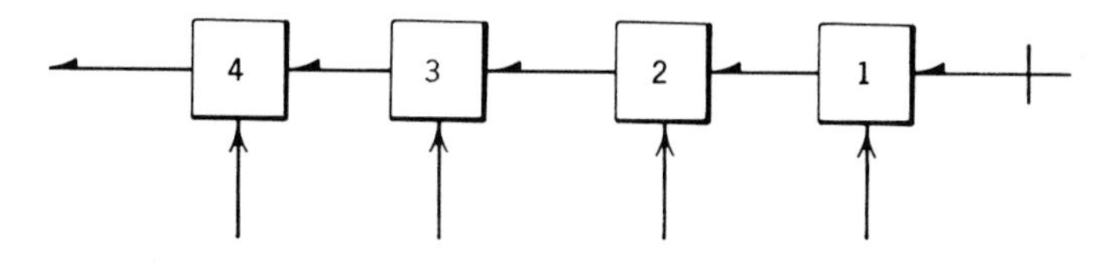

Fig. 9.3-2 Reversals in a terminal value problem.

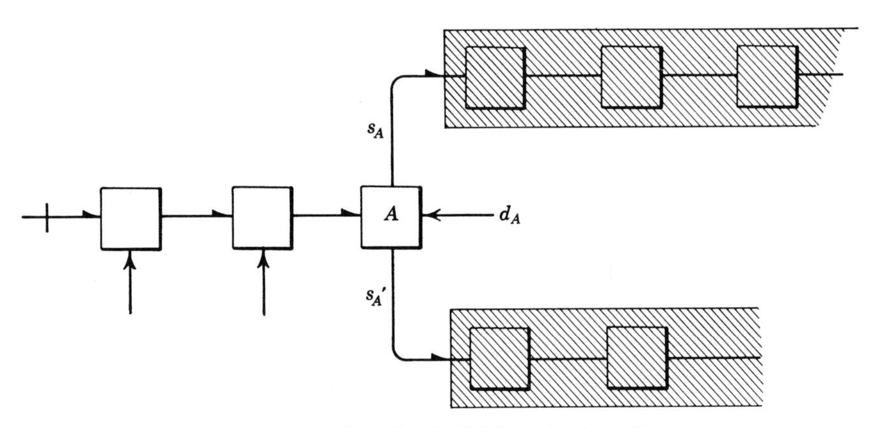

Fig. 9.3-3 Optimization of an initial diverging branch structure.

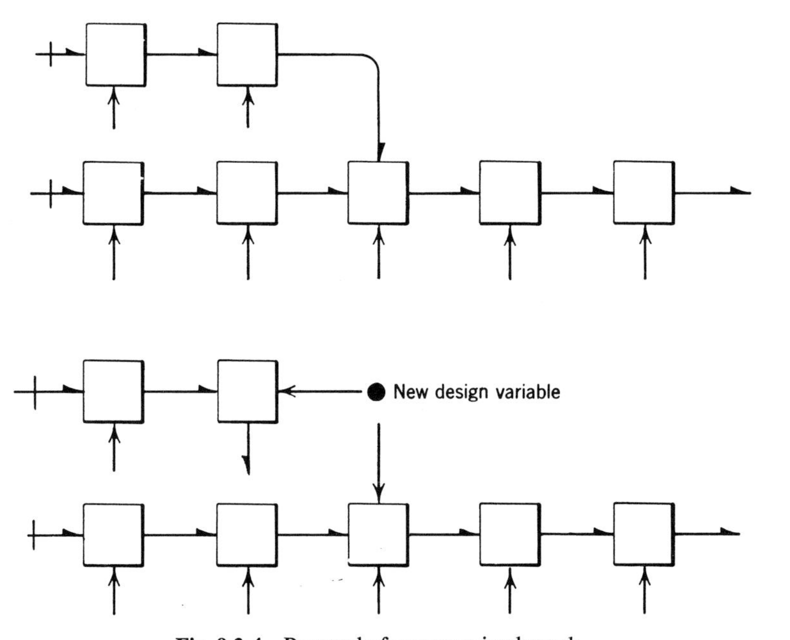

Fig. 9.3-4 Reversal of a converging branch.

9.4 THE STRATEGY OF STAGE COMBINATION

Once the dominant information flow structure is available for each component, attention is focused on the system of components. The ideas to be presented now are adapted from the work of D. J. Wilde of Stanford

University, R. Aris of the University of Minnesota, and G. L. Nemhauser of The Johns Hopkins University.[3]

In Chapter 8 we saw that under certain circumstances it was wise to consider an acyclic system as a collection of small stages and apply the principle of optimality thereto. On the other hand, frequently it was wise to combine all the stages into one larger stage to be optimized by direct search. In this section we look deeper into this idea of stage combination, developing strategies which ofttimes are better than the two extremes above.

Consider the three stage system in Fig. 9.4-1 with a number of output state variables and design variables at each stage. This system might be part of some larger macrosystem.

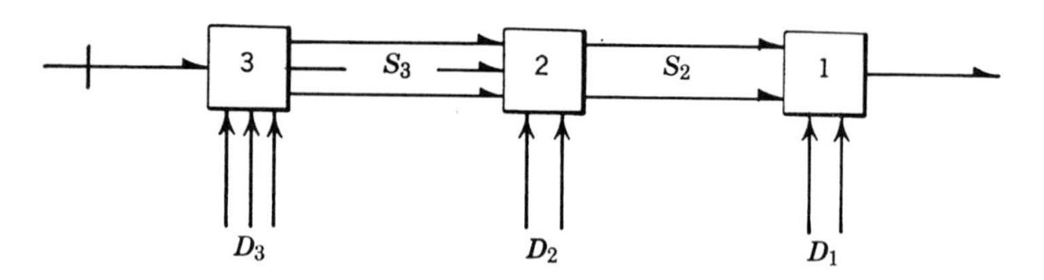

Fig. 9.4-1 A three stage serial structure.

There are four obvious optimization plans.

1. Combine all three stages into one large stage and use direct search.
2. Consider the stages separately and apply ordinary dynamic programming.
3. Combine stages 2 and 3 into one larger stage and apply dynamic programming to the remaining two stage system.
4. Combine stages 2 and 1 into one larger stage and apply dynamic programming.

In Chapter 6 it was mentioned that the direct search labor W_{DS} tends to increase exponentially with the number of variables D being searched

$$W_{DS} = a(b)^D$$

where the coefficients a and b depend on the particular direct search plan employed and the difficulty of the function being searched. Generally, b, the parameter of most interest now, ranged between two and, say, ten.

In Chapter 8 the concept of fixing the value of an input state variable at one of c different values during the suboptimization of the downstream

[3] R. Aris, G. L. Nemhauser, and D. J. Wilde, *A.I.Ch.E.J.*, **10**, 913 (1964). D. J. Wilde, *Chem. Eng. Progr.*, No. 3, **61** (1965).

stages was introduced, where the parameter c might range from, for example, two to ten or more.

We now evaluate the optimization efficiency of four plans in terms of the parameters a, b, and c, and number and kind of variables involved.

The number of calculations for the various plans are:

Plan 1. Direct Search

$$W_1 = a(b)^{D_1+D_2+D_3}$$

Plan 2. Ordinary Dynamic Programming

$$W_2 = a[(b)^{D_3} + (c)^{S_3}(b)^{D_2} + (c)^{S_2}(b)^{D_1}]$$

(9.4-1)

Plan 3. A Hybrid Plan

$$W_3 = a[(b)^{D_3+D_2} + (c)^{S_2}(b)^{D_1}]$$

Plan 4. A Hybrid Plan

$$W_4 = a[(b)^{D_3} + (c)^{S_3}(b)^{D_2+D_1}]$$

Now, it is rather difficult to extract any general stage combination rules from Eqs. 9.4-1. However, some interesting "rules of thumb" can be extracted by considering some limiting cases.

Case A: $b \approx c \gg 1$
Suppose now that a large number of points are needed to span a state variable during suboptimization, and that the labor of direct search is strongly dependent on the number of variables being searched. Then, in Eq. 9.4-1, the highest power will dominate. By replacing c with b, these equations can be simplified to

$$\begin{aligned} W_1 &= a(b)^{D_1+D_2+D_3} \\ W_2 &\approx a[(b)^{D_3} + (b)^{S_3+D_2} + (b)^{S_2+D_1}] \\ W_3 &\approx a[(b)^{D_3+D_2} + (b)^{S_2+D_1}] \\ W_4 &\approx a[(b)^{D_3} + (b)^{S_3+D_2+D_1}] \end{aligned} \tag{9.4-2}$$

When $D_1 = D_2 = D_3 = S_2 = S_3$, plans 2 and 3 are dominated by the lowest high power of b, with plan 3 being favored. However, should S_2 be much greater than the other equal variables, plans 1 and 4 would tend to be favored, with plan 1 winning when $S_3 + D_2 + D_1 > D_1 + D_2 + D_3$. As long as the coefficient b is high, say around 6 or 10, a check of the highest power of that coefficient is sufficient.

Figure 9.4-2 illustrates the application of these principles to a number of different three stage systems, showing a strategy of stage combination.

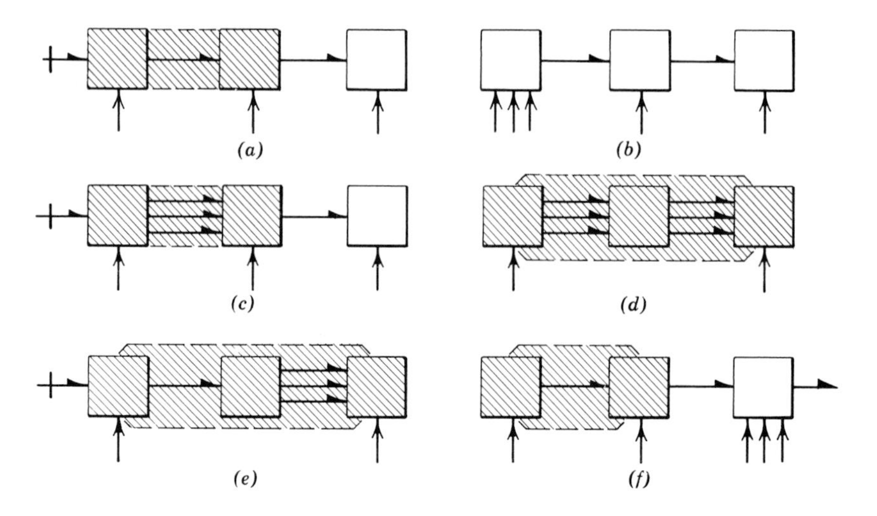

Fig. 9.4-2 Stage combination—Case A.

Case B: $b \gg c > 1$

Suppose now that a rather loose scatter of points is needed to span a state variable during dynamic programming, but still the work of direct search increases rapidly with dimension. Suppose, for example, that $c = 2$ and $b = 8$. It then takes three state variables to be equivalent to one design variable in Eq. 9.4-1, $(c)^3 = (b)^1$, and the equations reduce to

$$\begin{aligned} W_1 &= a(b)^{D_1 + D_2 + D_3} \\ W_2 &\approx a[(b)^{D_3} + (b)^{S_3/3 + D_2} + (b)^{S_2/3 + D_1}] \\ W_3 &\approx a[(b)^{D_3 + D_2} + (b)^{S_2/3 + D_1}] \\ W_4 &\approx a[(b)^{D_3} + (b)^{S_3/3 + D_2 + D_1}] \end{aligned} \tag{9.4-3}$$

The stage combinations which lead to the minimum optimization work in this case are shown in Fig. 9.4-3. Notice how the plans change between Figs. 9.4-2 and 9.4-3 as the numerical properties of the optimization tools change. In particular, notice in Case B (where the state variables are easily spanned) how a more complete fracturing of the system into subsystems is favored.

We might attempt to form some general rules to guide stage combination. Aris, Nemhauser, and Wilde proposed several rules of thumb for stage combination in the *usual* design problem, and these are quoted herewith.

Elimination of output state variables by stage combination should be considered for any stage having as many or more output variables than input variables.

Any stages without design variables should be combined with an adjacent stage. Where several combinations are possible, the one eliminating the most state variables is preferable.

We have seen that such rules of thumb can be approximately valid and useful guides to simplifying problems, and that they also can be misleading in unusual situations. It will be necessary to assess each new design problem quickly to establish the rule of stage combination which best fits the design. Several problems at the end of this chapter reinforce this idea.

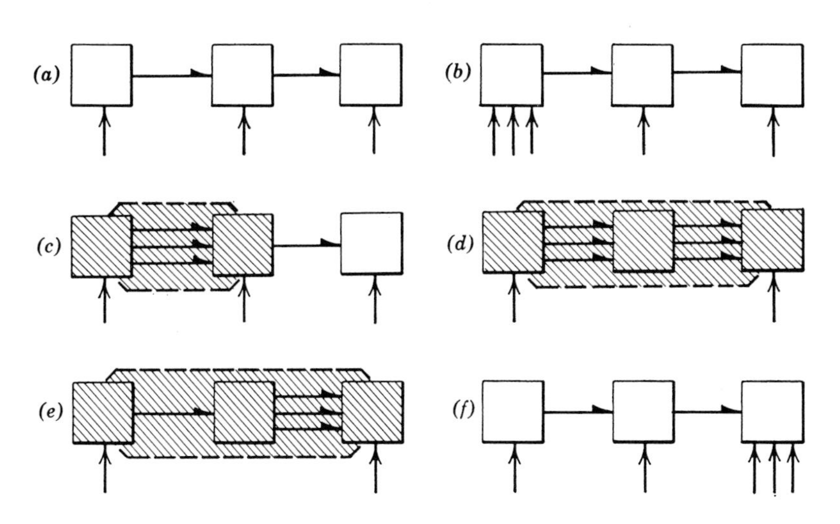

Fig. 9.4-3 Stage combination—Case B.

9.5 THE CUT STATE CONCEPT

Suppose now that the dominant structure which appears after a sensitivity test contains a recycle loop, as in Fig. 9.5-1. This structure is acyclic, save for the single recycle state variable.

Observe that changing the role of that recycle variable to a design variable is equivalent to cutting the recycle and feeding information into both ends of a serial structure. The solution of this interior serial optimization problem could be achieved by some best hybrid plan to yield the maximum objective for that fixed value of the recycle variable

$$\max_{\text{Interior}}(C) = \max_{\{d_1, d_2 \ldots\}} (U_1 + U_2 + \cdots) \tag{9.5-1}$$

Now, of course, we have no right to arbitrarily fix the value of the cut state variable C. Rather, we must search on C to find the maximum of this interior maximum objective.

$$\max_{\text{System}} = \max_{\{C\}} [\max (C)_{\text{Interior}}] \tag{9.5-2}$$

Notice that the direct application of the design variable selection algorithm of Chapter 3 leads to this plan also.

Thus, a strategy for attacking recycle problems might involve the following.

1. Fix the value of the recycle state variable C.
2. Solve the resulting serially structured problem for

$$\max_{\text{Interior}} (C)$$

3. Search on C for the

$$\max_{\{C\}} [\max_{\text{Interior}} (C)]$$

passing through Steps 1 and 2 for each search on C.

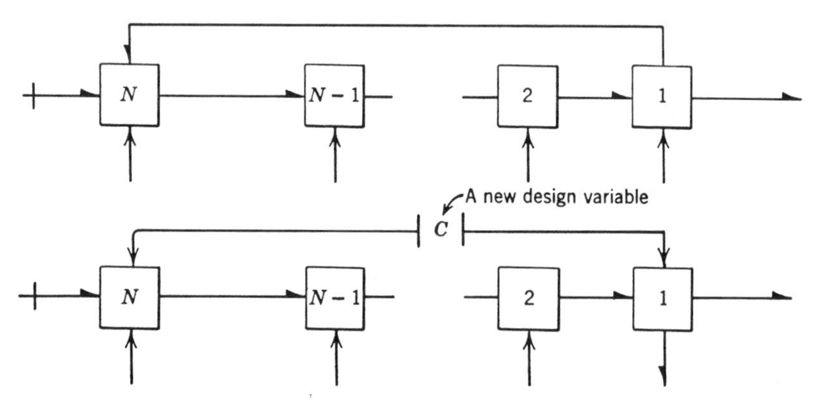

Fig. 9.5-1 Cutting the recycle state variable by a reassignment of design variable.

The conditions under which this *cut state* strategy is superior to, for instance, direct search can be determined quite easily for any special recycle situation. We illustrate this by analyzing the special case when there exists only one output state variable and one design variable per stage, and the interior problem becomes a simple serial structure after cutting, as in Fig. 9.5-1.

The structure of the interior problem and a reasonable stage combination are shown in Fig. 9.5-2. Stages 2 and 1, and N and $N-1$ are combined using the rule of thumb on stage combination, Section 9.4; hence, we are confining further discussion to situations under which that rule is valid.

$$W_{\text{Interior}} = a[(b)^2 + (c)^1(N-3)(b)^1]$$

The work to search on the single cut state variable is $a(b)$. Hence, the total work of optimization by the cut state strategy is

$$W_{CS} = a^2[(b)^3 + cb^2(N - 3)] \tag{9.5-1}$$

Comparing this to the work by direct search

$$W_{DS} = a(b)^N$$

reveals for the reasonable case, when $a = 2$, $b = 5$, and $c = 8$, that the cut state strategy is better when $N \geqslant 5$. Table 9.5-1 summarizes the computations which led to this observation.

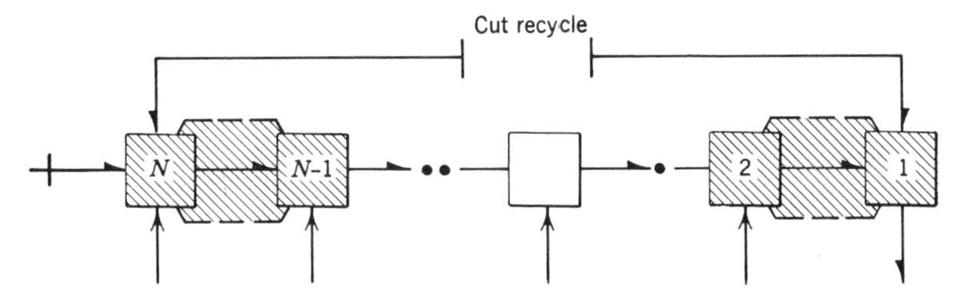

Fig. 9.5-2 Stage combination for the cut system.

Table 9.5-1 Comparing Cut State and Direct Search Strategies

N	3	4	5	10
$\frac{W_{CS}}{W_{DS}}$	2	1.05	0.3	0.00002

Aris, Nemhauser, and Wilde suggested a rule of thumb for cutting recycle loops.

Any loop with less than four design variables should be optimized with respect to all variables simultaneously (i.e., by direct search).

Again this rule is valid only for the usual case of $a \approx 2$, $b \approx c \gg 1$, and the unusual case should be analyzed separately.

This recycle rule is particularly interesting to chemical engineers, since chemical processing systems often contain tight little loops with only a few design variables contained therein. Direct search would then be the proper approach to these systems.[4]

[4] For a solution of the crosscurrent extraction problem of Section 3.3 with recycle using the cut state concept, see E. M. Rosen, *Chem. Eng. Sci.*, **19**, 999 (1964).

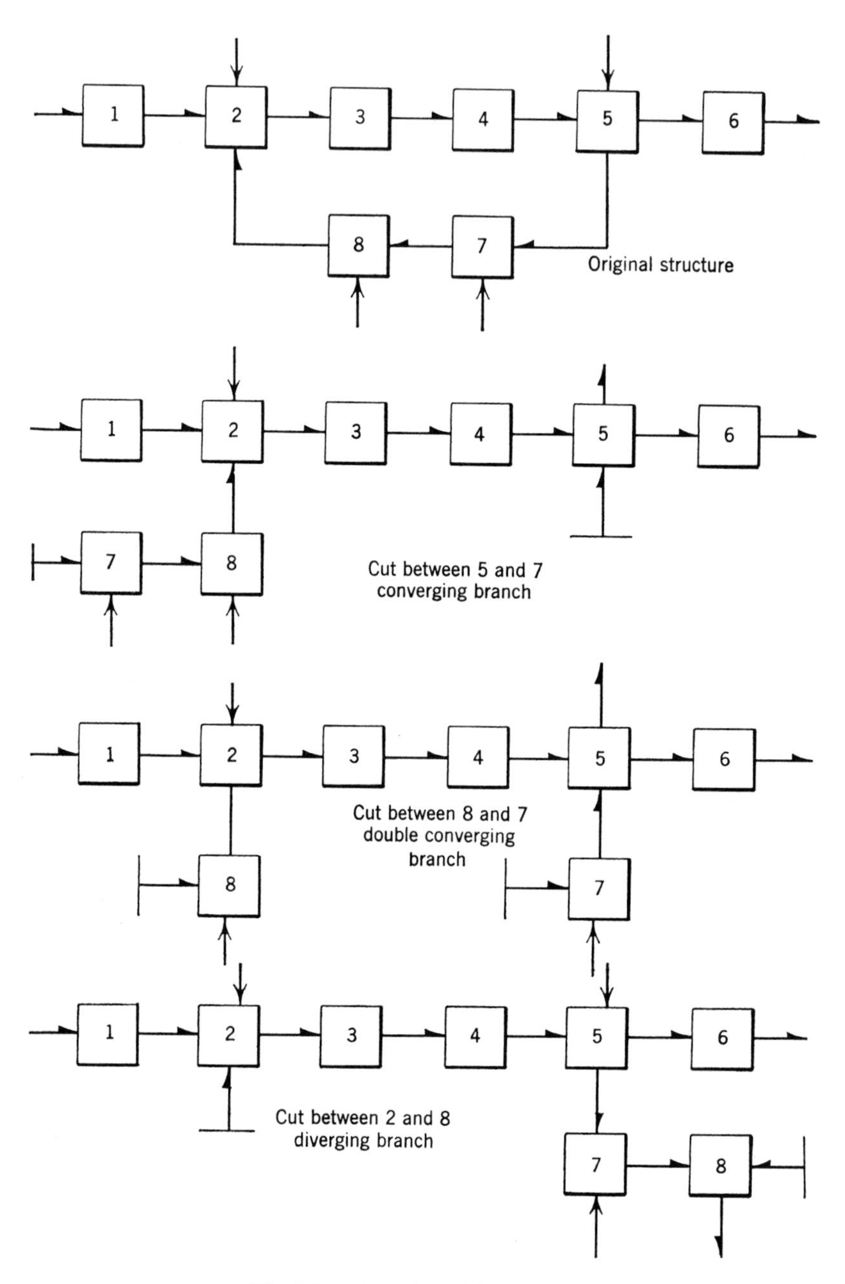

Fig. 9.5-3 Location of the cut state.

Quite often a large recycle loop can be cut in a number of places by electing to call an interstage variable a cut state variable (see Fig. 9.5-3). In Section 9.3 we saw that the diverging branch serial structure is relatively easy to suboptimize. The cut in a recycle problem should, therefore, be placed to generate a diverging branch interior problem whenever practical. We suggest:

The cut state should be inputs to multiple state input stages. When there is a choice it pays to put the most difficult stages at the head of a serial structure, for there they are only optimized once with known feed during suboptimization.

Observe how this rule applies in Fig. 9.5-3.

9.6 SUMMARY OF THE RULES OF THUMB

In this chapter and in the previous chapters we observed that the information flow structure of a processing system when properly interpreted provided a handle on the otherwise slippery optimization problems. These bits and fragments of observations have been combined into six rules for simplifying large complicated problems. These rules are now presented and discussed.

Rule 1. Fixed Output Constraints

"Any fixed output should be transformed into an input by decision or state reversal."

This concept of information flow reversal was introduced first in Chapter 3.

Rule 2. Irrelevant Stages

"If a stage has no objective function and if all of its outputs are system outputs, then that stage and its design variables should be eliminated."

Such stages would have been eliminated by the sensitivity study of Section 9.2.

Rule 3. Stage Combination

"Elimination of output state variables by stage combination should be considered for any stage having as many or more output variables than input variables.

"Any stages without design variables should be combined with an adjacent stage. Where several combinations are possible, the one eliminating the most state variables is preferable."

Rule 4. Small Loops

"Any loop with less than four design variables should be optimized with respect to all variables simultaneously."

Rule 5. Cut State Location

"The cut state variables should always be inputs to multiple input stages, to yield a diverging structure when possible. Try to place troublesome components at the head of serial information flow structures."

Rule 6. Order of Suboptimization

"Suboptimization should proceed in the direction opposite to that of the state arrows."

This is dynamic programming, discussed in Chapter 8.

The statement of the rules of thumb is only the beginning, since a complete understanding of their application is required before mastery can be claimed.

9.7 AN APPLICATION OF THE MACROSYSTEM METHODS

In this section we apply the six rules of thumb to a processing system whose information flow structure is shown in Fig. 9.7-1. Notice how the rules provide only a rough guide to the organization of an optimization plan, and that considerable flexibility remains. This is to be expected; it is only after we are deep into the details of the computations that sufficient information becomes available to select among the several approaches that might be taken. At this point no detailed calculations have been made. We wish to plan our first computational attack.

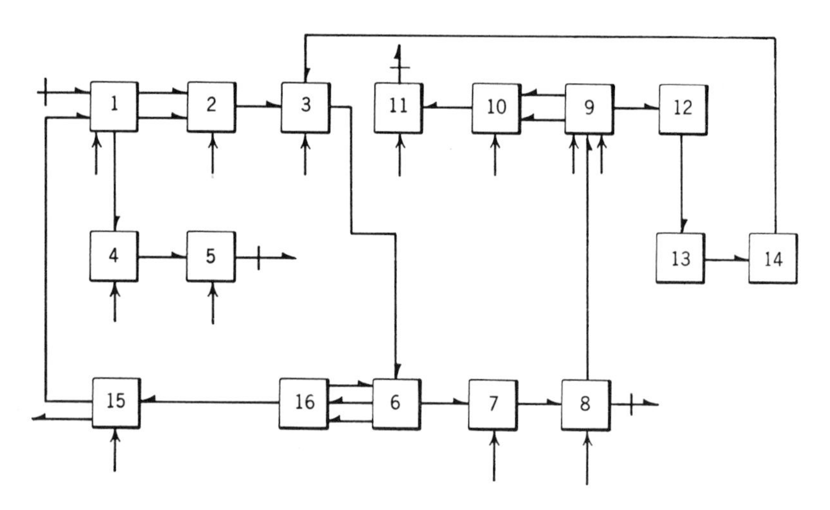

Fig. 9.7-1

Rule 1. Fixed Output Constraints

Fixed output constraints appear at units 5, 8, and 11. We elect to reverse the state variable from 4 to 5, the state variable from 1 to 4, and the design variable at 1 to meet the fixed constraint at unit 5. Decision reversals are made at units 8 and 11. Note the flexibility to adapt to special technical features of the units or the structure of the system which favors certain reversals.

Rule 2. Irrelevant Stages

We assume that the information flow structure describes the dominant problem, and that the components and streams pictured constitute the major features of the system, technically and economically. No irrelevant stages exist.

Rule 3. Stage Combination

A number of units have no design variables, either initially or after the reversals required by the fixed output constraints. Units 7 and 8 are combined. Units 12, 13, 14, and 3 are combined. Units 6, 16, and 15 are combined. Units 1 and 2 are combined. Units 11 and 10 are combined. Units 4 and 5 are combined, since 5 has as many outputs as inputs. The partially ordered system is shown in Fig. 9.7-2 after the application of the first three rules.

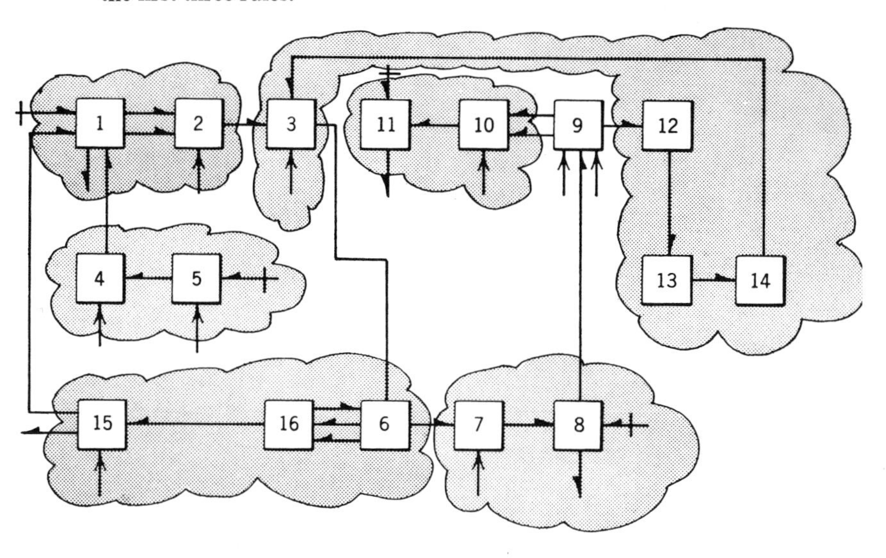

Fig. 9.7-2

Rule 4. Small Loops

There is one main loop with only three design variables in Fig. 9.7-2 consisting of the units 1-2, 3-14-13-12, and 15-16-6. This might be combined into one unit according to the small loop rule. This is a rather loose interpretation of this rule since that loop is connected to another loop involving units 9 and 7-8; no general rules were derived for loops within loops. Once these units are combined the structure involves one remaining loop encompassing units 1-2-3-14-13-12-6-16-15, 9, and 7-8 with various appendages. This loop involves six design variables.

Rule 5. The Location of Cut States

We now reach the point where we might consider the reassignment of design variables to place cut states within the loop structure. Notice that if we cut between unit 9 and unit 1-2-3-14-13-12-6-16-15 and make a decision reversal at unit 9, the structure will be acyclic and units 11-10 and 9 can be combined by the stage combination rule. The simple four component subsystem structure with a cut recycle loop shown in Fig. 9.7-3 results. This is just one of several locations that could have been considered for the cut state variable.

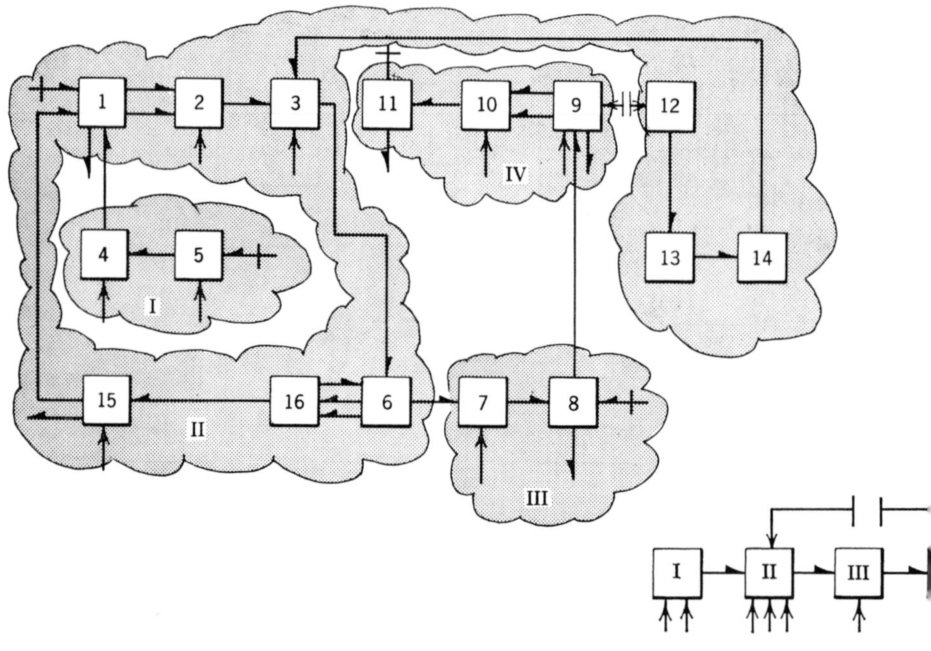

Fig. 9.7-3

Rule 6. The Order of Suboptimization

An attack on the suboptimization of this system is as follows:

Specify a value of the cut state design variable located on the recycle loop.

Suboptimize beginning at Subsystem IV and work up through Subsystem III and Subsystem II.

Return to the first and second steps and search on the cut state design variable until the loop is suboptimized.

Include Subsystem I as the last step in the suboptimization sequence.

We have seen that the initial structure, Fig. 9.7-1, contained far more detail than required, and that after an application of the several macrosystem rules, a rather simple subsystem structure appeared. This appears to be a common phenomenon, and many rather complex problems only appear so. However, the final structure, while simpler than the initial one, still demands a considerable expenditure of computational labor before the optimum design is exposed. This is also a common phenomenon in process engineering.

There are several thoughts to be derived from this example.

1. We have found some order in the attack on an extremely complex problem; imitating, in some sense, the effective intuitive manipulations of the experienced problem solver.
2. Our imitation of the abilities of the skilled problem solver is based on *six suggestive rules of thumb* which tend to lead towards a reduction of the complexity of the system.
3. These rules are only suggestive, giving the designer a latitude to impose his own preference on the problem. The information reversals, component groupings, cut state locations, and so forth are flexible; and this flexibility is essential since this system analysis is only part of the picture. A knowledge of the technology of the components would tend to favor certain of the several alternatives.
4. These system methods must work hand-in-hand with the detailed analysis of the technology within the blocks of the system. An analysis resting solely on system properties is as incomplete as an analysis which ignores the system structure focusing attention only upon the details of the reactors, stills, furnaces, and pumps. Systems analysis and component technology are two blades of a scissors; one is useless without the other for cutting through the complex problems of process engineering.

9.8 THE OPTIMIZATION OF A SULFURIC ACID PROCESS[5]

Lowry and Pike have illustrated the application of these methods by optimizing the design equations for a process which is to manufacture sulfuric acid using the contact method. This section is based exclusively on their work.

Figure 9.8-1 is the schematic flow diagram for a contact process to manufacture 98 per cent sulfuric acid and high pressure steam. We shall now briefly discuss the technology.

[5] I. Lowry and R. W. Pike, "A Dynamic Programming Study of the Contact Process," presented at the 61st National Meeting of the A.I.Ch.E., February 1967.

Burner. Molten sulfur at 423°K is injected into a burner for oxidation to sulfur dioxide, a reaction which is extremely rapid above 675°K. Previously dried air is used as the oxidizing agent. Heat losses from the burner are neglected.

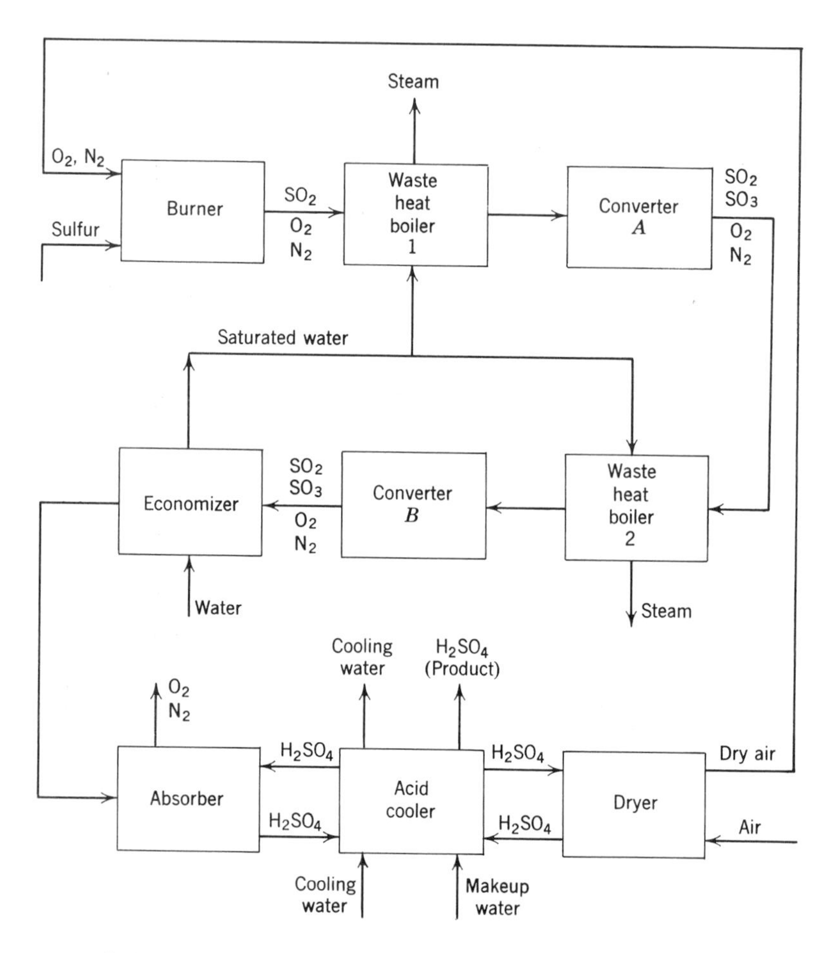

Fig. 9.8-1 Schematic diagram of the contact process for sulfuric acid.

Waste heat boilers. The hot gases leaving the burner must be cooled before further oxidation can be achieved efficiently. The cooling is done in a waste heat boiler from which the product is 225 psia steam superheated to 590°K. The recovery of heat as steam is very attractive because of the rising costs of fuels. A second waste heat boiler cools the gases after partial oxidation to sulfur trioxide takes place in the first converter.

Converters. The cooled sulfur dioxide is converted to sulfur trioxide in a two stage converter system. As the oxidation occurs the temperature of the gases rise thus lowering the equilibrium conversion. A second waste heat boiler is then used to lower the temperature of the gas to favor the higher conversion to sulfur dioxide in the second converter. Figure 9.8-2 shows the

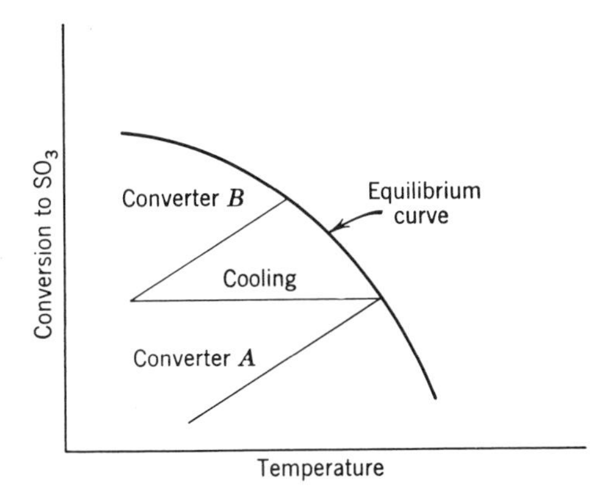

Fig. 9.8-2

temperature path the gases take in the excursion through converter *A*, waste heat boiler 2, and converter *B*.

Economizer. The economizer is a heat exchanger which takes water at 350°K and heats it to saturated water at 225 psia and 473°K to act as feed to the waste heat boilers.

Absorber. The gases leaving the economizer are passed countercurrent to 98 per cent sulfuric acid to absorb the sulfur trioxide, and the remaining gases are vented into the atmosphere.

Acid cooler. The acid circulating through the absorber and drier is cooled in a water cooled heat exchanger.

Dryer. It is important that there be no water in the air used in the burner, and the atmospheric moisture is removed by absorption in acid.

The detailed design equations for each of these processing components are not presented here for they are available elsewhere for examination.[6]

Figure 9.8-3 shows the information flow diagram obtained by examining the design equations. The flow rate of sulfur to the burner is fixed at $W_F =$ 10,000 pounds per hour. The flow rate of air to the burner appears as a design variable in the dryer, since the air is sent there first for moisture removal. The engineer is free to adjust the air rate W_a to achieve economy. The flow rate of water to the economizer W_{M1} is free to be adjusted as is

[6] Ivan Lowry, M.S. Thesis, Louisiana State University, 1966.

the split of that water, which when preheated is sent to the waste heat boilers as stream W_{M2} and W_{M3}. The other variable which is free to be adjusted is the rate of the coolant in the acid cooler W_{MO}. There are two major loops of information flow; one from the dryer to the burner and one from the economizer to each of the waste heat boilers.

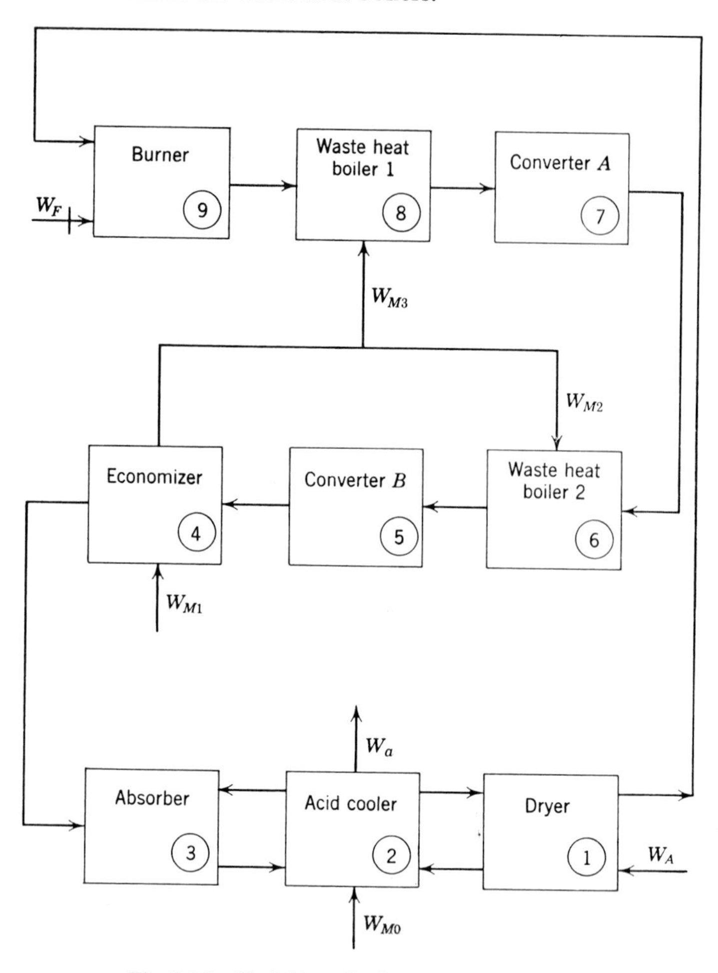

Fig. 9.8-3 Variables affecting the contact process.

The information flow loop involving the economizer, the waste heat boilers, and the converters was broken by a ruse which will be discussed in detail in Chapter 10, *Multilevel Attack on Very Large Problems.* An interim or fictitious price is assigned to the heated water leaving the economizer, and then it is required that the boilers purchase water at that price. This being done, it is no longer necessary to think of the boilers as being supplied by the economizer: the boilers may be thought of as purchasing the water from an auxiliary source, and the economizer may be thought of as being a supplier for that auxiliary source. This kind of price assignment ruse is very effective, and for the time being we shall assume it is a proper technique; later in Chapter 10

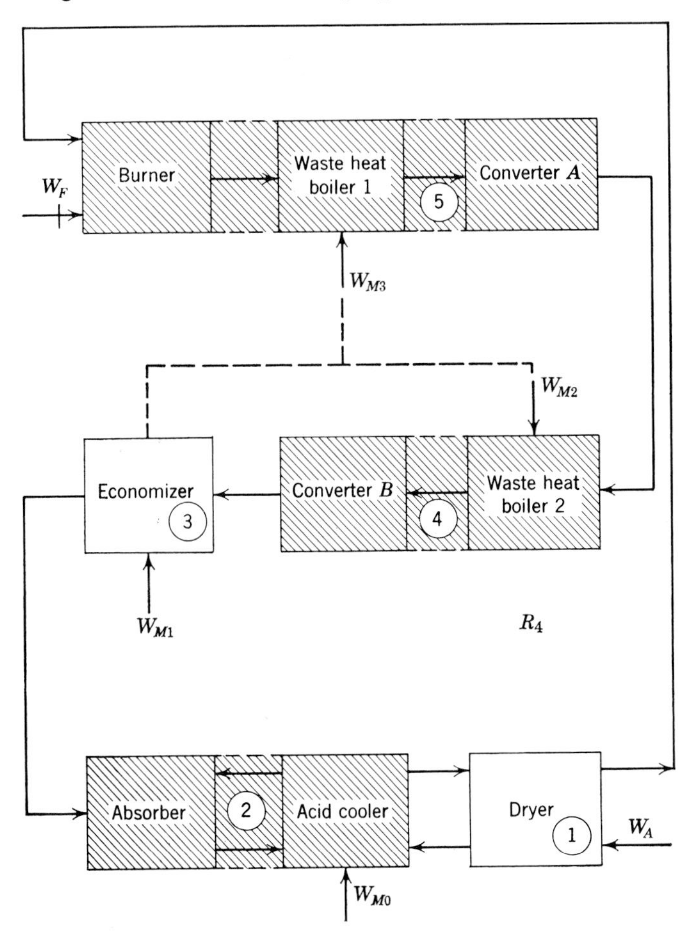

Fig. 9.8-4 Breaking recycle by a pricing ruse.

we shall examine the method more closely. There then results the simplified information flow structure of Fig. 9.8-4 with only one major recycle loop.

The burner, the converters, and the absorber have no design variables and should therefore be combined with adjacent stages so as to eliminate as many state variables as possible. There results the five stage system shown in Fig. 9.8-4. Applying the cut state concept to the major recycle loop requires that the flow rate and temperature of the dry air to the burner be fixed during optimization. This changes the design variable W_A at the drier to a state variable, and the dryer is combined with the absorber-acid cooler subsystem. Since the temperature of the dry air sent to the burner must be specified in

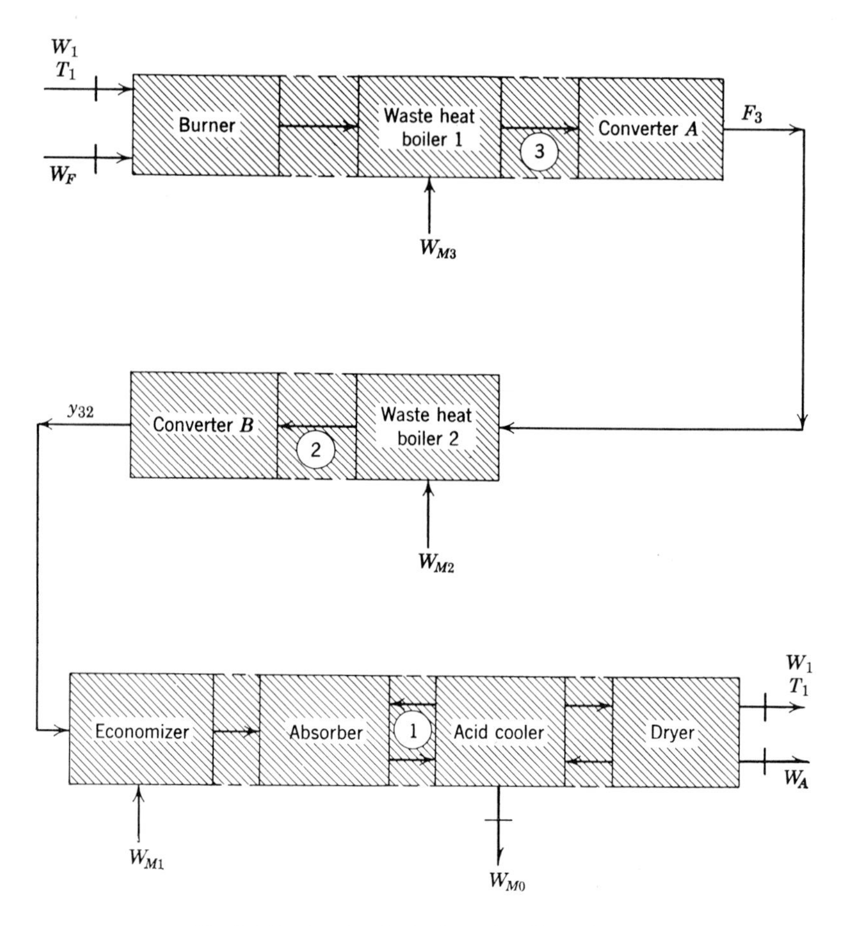

Fig. 9.8-5 Final optimization plan, three stages.

order to cut the recycle, the amount of cooling water sent to the acid cooler can be computed for any given dry air flow rate and temperature. That design variable becomes a state variable, and the absorber-acid cooler-dryer subsystem is then free of design variables and should be combined with the economizer to eliminate the state variable connecting the two. The three stage optimization problem in Fig. 9.8-5 results.

The optimization of the process then proceeds thus:

1. Reasonable values are assumed for the cut state variables W_1 and T_1, the flow rate and the temperature of the dry air entering the burner and leaving the dryer.
2. The economizer-absorber-acid cooler-dryer subsystem is suboptimized. The design variable used during the suboptimization was the temperature of the gases leaving the economizer, rather than the amount of cool water sent to the economizer as shown in the information flow diagram. This reselection of design variable was found to simplify calculations.
3. The economizer-absorber-acid cooler-dryer subsystem is combined with the converter B-waste heat boiler 2 subsystem as part of the next step in the sequential suboptimization of the process. The amount of water sent to waste heat boiler 2 is the major design variable during this suboptimization.
4. The entire system is now optimized by adjusting the rate at which water is sent to waste heat boiler 1.
5. There results the optimal design for the values assumed in step 1 for the dry air rate and temperature. It is then necessary to conduct a search over these variables to achieve the optimum design.

We have only sketched briefly through the very extensive work of Lowry and Pike to relate the concepts in this chapter to a realistic problem. However, we can benefit from the experience they gained. The following comments were offered.

> The optimization strategy described resulted from considerable trial and error, and must be developed by a combination of the methods presented formally and experience, intuition and luck.
>
> Adherence to the rules of thumb can simplify the task of problem formulation, although it is extremely important that the formulator be familiar with the technology of the process under study.

9.9 CONCLUDING REMARKS

The practicing chemical engineer stands to benefit greatly from the research in the areas of optimization and systems analysis. He must continually monitor the research literature and adopt any new methods as soon as they are shown to be useful. This is not an easy task, however, since the literature frequently reports developments of purely academic interest and little practical value alongside developments of vital practical importance. The references cited at the end of the chapter are selected to orient the reader in this confusing area. We have only sampled the field in the text material and the literature abounds with more methods of interest to the designer.

References

The basic paper in this area is:

R. Aris, G. L. Nemhauser, and D. J. Wilde, "Optimization of Multistage Cyclic and Branching Systems by Serial Procedures," *A.I.Ch.E. J.*, **10**, 913 (1964).

An interesting summary of this work can be found in:

D. J. Wilde, "Strategies for Optimizing Macrosystems," *Chem. Eng. Progr.*, No. 3, **61** (1965).

See also Chapter VI of:

G. Nemhauser, *Introduction to Dynamic Programming*, Wiley, New York, 1966.

and Chapter XX of:

R. Aris, *Discrete Dynamic Programming*, Blaisdell, Waltham, Mass., 1961.

We must realize that the material in this chapter is but an introduction to a vast area of optimization as described in:

D. J. Wilde and R. J. Beightler, "Foundations of Optimization," Prentice-Hall, Englewood Cliffs, N.J., 1967.

PROBLEMS

9.A. Consider an initial value serial optimization problem with N stages, each of which has a single design variable which can take on one of two values. Suppose further that each stage is connected to its neighbor by a single state variable which also can take on only one of two values. This kind of optimization problem might arise in the preliminary stages of process design.

(a) How many unique design alternatives exist?

Answer: 2^N

(b) How many comparisons of design alternatives must be made to find the optimal design by ordinary dynamic programming?

Answer: $4N - 2$

(c) How many comparisons must be made if alternate stages in the process are combined with their neighbor, to form a $N/2$ stage system to which dynamic programming is applied?

Answer: $4N - 4$

(d) How many comparisons must be made if the stages are combined into N/k subsystems of equal size, to which dynamic programming is applied?

Answer: $2^k(2N/k - 1)$

(e) Combine the first k stages into one subsystem and leave the remaining $N-k$ stages unaltered. How many design alternatives must be compared to optimize the process by dynamic programming?

$$\text{Answer: } 2^k + 4(N-k)$$

(f) Combine the last k stages into a subsystem and leave the remaining $N-k$ stages unaltered. How many comparisons must be made?

$$\text{Answer: } 2 + 4(N-k-1) + 2(2)^k$$

(g) Of the several optimization strategies suggested in this problem, which would be the best to apply to a ten stage process?

9.B. Repeat problem 9.A for an optimization problem differing only in that there are S state variables connecting each of the stages.

9.C. Suppose that the system in 9.A is modified to the extent that a single recycle stream connects the end of the process to the beginning. Under what conditions would you attempt to break the recycle by reselecting design variables?

9.D. Table 9.5-1 compares the cut state and the direct search optimization strategies for the special case of single state and design variables. This leads to the rule of thumb concerning small recycle loops. Derive a similar rule of thumb for the structure shown in Fig. 9.D-1.

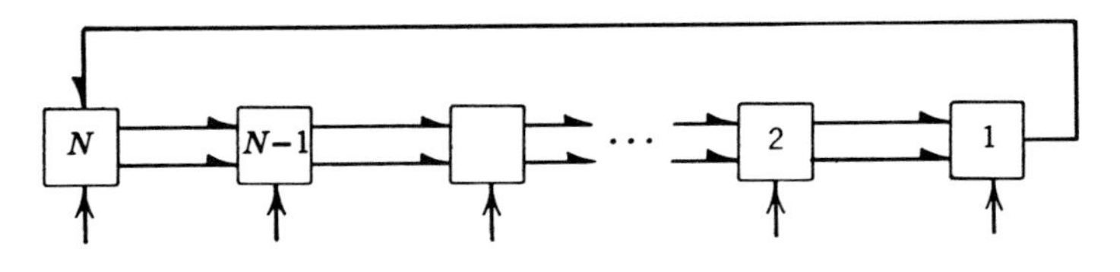

Fig. 9.D-1

9.E. Devise a macrosystem optimization strategy for the process shown in Fig. 3.8-3. How does this relate to the structure in Fig. 3.9-1?

9.F. Devise a macrosystem optimization strategy for the process whose information flow structure is shown in Fig. 3.10-2. Compare your plan to the structure in Fig. 3.10-4.

9.G. In the light of the developments of this chapter, suggest modifications in the example of Section 8.4.

9.H. Rudd et al. incorrectly extended the concept of suboptimization to the crosscurrent extraction problem with product recycle. Review the paper and find the error. The comments of Jackson prompted most of the later research on this recycle problem and related macrosystem problems.[7]

9.I. Discuss the relationship between the design variable selection algorithms of Chapter 3 and the cut state concepts of this chapter. Will the application of the design variable selection algorithm lead to the simplest optimization plan?

[7] D. F. Rudd et al., "On Optimum Crosscurrent Extraction with Product Recycle," *Chem. Eng. Sci.*, **17**, 277–281 (1962); R. Jackson, "Comments on Optimum Crosscurrent Extraction with Product Recycle," *Chem. Eng. Sci.*, **18**, 215 (1963).

10

MULTILEVEL ATTACK ON VERY LARGE PROBLEMS

A recurring dilemma in the design of processing systems has its origin in the size of the problems that must be attacked. The problems often are large enough to strain and exceed the abilities of a single team of engineers, and it is necessary to relegate the design responsibility to several more or less autonomous teams of experts, each responsible for a part of the system. How can the efforts of these teams be coordinated so that their designs will mesh together and form an optimal system? As in the preceding chapter, we now discuss methods at the forefront of research, which may eventually evolve into established practice.

Typical Problem

The design of a large processing complex with chemical plants around the country is to be divided into the design of a number of subprocesses which will purchase chemicals from each other and from the outside. What inplant prices should be set to force the several subdesigns to levels which maximize the efficiency of the large complex?

10.1 A LARGE SYSTEM DILEMMA

The engineer responsible for the design of very large processing systems often finds himself in a dilemma caused in part by enormous data handling problems. A practical limit exists on the number of details for which the single group of engineers can be held responsible, and the usual large processing system exceeds that limit. Thus, there is a desire to tear the large system into a number of smaller subsystems and allocate the responsibility for the design of the subsystems among a number of engineering groups.

Moreover, this modular approach has become a common business practice with talent available on contract for such special subsystems as inert gas generation, steam plants, and thermal cracking of hydrocarbons, for example. We see the responsibility for the design of a refinery system distributed among the crude desulfurization design group, the waste treatment systems design group, the fractionation system design group, and so forth, the original system being too large for one group of engineers to be responsible for all of the details.

The source of a dilemma is the realization that tearing a large system design problem into a number of smaller design problems, using the macrosystem methods of Chapter 9, for example, may unduly restrict the original problem, and that optimization can be achieved only if free and unrestricted interaction is allowed among all portions of the system and its surroundings. Tearing may inadvertently sever some critical interaction, and the suboptimization of a carelessly torn subsystem may not lead to an optimal design. In an extremely large problem, the engineer is caught between an absolute need for reducing the size of the individual units of responsibility and the realization that the careless optimization of any of the smaller subsystems may not be in the best interests of the overall system. How then can the efforts of those responsible for the subsystems be coordinated to achieve the system goals?

First we discuss a multilevel approach in which the responsibility for the engineering of each of the subsystems is assigned to individual engineering groups. These groups, at what is called level I, are only responsible for the design of their subsystem and need not directly concern themselves with the design of any other portion of the process. The responsibility for the overall system behavior is delegated to a coordinating group, which is said to be at level II. The second level is divorced from the need for considering all the details of the subsystems and is charged only with assuring the cooperation of the other groups in achieving the systems goal. Each group then has a domain of responsibility within its limited but still considerable abilities.

We describe a technique suggested by L. Lasdon of the Case Institute of Technology[1] for coordinating this two level attack in which communication is maintained between the two levels by supply, demand, and price parameters. It should be realized that the two level attack is just one of a number of possible approaches that might be taken, and that the emphasis placed on this method is for pedagogical reasons, to give the initiate an introduction to the kinds of thinking involved. In the last section of this chapter we examine the more extensive decomposition principles used in SYMROS, a computer program for scheduling the activities of the Shell Oil Company.

[1] Leon Lasdon, "A Multilevel Technique for Optimization," Systems Research Center report SRC50-C-64-19, Case Institute of Technology, Cleveland, Ohio, 1964.

10.2 THE SYSTEM AND ITS SUBSYSTEMS

Focus attention now on the large system in Fig. 10.2-1. Notice that the system has been divided into three subsystems which interact via the state variables s, which might be the flow rates of intermediate products. The macrosystem methods of Chapter 9 might have been employed to group the system thus, or the structure might well have appeared naturally from the geographical position of the subsystems.

The design goal is to select the detailed designs for the subsystems which maximize the overall system objective function, U, consisting of the contributions U_A, U_B, and U_C from the three subsystems. The contribution from subsystem A might be, for example, the profit calculated from the value of goods sold by that subsystem to the surroundings, less the manufacturing costs and the appropriate charges for capital invested therein. *No charge is assigned, at this time, to material or energy obtained from or given to any other subsystem*, since there is no way of assessing the value of such *inplant* transfer. Internal transfers are considered to be unlimited and free during the development of the objective function for each subsystem. In this way the subsystems are torn free from the system.

Each group at level I then has within its domain of responsibility the following:

(*a*) State variables s which connect the system with other subsystems.

(*b*) The design details d of the subsystem about which each group is the *sole* authority.

(*c*) The contribution that the subsystem makes to the overall system objective.

The large system design problem may then be written as

$$\max_{(d_A,\, d_B,\, d_C \cdots)} \quad (U_A + U_B + U_C \ldots) \tag{10.2-1}$$

with the responsibility assigned to the subsystem groups as follows.

	Objective Function	*Design Detail*
Group I_A:	$U_A(s_3, s_1, d_A)$	$f_A(s_3, s_1, d_A) = 0$
Group I_B:	$U_B(s_1, s_2, d_B)$	$f_B(s_1, s_2, d_B) = 0$
Group I_C:	$U_C(s_2, s_3, d_C)$	$f_C(s_2, s_3, d_C) = 0$

The simple expression $f_A(s_3, s_1, d_A) = 0$ for the design detail of subsystem A is a shorthand expression of all the tools that the group may elect to bear, any and all of the methods discussed in this text. The design relation merely states that for a given input state variable s_3 and for a given design detail d_A, the group responsible for this area of the system can predict its behavior and assign a value to the output s_1. This may well require the efforts of a team of engineers aided by the most extensive of computing facilities.

Suppose now that the subsystems are of a complexity which is at the limits of responsibility for a single design team and, hence, the systems design problem, which necessarily involves the consideration of more than one subsystem, is beyond attack by usual methods. How do we devise a strategy to attack this problem?

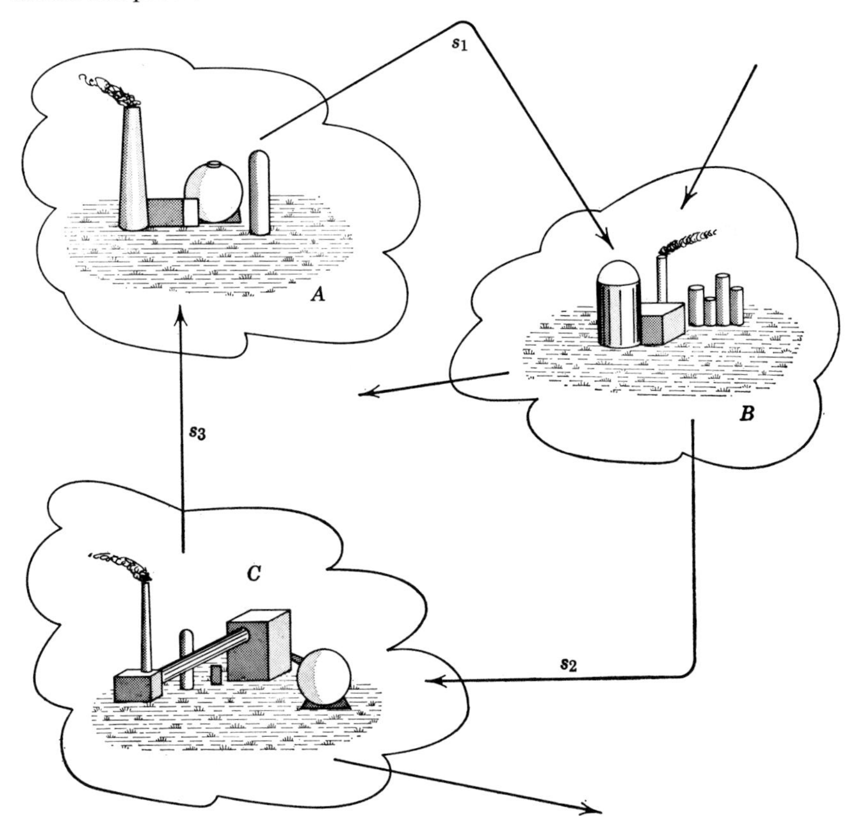

Fig. 10.2-1 The system and its subsystems.

10.3 SUBOPTIMIZATION AND FORCED COOPERATION

At the outset, we know that the optimization of any subsystem by itself without regard for interactions with other parts of the system is not part of an optimal strategy. A gain achieved by the optimization of one subsystem might be at the expense of a more than compensating loss in some other subsystem.

We are then led to the problem of forcing cooperation among the groups of subsystem designers. There is the need for a new coordinating group to assign proper suboptimization goals to the subsystems to force their cooperation.

Such cooperation can be achieved by *requiring* that each subsystem buy its input state variables from and sell its output state variables to the other subsystems *at assigned prices*. The level II coordinating group then has the task of assigning and adjusting the inplant prices of these state variables until the level I groups have cooperated in solving the systems problem while attempting to solve their own suboptimization problems.

Thus, being required to buy and sell from within, the level I groups will strive to maximize the new objective functions.

$$\begin{aligned} &\text{Group } \mathrm{I}_A: && \max_{\{s_3, d_A\}} (U_A + p_1 s_1 - p_3 s_3) \\ &\text{Group } \mathrm{I}_B: && \max_{\{s_1, d_B\}} (U_B + p_2 s_2 - p_1 s_1) \\ &\text{Group } \mathrm{I}_C: && \max_{\{s_2, d_C\}} (U_C + p_3 s_3 - p_2 s_2) \end{aligned} \tag{10.3-1}$$

The p_1, p_2, and p_3 inplant prices are to be prescribed by the level II coordinating group.

We pause now to make a few observations. Notice that the level I groups have additional variables to manipulate during suboptimization, namely, the amount of the input state variable demanded. Thus, for given prices, each subsystem will demand a certain amount of its input state variable in an attempt to achieve its local optimization goal. But that demanded by one subsystem must be supplied by another, and the prices which force equality of supply and demand are yet unknown.

We feel intuitively that the system will operate most efficiently when all of the inplant supplies and demands are equal. And this is so for a large class of optimization problems which exhibit certain properties of continuity and differentiability. The subproblems are extracted from the Lagrangian of the original systems problem and the theory of Lagrange multipliers can be used to validate the suboptimization design criteria.[2]

The multilevel attack might proceed thus: The level I subsystem groups set forth a tentative design based on the assumption that the inplant material is available free of charge. The level II coordinating group compares the supplies and demands and attempts to assign artificial inplant prices for the state

[2] See for example H. Everett, "Generalized Lagrange Multiplier Method for Solving Problems of Optimum Allocation of Resources," *Operations Res.*, **11**, 399–417 (1963).

variables which will cause the groups at level I to readjust their tentative designs towards the point where the demands of each subsystem equal the supplies afforded by the other subsystems. The level I groups communicate with the level II via supply and demand variables, and the level II communicates via the price parameters as illustrated in Figs. 10.3-1 and 10.3-2.

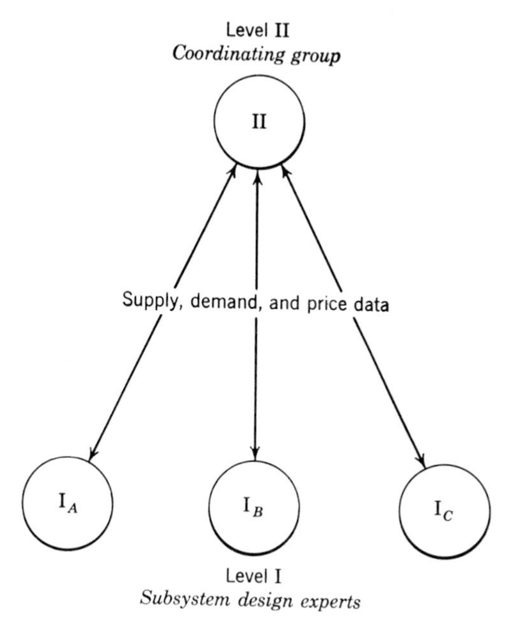

Fig. 10.3-1 A two level attack structure.

Upon receipt of supply and demand data, the coordinating group is faced with the problem of adjusting the price parameters to more nearly equate the supplies and demands on the next round. The data are of the form:

Demanded by one subsystem	*Supplied by another subsystem*
s_1^{demand}	s_1^{supply}
s_2^{demand}	s_2^{supply}
s_3^{demand}	s_3^{supply}

The coordinating group must analyze these data and find a new set of prices for the next go around. One attack might be to solve the minimization problem,

$$\min_{\{p_1,p_2,p_3\}} \left(\sum_i |s_i^{\text{demand}} - s_i^{\text{supply}}| \right) \tag{10.3-2}$$

using the direct search approach discussed in Chapter 6. However, at this point we elect to introduce some intuitive reasoning. When the demands for a commodity exceed the supply, the value of that commodity increases in the marketplace. This provides more incentive to produce and less incentive to consume, thus tending to drive the supply towards the demand in the future. This suggests a method of price adjustment for our systems coordination problem. The new price might be estimated by

$$p_i^{\text{new}} - p_i^{\text{old}} = K_i(s_i^{\text{demand}} - s_i^{\text{supply}}) \tag{10.3-3}$$

for

$$i = 1, 2, 3 \ldots$$

where K_i are constants of proportionality which determine the rate of convergence of the price adjustment scheme. The prices tend to rise when the demand exceeds the supply and tend to fall when the supply exceeds the demand.

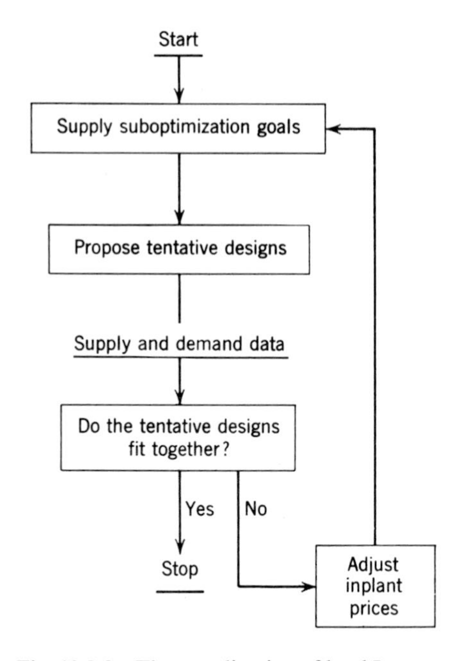

Fig. 10.3-2 The coordination of level I groups.

10.4 AN APPLICATION OF THE TWO LEVEL METHOD

In this section this simple two level method of subsystem coordination is applied to the optimization of the system shown in Fig. 10.4-1. In Fig. 10.4-1 the design and economic data are presented graphically. For example, the graph for subsystem *A* consists of the *maximum profits* this subsystem can earn

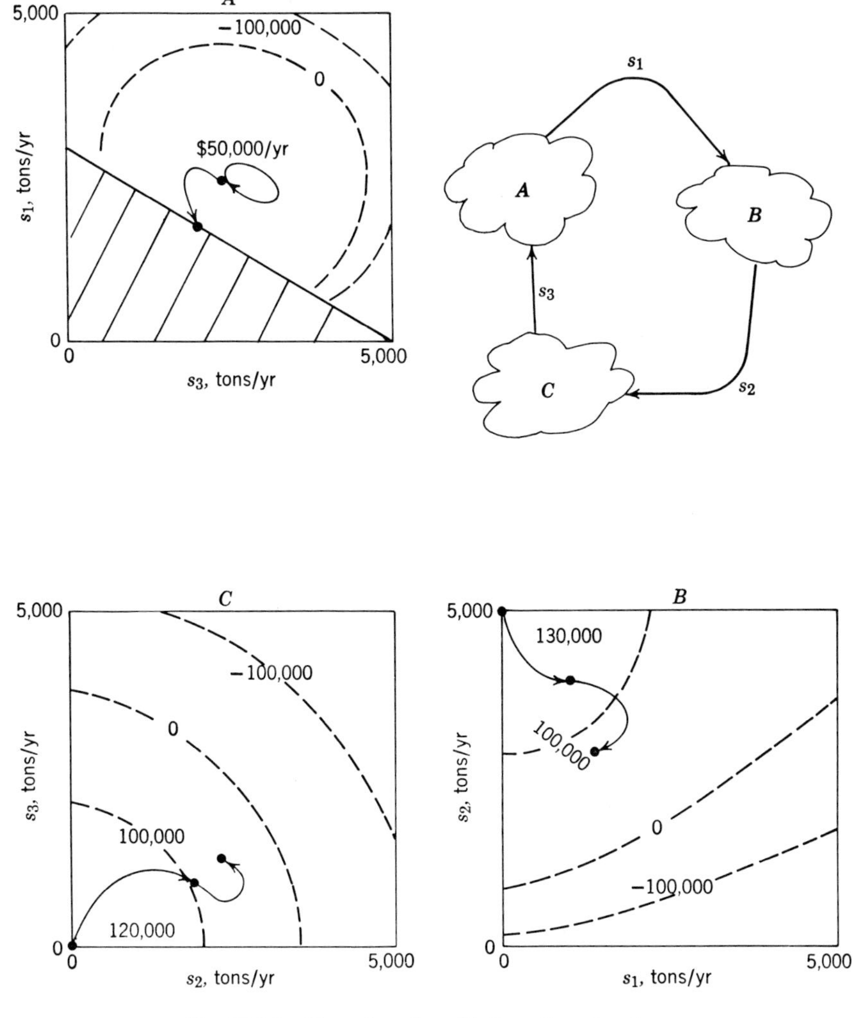

Fig. 10.4-1 The external profits from the subsystems.

from the *outside* by demanding a certain amount of s_3 from subsystem C and supplying an amount of s_1 to subsystem B. The subsystems are sufficiently large so that the estimation of the profit for one such point demands the full efforts of a team of specialists. We can just refer to the graph for that information, a luxury not available to the subsystem design groups. The design data are placed in this unusual but convenient form to focus attention on the methods of coordination and to suppress dealings with detailed numerical manipulations.

STRONG COMPONENT DOMINANCE. One erroneous approach to the design of this system involves the yielding of precedence to the strongest subsystem. Subsystem B has a potential profit of \$130,000/yr, whereas subsystems A and C have a potential of only \$50,000/yr and \$120,000/yr of profit. The strong component dominance principle would then force subsystems A and C to yield to those demands.

Subsystem B would demand $s_1 = 0$ and $s_2 = 5{,}000$ tons/yr to achieve its maximum profit $U_B = 130{,}000$. To provide the feed of $s_1 = 0$, subsystem A must demand a feed of $s_3 = 5{,}000$ from subsystem C.

The design of the system based on the dominance of the strong component is shown below, a decidedly uneconomical design.

Subsystem	*Demand*	*Supply*	*Profit*
A	$s_3 = 5{,}000$	$s_1 = 0$	$-250{,}000$
B	$s_1 = 0$	$s_2 = 5{,}000$	$+130{,}000$
C	$s_2 = 5{,}000$	$s_3 = 5{,}000$	$-200{,}000$
		Total Profit \$/yr	$-320{,}000$

The dominance of the strongest component in a system leads to an unprofitable design.

Forced cooperation. The two level method of optimization is now applied to this problem to force the cooperation of the subsystem in achieving the maximum systems profit. The convergence is illustrated in Fig. 10.4-1.

Action-Level II: Allow each subsystem complete freedom from the actions of the other subsystems by the arbitrary assignment of internal prices at zero.

$$p_1 = 0, \qquad p_2 = 0, \qquad p_3 = 0$$

Reaction-Level I: Each subsystem responds by striving for maximum profit.

Subsystem	*Demand*	*Supply*
A	$s_3 = 2{,}500$	$s_1 = 2{,}500$
B	$s_1 = 0$	$s_2 = 5{,}000$
C	$s_2 = 0$	$s_3 = 0$

Reaction-Level II: These designs do not fit together, the optimum has not been achieved.

Variable	*Demand*	*Supply*	*Excess Demand*
s_1	0	2,500	$-2{,}500$
s_2	0	5,000	$-5{,}000$
s_3	2,500	0	2,500

The prices must be adjusted to drive the excess demand to zero. Using the economic analogy, the argument might be made that the price is too low for a commodity in excess demand. The new set of prices reflects this thinking.

$$p_1 = -1, \quad p_2 = -2, \quad p_3 = 1$$

Reaction-Level I: The subsystems are now required to purchase their demanded inputs and sell their outputs according to the new price schedule dictated by level II. This leads to the following designs:

Subsystem	*Demand*	*Supply*
A	$s_3 = 2{,}500$	$s_1 = 2{,}500$
B	$s_1 = 1{,}000$	$s_2 = 4{,}000$
C	$s_2 = 2{,}000$	$s_3 = 1{,}000$

Reaction-Level II: The situation has been improved by this new set of prices, but excess demand still exists.

Variable	*Demand*	*Supply*	*Excess*
s_1	1,000	2,500	–1,500
s_2	2,000	4,000	–2,000
s_3	2,500	1,000	1,500

The prices p_1 and p_2 should be decreased and p_3 increased. Try

$$p_1 = -2, \quad p_2 = -4, \quad p_3 = 3$$

Reaction-Level I:

Subsystem	*Demand*	*Supply*
A	$s_3 = 2{,}000$	$s_1 = 1{,}800$
B	$s_1 = 1{,}400$	$s_2 = 3{,}000$
C	$s_2 = 2{,}300$	$s_3 = 1{,}300$

Reaction-Level II: The demands and supplies have been nearly met by this round of suboptimizations.

Variable	*Demand*	*Supply*	*Excess*
s_1	1,400	1,800	–400
s_2	2,300	3,000	–700
s_3	2,000	1,300	+700

Considering the accuracy of the graphs, closer convergence might well be meaningless. The design teams have achieved a sufficient degree of cooperation on the last round.

If the final design is patched together on the basis of this rough solution, the following system design might result.

Subsystem	*Demand*	*Supply*	*Profit*
A	$s_3 = 2{,}000$	$s_1 = 1{,}800$	35,000
B	$s_1 = 1{,}800$	$s_2 = 3{,}000$	70,000
C	$s_2 = 3{,}000$	$s_3 = 2{,}000$	0
		Total Profit	\$105,000/yr

More profitable designs exist, only three steps were performed and convergence was not achieved. Even so, the point has been illustrated that cooperation among the subsystems in any system is essential. Here, the profit from the dominant subsystem *B* has been lowered from its maximum of \$130,000/yr to the value of \$70,000/yr gaining more from subsystems *A* and *C* in exchange.

10.5 OPTIMIZATION OF A THERMOFOR CATALYTIC CRACKING PROCESS[3]

Nunez and Brosilow report the optimization of the design equations of one of Mobil Oil Company's catalytic cracking plants shown in Fig. 10.5-1 by the multilevel methods. This will be used as a bridge between the simple illustration of Section 10.4 and industrial problems.

The Airlift Thermofor Catalytic Cracking process is used to produce high quality gasolines, increase the yield of fuels, and reduce the yield of lower value residual oils. Oil is preheated, mixed with a recycle stream and heated in a furnace to near the cracking temperature. The vapor-liquid mixture then is mixed with a flowing solid catalyst in the reactor where the catalytic cracking occurs. The vapor products from the reactor are separated by fractional distillation. The catalyst gravitates to a kiln where it is regenerated by burning off a coke deposit. The catalyst then flows to a lift pot where air carries it to a surge separator, and the catalyst flows down the seal leg to the reactor.

The Mobil Oil Company has found it worthwhile to simulate the performance of such TCC processes on the digital computer, and in this case the design equations then take the form of computer subroutines which describe the performance of various parts of the process. For the principles which lead to the computer simulation of processes see Weekman[4] and Crowe et al.[5]

[3] E. Nunez and C. Brosilow, "Multilevel Optimization Applied to a Catalytic Cracking Plant," CIC meeting, October 1966.

[4] V. Weekman et. al., " Computer Simulation of a Moving Bed Catalyst Regenerator," *Ind. Eng. Chem.* No. 1, **59** (1967).

[5] C. Crowe et al., *Chemical Plant Simulation*, Mc Master University (1969).

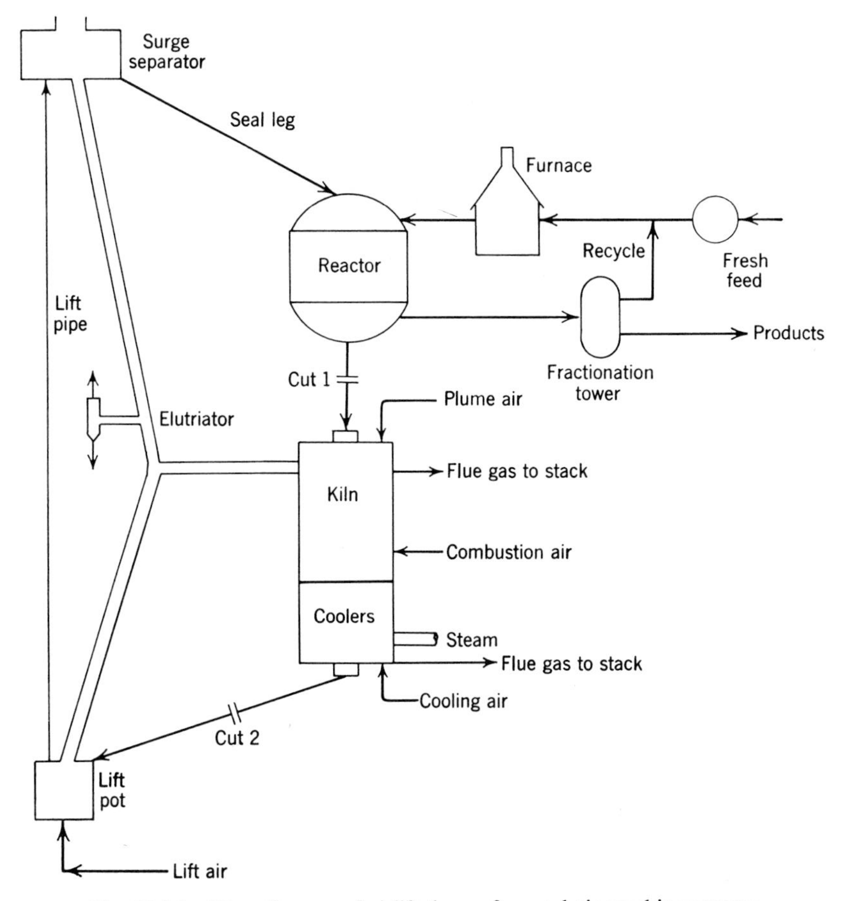

Fig. 10.5-1 Flow diagram of airlift thermofor catalytic cracking process.

The optimization problem is to maximize the profit rate which is a function of primarily the product value, processing cost, and catalyst processing cost. The major design variables and the range of variation are as follows.

Variable	*Nomenclature*	*Range*
Recycle ratio	RR	0–0.35
Total feed enthalpy (Btu/lb)	TFH	450–650
Lift air temperature (°F)	LAT	120–450
Combustion air temperature (°F)	CAT	120–1000
Total combustion air rate (SCFM)	CAR	30,000–42,000
Plume air rate (SCFM)	PAR	1,500–3,500
Cooling air rate (SCFM)	KAR	2,000–12,000
Fraction of combustion air to the top zone	XTOP	0.55–0.70

The dependent state variables include these which are constrained:

Wet gas rate (WETGAS) $\leq$ 8550 SCFM
Unstabilized gasoline (UGASO) $\leq$ 7500 B/D
Catalyst temperature above air inlet (TKA) $\leq$ 1275°F
Catalyst temperature above cooling coils (TKC) $\leq$ 1275°F
Mole fraction of O_2 from bottom zone (XO2) ≤ 0.05
Weight fraction of residual coke on catalyst (XRC) ≤ 0.01
Catalyst seal leg temperature (TCIN) 925–1075°F

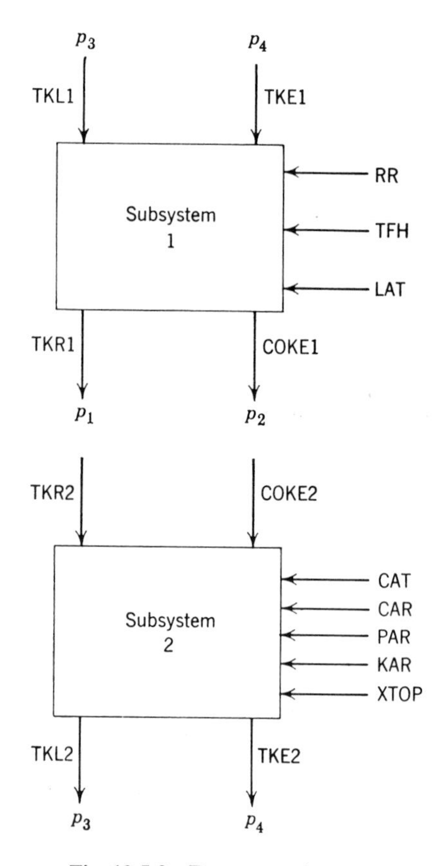

Fig. 10.5-2 Decomposed system.

In Fig. 10.5-2 the TCC process is decomposed into two subsystems, the kiln regenerator section and the reactor section. This is accomplished by tearing the following variables.

	Nomenclature
Temperature of the catalyst in the lift pot	TKL
Temperature of the catalyst in the elutriator	TKE
Temperature of the catalyst leaving the reactor	TKR
Amount of coke leaving the reactor	COKE

The two subproblems which arise are then:

Subproblem Number 1. Given prices p_1, p_2, p_3, p_4, maximize

$$[U_1 + p_1(\text{TKR1}) + p_2(\text{COKE1}) \\ - p_3(\text{TKL1}) - p_4(\text{TKE1})]$$

by manipulating RR, TFH, LAT, TKL 1, TKE1 subject to

$$0 < \text{RR} < 0.35$$
$$450 \leq \text{TFH} \leq 650$$
$$120 < \text{LAT} < 450$$
$$\text{WETGAS} \leq 8550$$
$$\text{UGASO} \leq 7500$$
$$925 \leq \text{TCIN} \leq 1075$$

Subproblem Number 2. Given prices p_1, p_2, p_3, p_4, maximize

$$[U_2 + p_3(\text{TKL2}) + p_4(\text{TKE2}) \\ - p_1(\text{TKR2}) - p_2(\text{COKE2})]$$

by manipulating CAT, CAR, PAR, KAR, XTOP, TKR2, COKE2 subject to

$$120 < \text{CAT} < 1000$$
$$30{,}000 \leq \text{CAR} \leq 42{,}000$$
$$1{,}500 \leq \text{PAR} \leq 3{,}500$$
$$2{,}000 \leq \text{KAR} \leq 12{,}000$$
$$0.55 \leq \text{XTOP} \leq 0.70$$
$$\text{TKA} \leq 1275$$
$$\text{TKC} \leq 1275$$
$$\text{XO2} \leq 0.05$$
$$\text{XRC} \leq 0.01$$

These suboptimization problems were solved by Nunez and Brosilow using the methods of direct search mentioned in Chapter 6.

The prices p_1, p_2, p_3, and p_4 were adjusted during the coordinating activity by minimizing the *dual function* γ

$$\gamma = U^\circ_1 + U^\circ_2 + p_1(\text{TKR1}^\circ - \text{TKR2}^\circ) + p_2(\text{COKE1}^\circ - \text{COKE2}^\circ) \\ + p_3(\text{TKL2}^\circ - \text{TKL1}^\circ) + p_4(\text{TKE2}^\circ - \text{TKE1}^\circ)$$

The $^\circ$ in the variables indicates the optimum values obtained from the subproblems.

The dual function γ is recognized as the sum of the suboptimization goals and can be shown to be not less than the optimum profit for the system. The dual function γ is equal to the optimum profit when the prices which force equality of the supply and demand data have been found. Since the gradients of the dual function can be computed easily as

$$\frac{\partial \gamma}{\partial p_1} = \text{TKR1}^\circ - \text{TKR2}^\circ$$

$$\frac{\partial \gamma}{\partial p_2} = \text{COKE1}^\circ - \text{COKE2}^\circ$$

$$\frac{\partial \gamma}{\partial p_3} = \text{TKL2}^\circ - \text{TKL1}^\circ$$

$$\frac{\partial \gamma}{\partial p_4} = \text{TKE2}^\circ - \text{TKE1}^\circ$$

methods of direct search which use the gradient can be employed during coordination.

Table 10.5-1 shows the progress of the price adjustment. The dual function decreases as the optimum conditions are approached until iteration 8 is reached. There divergence occurs, and this was traced by Nunez and Brosilow to an inaccurate suboptimization by one of the level I optimizers. The coordinating level had received erroneous information. We shall not delve further into the details of the optimization but point to the final results in Table 10.5-2. Reported are the optimum conditions found by Mobil Oil Company's engineers by other optimization methods and two conditions found during multilevel optimization. It is interesting to notice the fictitious prices which correspond to these solutions. Relatively small differences in the operating conditions correspond to large changes in the prices, suggesting that internal prices employed within a firm to best coordinate the activities of the firm may bear little relationship to the actual prices of those commodities in the marketplace.

Table 10.5-1 Run Using Hooke and Jeeves Method for Suboptimization and Fletcher-Powell Method for coordination

Iteration[a]	p_1	p_2	p_3	p_4	Dual Function
1	0.0000	0.0000	−0.0000	0.0000	65,093
2	0.0199	0.0023	−0.1000	0.0682	65,071
3	0.0398	0.0046	−0.2000	0.1365	65,049
4	0.0797	0.0092	−0.4000	0.2729	65,017
5	0.1593	0.0185	−0.8000	0.5459	64,973
6	0.3187	0.0369	−1.6000	1.0917	64,886
7	0.6373	0.0739	−3.2000	2.1834	64,710
8	1.2746	0.1478	−6.4000	4.3669	65,452
9	0.6875	0.0797	−3.4521	2.3555	64,947
10	0.6383	0.0740	−3.2049	2.1855	64,905
11	0.6373	0.0739	−3.2000	2.1834	64,904

[a] An iteration is defined as every time a new set of prices is assigned by the coordination algorithm

Table 10.5-2

		Mobil's Optimum	Feasible Solution 1	Feasible Solution 2
Feasible operating values:	RR	0.25	0.25	0.25
	TFH	630.6	629.	629.
	LAT	713.7	750.	750.
	TKL	1082.2	1045.45	1014.1
	TKE	941.5	927.6	890.91
	CAT	120.	120.	120.
	CAR	34668.	41594.	34432.
	PAR	2456.	3500.	1521.
	KAR	2000.	2000.	12000.
	XTOP	0.55	0.55	0.60
	TKR	913.7	892.19	873.44
	COKE	5.29	5.13	4.97
Objective function:		64,645	63,644	62,791
Calculated prices:	p_1	9.2612	3.8121	36.95
	p_2	3299.	2948.	4534.
	p_3	44.69	42.18	71.10
	p_4	0.4147	0.4012	0.6702

10.6 SYMROS, SYSTEM FOR MULTI-REFINERY OPERATIONS SCHEDULING

Having presented a brief introduction to the strategy of solving extremely large engineering problems, we close our study of optimization in process engineering by relating a specific problem that is now solved regularly as part of the normal operation of an integrated oil refining company. This is perhaps the largest numerical problem reported by the process industry.

J. C. Ornea and G. G. Eldredge of the Shell Development Company[6] have discussed the capability of SYMROS, an extensive computer *SY*stem for *M*ulti-*R*efinery *O*perations *S*cheduling. This computer program is used to coordinate the operations of refineries scattered over the world, expressly the size of the problem discussed in this chapter.

Figure 10.6-1 shows part of the layout of a typical refinery complex consisting of several manufacturing centers servicing a wide market and drawing

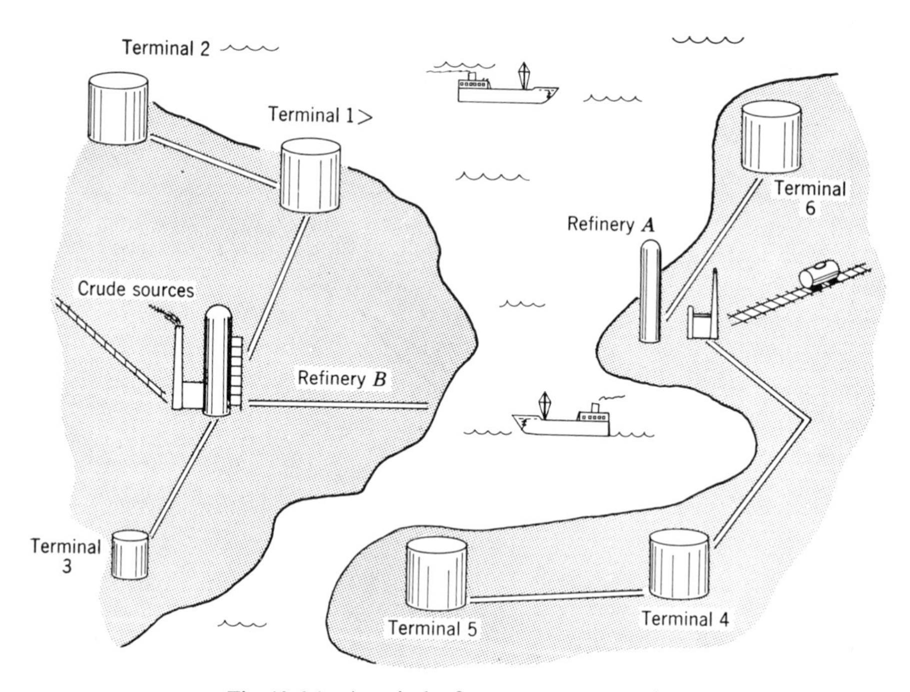

Fig. 10.6-1 A typical refinery-transport complex.

[6] J. C. Ornea and G. G. Eldredge, "Nonlinear Partitioned Models for Plant Scheduling and Economic Evaluations," A.I.Ch.E.–I.Chem.E. Joint Meeting, London, June 1965, Paper 4.15.

from a number of sources of raw materials. This complex is connected by the movement of materials via pipeline, deep-water tanker, truck, barge, and railroad.

The problem of coordinating this system is extremely difficult. The major questions that must be answered include the following.

1. How much crude oil should be allocated to each refinery?
2. How should the feeds be selected by each processing unit at each refinery? These feeds may consist of the crude oil, intermediate feeds from other units, or recycled streams. Intermediate feeds may even be shipped in from another refinery; for example, a catalytic reforming feed from refinery *A* from a particular crude oil may be sent to refinery *B* for special processing.
3. What processing conditions should be maintained at each refinery? These include temperatures, pressures, recycle rates, space velocities, catalyst activities, and so forth.
4. How should the products from the process units be blended to meet specifications? Should special products be sent to another refinery to be used more efficiently there?
5. Which terminal facility should be supplied from which refinery? This depends on the availability and cost of transportation as well as the ability of the refineries to manufacture the products.

All these questions are complicated by the fact that the refineries differ in age and design, local costs of labor and utilities vary widely, the transportation system is complex, the demands for the products vary, and there are nonuniformities in local governmental regulations.

Needless to say, we have outlined a very important but extremely difficult problem. Table 10.6-1 contains a *very abbreviated* schedule for a day's operation of the two-refinery complex shown in Fig. 10.6-1. These are the kinds of answers required of the process engineer.

SYMROS translates such refinery scheduling problems into numerical form, solves the optimization problem, and presents the results in a form usable by the technologists. We now outline the working of SYMROS.

The variables in the problems are divided into two classes: *X*-variables and *Y*-variables.

1. The *Y*-variables represent key decisions which when made reduce the remaining problem to a set of *linear programs*.
2. The *X*-variables are the remaining decisions which may be treated by linear programming methods.

Thus, the efficient methods of linear programming are brought to bear on *a portion* of the large problem. Selected *Y*-variables are shown in Table 10.6-2.

Table 10.6-1 Simplified Outline of a Refinery Problem (Basis: One Calendar-Day)

Crudes and Raw Materials, Available and Used	Crudes							
	Sweet Domestic	Sour Domestic	Imported	Natural Gasoline	Butane Purchase	Total Bbl	Crude Capacity	Raw Material Cost (M$)
Available (M bbl)	200	300	100	15	Open			
Cost ($/bbl)	3.00	2.75	2.75	3.00	2.50			
Used								
Ref. A (M bbl)	150	0	50	15	9	224	250	655
Ref. B (M bbl)	50	200	50	0	5	305	300	850
Total (M bbl)	200	200	100	15	14	529	550	
Cost (M $)	600	550	275	45	35			1505
Manufacture at the Refineries								
	LPG+ Refy. Gases	Gasoline	Jet Fuel	Diesel Fuel	Furnace Oil	Residual Fuel	Total	Marginal Cost (M $)
Manufactured								
Ref. A (M bbl)	20	150	20	20	15	10	235	250
Ref. B (M bbl)	40	155	30	21	27	25	298	275
Total (M bbl)	60	305	50	41	42	35	533	525
Market Value at Terminals ($/bbl)								
Terminals								
1–3	3.00	5.05	4.00	3.00	3.25	2.75		
4–6	3.00	5.00	4.00	3.05	3.25	3.00		
Surplus	3.00	5.00	—	3.00	—	2.75		
Market Demands at Terminals (M bbl)								Total Sale Value (M $)
Terminals								
1	10	30	10	10	—	6	66	268
2	0	70	20	10	5	3	108	488
3	10	40	—	—	11	7	68	287
4	10	35	—	—	10	—	55	240
5	0	30	—	—	10	10	50	215
6	20	80	20	16	6	5	147	625
Saleable Surplus	10	20	—	5	—	4	39	154
Total	60	305	50	41	42	35	533	
Value (M $)	180	1532	200	122	143	100		2277

TRANSPORTATION

	Refinery *A*		Refinery *B*		Total	
	(M bbl)	($/bbl)	(M bbl)	($/bbl)	(M bbl)	Cost ($)
Crudes to Refineries						
Sweet Domestic	150	0.05	50	0.05	200	10 000
Sour Domestic	0	0.10	200	0.05	200	10 000
Imported	50	0.10	50	0.10	100	10 000
Natural Gasoline	15	0.05	0	0.05	15	750
Butane Purchased	9	0.10	5	0.10	14	1 400
Total	224		305		529	
Transportation Cost						32 150
Exchanges Between Refineries						
Gasoline Components (B to A)	20	0.10	—	—	20	2 000
Reforming Feeds (A to B)	—	—	10	0.10	10	1 000
Total					30	
Transportation Cost						3 000
Products to Terminals						
1	66	0.02			66	1 320
2	98	0.02	10	0.12	108	3 160
3	61	0.03	7	0.13	68	2 740
4	45	0.12	10	0.02	55	5 600
5	—	—	50	0.03	50	1 500
6	—	—	147	0.04	147	5 800
Surplus			39	0.00	39	0
Total	270		263		533	
Transportation Cost						20 120

NET RETURN SUMMARY	Net Return ($/day)
Crudes and Raw Materials Cost	−1 505 000
Manufacturing Cost, Marginal	−525 000
Market Value at Terminals	2 277 000
Transportation Cost	
Crudes	−32 150
Exchanges	−3 000
Products	−20 120
Total Net Return	191 730

Table 10.6-2 Selected Y-Variables

	Minimum Value	Maximum Value	Current Value
Refinery A variables			
1. Sweet domestic crude oil used (M bbl/day)	0	250	150
2. Imported crude oil used (M bbl/day)	0	100	50
3. Fraction straight run naphtha to blending	0	1	0.2
4. Tetraethyl lead in premium gasoline (cm^3/gal)	0	4	2.6
5. Tetraethyl lead in regular gasoline (cm^3/gal)	0	4	1.6
Interrefinery shipments			
6. Reforming feed from refinery A to B (M bbl/day)	0	20	10
Coupling variables between blending and shipping subproblems			
7. Refinery A premium gasoline production (M bbl/day)	0	1000	100
8. Refinery A residual fuel production (M bbl/day)	0	1000	10

Of the five classes of decisions mentioned above, the first three involve nonlinearities, and thus must be treated as Y-variables. Blending and transportation are normally treated as X-variables.

The performance of each refinery is described by computer simulations to analyze the effects of the Y-variables such as temperatures, pressures, fractions bypassed, or degree of chemical conversion. These simulators are massive computer programs in themselves.

The following data must be supplied to the program:

Crude allocation. The number, amounts, and availabilities of crude oil and other raw materials at each refinery.

Marketing terminals. The number of deep-water terminals and the product demand at each. The demand may be specified to be met exactly, or as a maximum or minimum.

Refinery production. Products to be made at each refinery. Product specifications to be met by blending or by percentage recipes.

Components. Properties of product components.

Interrefinery shipments. The amounts of interrefinery shipments. Components from any refinery may be blended with components from any other refinery.

Refinery simulator data. The input to the refinery simulators. This includes initial settings on the operating variables for the refineries.

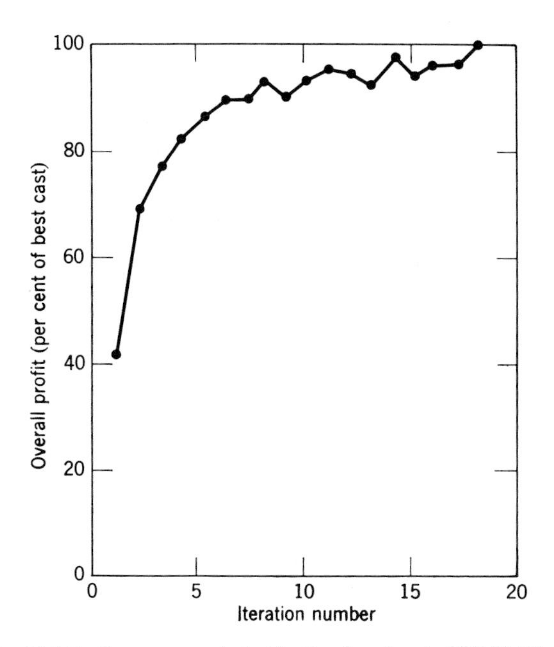

Fig. 10.6-2 Improvement of objective function in SYMROS run.

SYMROS operates in two phases:

Setup phase. The basic data for a new problem is set up and translated by the computer into a numerical form. This includes data input, making program library tape, and organization and formulation of constraints in matrix form. Errors and inconsistencies in problem formulation are checked for.

Optimizing phase. The method of partition programming aided by a gradient projection optimizer is brought to bear on the iterative solution of this problem.[7] Successive solutions are generated until no further improvement is observed. Figure 10.6-2 shows the improvement in the overall profit from a typical refinery complex. Table 10.6-3 shows the size of refinery models normally used by SYMROS.

[7] J. B. Rosen and J. C. Ornea, *Management Sci.*, **10** (1963).

Table 10.6-3 Sizes of Refinery Models Commonly Used

	Three-Refinery Model	Single-Refinery Model
Refineries	3	1
Simulator core loads	7	3
Minutes per cycle (IBM 7094)	30	10
Subproblem matrices	6	3

Matrix Name	Matrix Number	Variables	Constraints	Matrix Number	Variables	Constraints
Gasoline products	1	60	60	1	20	50
Fuels	2	55	50	2	20	50
Distillates	3	50	35			
Special products	4	70	60			
Capacities	5	45	50	3	10	20
Shipping	6	45	20			
Total		325	275		50	120
Y-variable matrix	—	50	50	—	15	15
Total		375	325		65	135

10.7 CONCLUDING REMARKS

The main purpose of this chapter is to acquaint the newcomer to optimization with the kinds of thinking which are necessary to attack large industrial problems. In practice, industrial problems have been decomposed into subsystems and then optimized empirically without the benefit of a rigorous mathematical base. It is now recognized that the tools of optimization are reaching the level of sophistication to handle problems which were beyond mathematics and in the area of empirical optimization.

Needless to say, we have only attempted to be suggestive in this chapter and not exhaustive. A study of the following literature is recommended.

References

See the following for basic concepts of decomposition, decentralization, and coordination.

K. J. Arrow and L. Hurwicz, "Decentralization and Computation in Resource Allocation," *in Essays in Economics and Econometrics*, R. Pfouts, ed., University of North Carolina Press, Chapel Hill.

Marschak, "Centralization and Decentralization in Economic Organization," *Econometrica*, No. 3, **27** (1959).
L. Lasdon and Schoeffler, "Decentralized Plant Control," *ISA J.*, October 1964.
C. Brosilow and L. Lasdon, "A Two Level Optimization Technique for Recycle Processes," A.I.Ch.E.–I.Ch.E. Symposium Series, No. 4, London (1965).

Fundamental work on the decomposition of processes is presented in:

R. Jackson, "Some Algebraic Properties of Optimization Problems in Complex Chemical Plants," *Chem. Eng. Sci.*, **19**, 19–31 (1964).

Linear programming problems can be approached by decomposition.

G. Dantzig and P. Wolfe, "The Decomposition Principle of Linear Programming," *Operations Research*, **8**, 101 (1960).
D. J. Wilde, "Production Planning of Large Systems," *Chem. Eng. Progr.*, **59**, 1 (1963).

See also:

J. B. Rosen and J. C. Ornea, "Solution of Nonlinear Programming Problems by Partitioning," *Management Sci.*, No. 1, **10** (1963).
E. Singer, "Simulation and Optimization of Oil Refineries," *Chem. Eng. Progr., Symp. Ser.* No. 37, **58**, 62–74 (1964).
L. K. Kirchmayer, *Economic Control of Interconnected Systems*, Wiley, New York, 1959.
G. Kron, *Diakoptics: The Piecewise Solution of Large Scale Systems*, MacDonald, London, 1963.

PART III

Engineering in the Presence of Uncertainty

It was tacitly assumed in the first two parts of this text that all of the information needed for process design was available, and that we needed only to analyze the information properly to obtain the best design. Unfortunately this is not true, and the engineer must function in the rather uncertain environment to be described in Part III of this text.

Chapter 11, *Accommodating to Future Developments*, deals with methods for determining characteristics which best fit the process into the changing future.

Chapter 12, *Accounting for Uncertainty in Data*, is concerned with techniques for making the best gamble when design data are known to be in error.

Chapter 13, *Failure Tolerance*, is concerned with the complete failure of system components and with methods for designing the system to possess a tolerance thereto.

Chapter 14, *Engineering around Variations*, deals with the design of systems to respond best to a variable environment.

Chapter 15, *Simulation*, describes the use of process simulation to account for the variety of uncertainties mentioned above.

11

ACCOMMODATING TO FUTURE DEVELOPMENTS

The useful life of a processing system necessarily extends into the future, and the optimum design attempts best to meet that future. By the same measure that any forecast is imprecise and uncertain, so is the optimum design imprecise and uncertain. While, characteristically, we are committed only to the initial step in a larger production campaign which may embrace the construction of a number of systems over a number of years, there is, however, an element of long range commitment which exacts a penalty for an error in accommodating to the future. We are concerned with determining a suitable initial size for a processing system.

Typical Problem

The demands for fertilizer in Iceland have been increasing over the years. What initial steps should be taken in a larger campaign to meet the existing and forecast demand?

11.1 ANTICIPATING THE FUTURE

Forecasting is uncertain and imprecise, yet unavoidable; since processes necessarily must respond to future demands. Ignorance of the future is one of the most confounding factors in process engineering, obscuring the view of the *best* process. Somehow, the effects of this ignorance must be minimized, and the view of the best system clarified to the point where the engineer can confidently create sufficiently optimal systems.

One approach to this problem involves the preparation of *reasonable* forecasts of the future, followed by an accommodation to those forecasts.

The engineer must ask:

What has happened in the past?
What is the best that may happen in the future?
What is the worst?
What is most likely to happen?

The following question has *no* certain answer.

What is going to happen?

In this chapter we confine our attention to accommodating to the most probable forecasts. In Chapter 12 we examine quantitative methods for dealing with uncertainty in such forecasts.

Forecasts of interest to the process engineer fall into two main groups: business forecasting, and technical or environmental forecasting.

Business forecasting comprises the prediction of market demands and prices of materials, energy, and finished products; the cost and availability of labor; the extent of competition for markets and talent, and the like. A prominent characteristic of business is the influence of directed acts of entities, such as competitors, who move in opposition to our goals. Thus, the theory of games *should* play a vital role in the interpretation of business forecasts. However, the mathematics of game theory has not advanced to the point where it is generally useful in interpreting industrial situations,[1] and adjustments for these directed entities are commonly made intuitively by those trained in economics.

Even so, the engineer cannot ignore business forecasts: the processing system must respond to these forecasts. For example, Pennsalt, a large chemical manufacturer, requires that a ten year forecast of sales volume, prices, costs, future capital requirements, and return on investments accompany each capital appropriation of more than $100,000. In that environment every major engineering decision is influenced by business forecasting.

To illustrate the kinds of business forecasts which greatly influence the engineering of a process we cite an example. The sales price for a commodity will not remain static unless the available production capacity for that commodity maintains the same position with respect to the changing demand. Excess demand tends to increase prices, and excess supply tends to decrease prices. But, when profit margins are high, new capacity will be built at a fast pace as more and more competitors enter the market. Prices, therefore, tend to fall with time as long as some producer can earn a return on his capital by operating a modern, efficient and well-designed process.

The floor under prices is the sales price at which the most efficient operator

[1] However, see R. R. Hughes and J. C. Ornea, "Decision-Making in Competitive Situations," *World Petrol. Conf.*, April 1967.

can make a reasonable profit with a technically and economically superb process. Sales price often can thus be forecast as a sum of a fairly stable floor price F plus a margin over this floor price which decays rapidly as more competitors attempt to capture the market.

$$\text{Price} = M_0 e^{-K_M \theta} + F \tag{11.1-1}$$

The floor price is based on cost estimates for the largest and most efficient plant that is likely to be built over the years and must allow for technological improvements as well as changes in fixed and construction costs. The floor price may decay much less rapidly than the margin prices, for example, on the order of 1 to 4 per cent per year, as compared with 10 to 20 per cent per year for the margin.

Clearly, these and other purely business factors must be accounted for in the design of a process, if the process is to be useful to the parent firm and to society.

Technical forecasting includes predictions of how nature may be expected to respond to our actions. Included, for example, are predictions of the activity and selectivity of catalysts, the durability of materials of construction, the need for maintenance and replacement of equipment, variations in environmental conditions, such as water purity and air temperature. These factors tend to be neutral to or even partially under the control of the engineer, rather than in opposition. For example, the life of a reactor, forecast at ten years, might be extended indefinitely by maintenance should this fit into our plans.

An important factor in technical forecasting is the ever-increasing competence of the technical society. Hirschmann[2] has made a study of this factor and suggests that the actual capacity Q_a of a given process might be expected to increase according to the learning curve, Eq. 11.1-2

$$\frac{Q_a}{Q_d} = 1 + \left(\frac{Q_\infty}{Q_d} - 1\right)[1 - \exp(-K\theta)] \tag{11.1-2}$$

where Q_d is the design capacity, Q_∞ the ultimate capacity. For catalytic cracking units, Hirschmann found $Q_\infty/Q_d \approx 2$ and $K = 0.1$.

For example, in 1951, the worldwide installed design capacity of fluid catalytic cracking units was 1,200,000 barrels per stream day, but the actual throughput was 1,600,000 barrels per stream day. The capacity of the fluid cat-cracking industry had been increased by one third over the design capacity. Such increased capacity is obtained in two ways (1) by a more nearly optimum adjustment of the operating variables, and, (2) by minor equipment revisions to remove bottlenecks found through operating experience.

[2] W. B. Hirschmann, *Harvard Business Review*, No. 1, **42**, 125 (1964); *Chem. Eng.*, March 30, 1964.

The increasing technical competence of an engineering group also influences the eventual costs of a processing system, as Hirschmann has documented. For example, over a period of ten years, the time required to put a Whiting refinery fluid cracking unit of the American Oil Company on-stream was reduced by more than one half, thus materially reducing the investment in such processes. In a second example, an oil refinery at Fawley, England built in the early 1950's could have been duplicated five years later at a construction cost 70 per cent of the original. Finally, the investment required to duplicate, in 1955, a downflow fluid catalyst oil processing plant built in 1942 is only one third the original, as a result of increased technical competence.

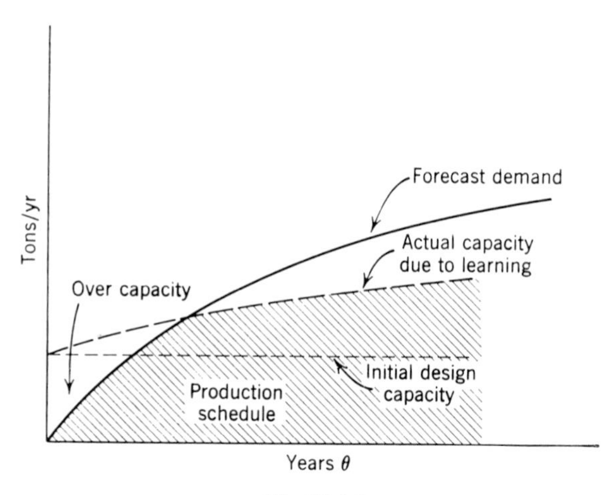

Fig. 11.1-1

Thus, the basic problem we address in this chapter is that of estimating the best initial size of a processing system in a dynamic economic and technical environment. Figure 11.1-1 illustrates this problem by showing a forecast of the market available to a process, an initial design capacity followed by the improvements to be expected from the learning curve effect, and the resulting production schedule for the process. The basic problem in the preliminary phases of process planning is that of estimating the initial design capacity which economically balances the initially unproductive period of overcapacity against the savings in investment per unit capacity which arise from larger processes.

11.2 ACCOMMODATING TO A LINEAR DEMAND FORECAST

In this section we determine a suitable initial capacity for a processing system which is obligated to meet an ever-increasing known linear demand in an otherwise static environment. The goal is to meet this demand by a sequence of plant additions timed to involve a minimum present value for the sum of all investments, initial and subsequent.

The simple picture envisioned is as follows.

The demand forecast is linear, increasing on into the indefinite future.

$$D = a\theta \tag{11.2-1}$$

This demand must be met.

The investments in the initial plant and any future expansions follow the power law.

$$I = I_B\left(\frac{Q}{Q_B}\right)^M \tag{11.2-2}$$

The economic and technical environment is static.

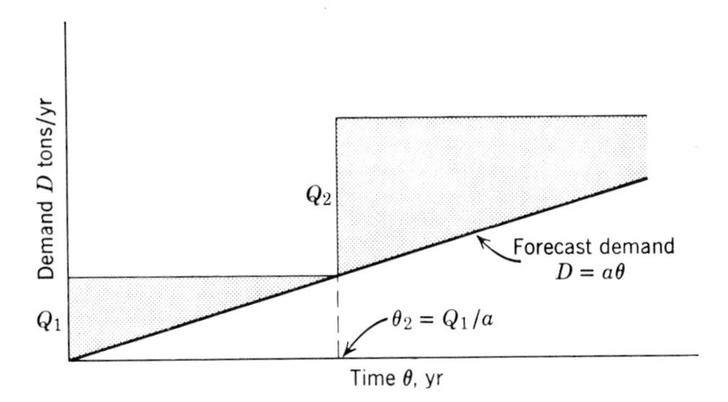

Fig. 11.2-1 A feasible but not optimal expansion plan.

While this simple picture will be made more realistic later, as the conditions are relaxed, an example of a reasonable accommodation to this environment is most instructive.

The present value of an investment of I dollars to be made θ years in the future is

$$Ie^{-i\theta}$$

where i is the expected rate of return from capital invested in the firm's activities. The present value of a sequence of investments I_j made at years θ_j is then

$$\text{Present value} = \sum_{j=1}^{\infty} I_j e^{-i\theta_j} \tag{11.2-3}$$

Now if our goal is to meet the demand with a sequence of investments designed to result in the minimum present value of the total investments, Eq. 11.2-3 becomes our design objective function. This function may be regarded as the fund that must be set aside to finance the series of investments.

Figure 11.2-1 illustrates a *feasible* set of investments in plant capacity; feasible in the sense that the demand is met, but not necessarily optimally. Notice, how the time θ_j at which the additional capacity Q_j is to be added is

$$\theta_j = \frac{1}{a} \sum_{k=1}^{j-1} Q_k \tag{11.2-4}$$

Every time the demand reaches the existing plant capacity, the same problem appears, namely, the accommodation to a linear forecast which extends on into the indefinite future. If the same problem appears each time, the solution must be the same. Consequently, the optimal expansion plan necessarily involves expansion in identical increments.

The time for the jth addition is then

$$\theta_j = \frac{(j-1)Q^*}{a}$$

where Q^* is the size of the optimal identical expansions, yet to be determined.

Equation 11.2-3, the present value, reduces to

$$\begin{aligned}\text{Present value} &= \sum_{j=1}^{\infty} I_B \left(\frac{Q}{Q_B}\right)^M e^{-i(j-1)Q/a} \\ &= I_B \left(\frac{Q}{Q_B}\right)^M \sum_{j=1}^{\infty} e^{-i(j-1)Q/a}\end{aligned}$$

Evaluating the sum, using the identity

$$\sum_{j=0}^{\infty} X^j = \frac{1}{1-X}$$

gives

$$\text{Present value} = I_B \left(\frac{Q}{Q_B}\right)^M \Big/ (1 - e^{-iQ/a}) \tag{11.2-5}$$

Differentiating Eq. 11.2-5 with respect to Q and setting the derivative to zero gives the following expression for the optimal plant addition

$$M(1 - e^{-iQ^*/a}) - \frac{iQ^*}{a} e^{-iQ^*/a} = 0 \tag{11.2-6}$$

Equation 11.2-6 is plotted in Fig. 11.2-2, and is subject to some interesting interpretations.

For a given M, the term Q^*i/a is constant. For example, the following *rule of thumb* might be established.

Should we suspect that the investment in a processing system will increase with capacity according to the "six-tenths" rule, the recommended initial capacity is approximately

$$Q^* = \frac{a}{i} \tag{11.2-7}$$

Having witnessed the logical development of this rule of thumb, we are aware of its weaknesses and its limited range of applicability. How often are other rules accepted when their underlying principles are unavailable for examination or perhaps even nonexistent?

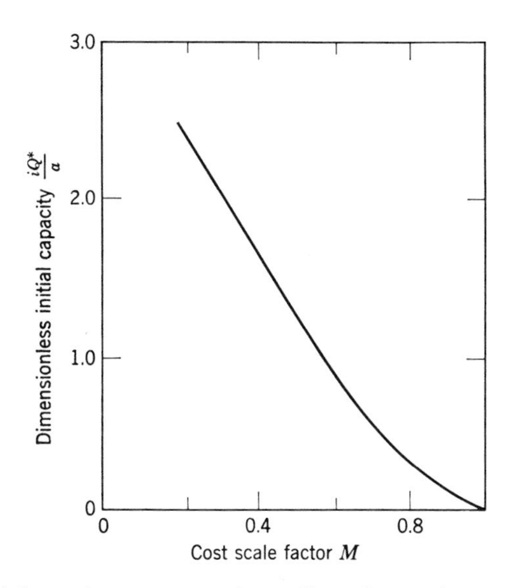

Fig. 11.2-2 Initial capacity to accommodate to linear forecast in a static environment.

Observe further, in Fig. 11.2-2, that as M tends to one, smaller expansion steps are recommended. This is reasonable since when M equals one there is no cost advantage attending the larger system. However, as M tends to zero the expansion steps should be larger, expressing the fact that there is an investment saving associated with building single large facilities.

Example:
The demands for steam in a proposed processing system are expected to rise by 20,000 pounds steam per hour each year on into the indefinite future. Money invested in the firm is returning \$0.20 per year per dollar invested, and the installed cost of boilers is

$$I = \$600{,}000\left(\frac{Q}{100{,}000}\right)^{0.6}$$

where Q is the capacity in pounds steam per hour. What is the best initial capacity, following the tentative analysis in this section?

$$Q^* = \frac{a}{i} = \frac{20{,}000}{0.20} = 100{,}000 \text{ pounds per hour}$$

11.3 NONZERO INITIAL DEMAND

Frequently a finite demand exists for the services of a system when the engineer arrives on the scene. This existing demand might have been satisfied by some other means, such as the purchase of material from a more costly source, or it might have remained unsatisfied, hindering further growth of some larger complex. We now consider the problem of accommodating to a forecast which consists of an unsatisfied initial demand D_0 and a linear growth of demand with time.

$$D = D_0 + a\theta \qquad (11.3\text{-}1)$$

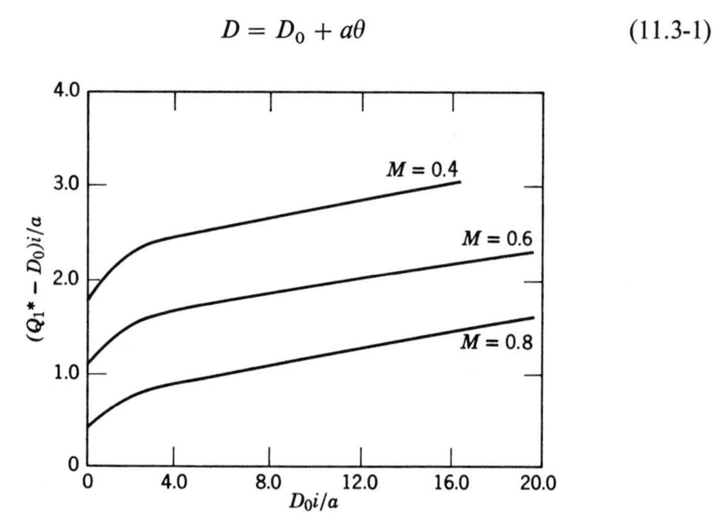

Fig. 11.3-1 Overdesign for initial and linearly increasing demand.

Suppose that an initial system with a capacity $Q_1 > D_0$ has been put in operation, and we consider the situation when, at time θ_2, a second system is to be built.

$$\theta_2 = \frac{Q_1 - D_0}{a} \qquad (11.3\text{-}2)$$

At that time there exists a zero initial demand and a linear forecast. Thus, the problem which confronts the engineer at the time of the second expansion of capacity is that of Section 11.2. Conclusion: All system expansions, after the initial system is no longer able to meet the demands, should be of a capacity given by Eq. 11.2-6 and Fig. 11.2-2, here denoted by Q^*.

We have determined the optimal expansion plan and now focus attention on the assignment of the optimal initial system capacity $Q_1{}^*$.

The present value of all investments in the sequence of system additions is

$$I_B\left(\frac{Q_1}{Q_B}\right)^M + e^{-(Q_1 - D_0)i/a}\left[\frac{I_B(Q^*/Q_B)^M}{1 - e^{-iQ^*/a}}\right] \tag{11.3-3}$$

The term on the left is the investment in the initial process of capacity Q_1, and the term on the right is the present value of all future expansions, following Eq. 11.2-5.

Setting the derivative of Eq. 11.3-3 with respect to Q_1 to zero gives the following expression for the optimal initial capacity.

$$\left(\frac{Q_1{}^*i}{a}\right)^{M-1} e^{-Qi^*/a} = \left(\frac{Q^*i}{a}\right)^{M-1} e^{(D_0 - Q_1^*)i/a} \tag{11.3-4}$$

We might define an overdesign factor as $(Q_1{}^* - D_0)i/a$ and a dimensionless initial demand as $D_0 i/a$. The only other parameter is then M, as Fig. 11.3-1 shows.

In the next section we discuss an application of these concepts.

11.4 ACCOMMODATING TO ICELAND'S FERTILIZER NEEDS (1950)[3]

We are now in a position to attack some realistic problems, with the hope that our tools *may* fit the task. An excellent case history upon which to test the developments is afforded by the design and construction of Iceland's first chemical fertilizer manufacturing system in the early 1950's.

The demand of Iceland for nitrogenous fertilizers had been increasing as shown in Fig. 11.4-1, during the period 1940 to 1950 and had been met by

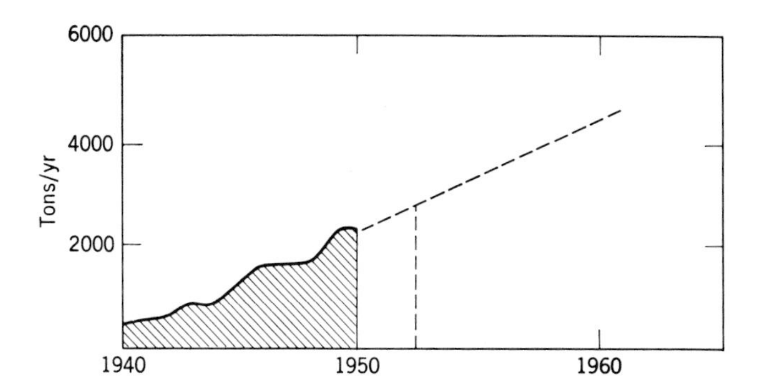

Fig. 11.4-1 The demand for fertilizer in Iceland 1940 to 1950.

[3] Private Communication with R. Thordarson, Manager, Iceland Fertilizer Company.

importing ammonium nitrate from overseas. The assignment is that of recommending the initial capacity of a system for manufacturing this material in Iceland and planning for future expansions.

The first problem is that of forecasting the demand for fertilizers. This properly would be done after extensive discussion with local agricultural experts using forecasts of possible increases of agricultural activity which might accompany the expected population increases. Perhaps, speculations of the probable effects of the then recent termination of World War II are in order. Unfortunately, these necessary considerations are beyond the scope of this text.

A linear extension of the demands observed over the period of 1940 to 1950 gives the following forecast.

$$\begin{aligned} D &= 2{,}700 + 230\,\theta \text{ tons/yr} \\ D_0 &= 2{,}700 \text{ tons/yr} \\ a &= 230 \text{ tons/yr}^2 \end{aligned} \tag{11.4-1}$$

where $\theta = 0$ at 1952, the time when the system is expected to begin operation.

Assuming that the investment in an ammonium nitrate production facility follows the "six tenths rule," and that the value of money within the firm is that within an average chemical company

$$M = 0.6$$
$$i = 0.12$$

the dimensionless initial capacity, to be used in Fig. 11.3-1, is

$$\frac{D_0\, i}{a} = \frac{(2{,}700)(0.12)}{(230)} \approx 1.4$$

The dimensionless overdesign factor is then, using Fig. 11.3-1,

$$\frac{(Q_1{}^* - D_0)i}{a} \approx 1.4,$$

an accidental similarity of values. The recommended initial capacity is then

$$\begin{aligned} Q_1{}^* &\approx D_0 + \frac{1.4a}{i} \\ &= 2{,}700 + \frac{(1.4)(230)}{(0.12)} = 5{,}400 \text{ tons per year} \end{aligned}$$

The first expansion of this system is forecast for

$$\theta_2 = \frac{Q_1 - D_0}{a} = \frac{2{,}700}{230} \approx 12 \text{ years (1964)}$$

The recommended expansion, Fig. 11.2-2, is

$$Q^* = \frac{1.0a}{i} = \frac{(1.0)(230)}{(0.12)} \approx 1{,}900 \text{ tons per year}$$

This particular plan for accommodating to Iceland's future needs for fertilizer is illustrated in Fig. 11.4-2.

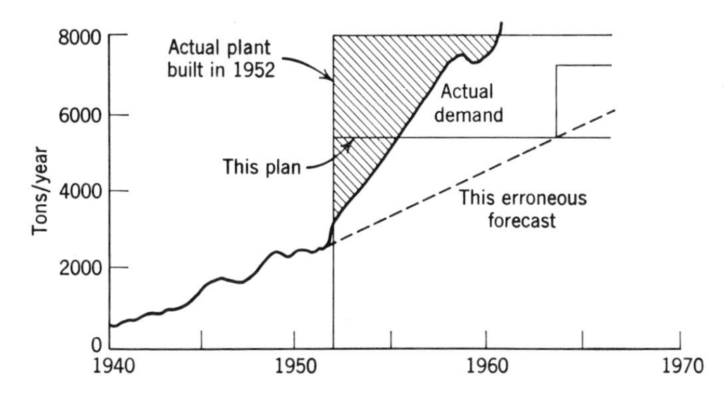

Fig. 11.4-2 What actually happened.

This recommendation should be taken only as suggestive, since a number of important factors have been omitted. For example, might a larger system be in order to respond better to annual variations in weather conditions? To the occasional shutdown caused by equipment failure? To uncertainties in the availability of such resources as electric power? Methods to account for such effects are discussed in Chapters 12 through 15.

Now let us observe what actually happened. The Icelandic Government authorized the construction of an ammonium nitrate manufacturing system with a capacity in the range of 5,000 to 10,000 tons per year. On the basis of experience, the engineers recommended an initial plant with a capacity of 8,000 tons per year and with a provision for doubling the capacity in the future. Notice in Fig. 11.4-2 how the demand for fertilizer increased markedly when the Icelandic fertilizer became available in 1952. The unforeseen increase in demand has an interesting explanation. Prior to the construction of the Iceland facility, ammonium nitrate was shipped in, and orders had to be made months in advance. It was impossible to reorder after the original orders went in; the farmers made do with their original estimates of fertilizer need. However, once the Icelandic plant began manufacturing, the fertilizer orders could be made at any time, and the demands, therefore, increased. This facility had an autocatalytic effect. The availability of the fertilizer increased the demand, a factor not included in our simple extrapolation.

This example has illustrated the direct application of the theoretical analysis to a practical engineering problem indicating why a complete understanding of the circumstances surrounding the origin of the primitive problem is essential. The unforeseen, but perhaps predictable, autocatalytic effect invalidated an otherwise suitable initial processing plan. Clearly, in actual engineering problems a more detailed and involved analysis must be made.

11.5 SIZING NEW CHEMICAL PLANTS IN A DYNAMIC ECONOMY

The problems considered in the several preceding sections are overly simplified to illustrate the kind of thinking involved in accommodating to the future. Real life problems in process engineering are rarely so simple. Markets grow, prices decay, costs increase, the capacities of processes change as experience is gained through operation, and processes accumulate cost disadvantages through technical obsolescence. Even assuming that the sales price of the products and the purchase price of the raw materials will remain unchanged over the life of a process is a gross oversimplification.

W. W. Twaddle and J. B. Malloy of Amoco Chemicals Corporation, Chicago, have illustrated the detailed analysis which must attend any critical industrial problem. We base this section on their report.[4]

Consider the problem of sizing a process to produce a monomer for a plastic which is experiencing a rapidly growing demand.

Market forecast. The market is currently 500 MM lb/yr and is growing at 30%/yr. The growth rate is expected to smooth out to 15%/yr in four years and then to 4%/yr in the distant future.

$$D = 500 \exp\{(0.04\theta) + 1.210[1 - \exp(-0.215\theta)]\} \qquad (11.5\text{-}1)$$

The firm's share of the market is forecast at 4 per cent initially, rising to 10 per cent according to the following equation:

$$f = 0.04 + 0.06[1 - \exp(-0.278\theta)] \qquad (11.5\text{-}2)$$

Price forecast. The price floor during the 15-year process life is calculated in Table 11.5-1. The largest process presently in the industry is 200 MM lb/yr, and a 1,000 MM lb/yr process is the largest expected in 15 years. Accounting for expected changes in raw materials, costs, improvements in yields, and so forth, gives the following floor price forecast.

$$\text{Floor price} = 12.08 \exp(-0.0329\theta) \qquad (11.5\text{-}3)$$

[4] W. W. Twaddle and J. B. Malloy, "Evaluating and Sizing New Chemical Plants in a Dynamic Economy," *Chem. Eng. Progr.*, July 1966.

Table 11.5-1

Estimating the Price Floor

	Now	*15 Years Later*
Plant size, MM lb/yr	200	1,000
Investment, $MM	40	118
VARIABLE COST (¢/lb)		
Raw material	5.63	3.34
Other	1.50	1.40
	7.13	4.74
FIXED COSTS (¢/lb)[a]		
Labor	0.40	0.13
Maintenance, insurance, taxes	1.26	0.75
Selling, administration, research and development	0.70	0.20
	2.36	1.08
CAPITAL CHARGES (¢/lb)		
6 Per cent cost of capital	2.59	1.52
	12.08	7.34

$$\text{Price Floor} = 12.08e^{-0.0329\theta}$$

Allowance for Working Capital

CURRENT ASSETS	
Accounts receivable	1 month sales
Raw material inventory	1 month raw material cost
Product inventory	1 month processing cost
Cash	1 month processing cost, ex raw material
Supplies	2 per cent of investment
CURRENT LIABILITIES	
Accounts payable	1 month variable cost
Wages payable	½ month labor cost
Taxes payable	6 month tax cost

[a] Equivalent ¢/lb charge at a 6% discount rate. Labor at $700M/yr now, growing at 3.5%/yr. Maintenance, insurance, and taxes at 5.5% of investment, growing at 3.7%/yr. Selling, administrative, and R & D at $600M/yr now growing 3.5%/yr.

It is assumed that the margin price over the floor price decays at 18%/yr, and that the current price for the monomer is 23¢/lb. Thus, the price projection is

$$\text{Price, ¢/lb} = 10.92e^{-0.18\theta} + 12.08e^{-0.0329\theta} \tag{11.5-4}$$

The price drops from 23 ¢/lb to 8.1 ¢/lb in fifteen years.

Variable manufacturing costs are expected to decrease from 7.13¢/lb in Table 11.5-1 according to Eq. 11.5-5

$$\text{Variable costs, ¢/lb} = 7.13e^{-0.0272\theta} \tag{11.5-5}$$

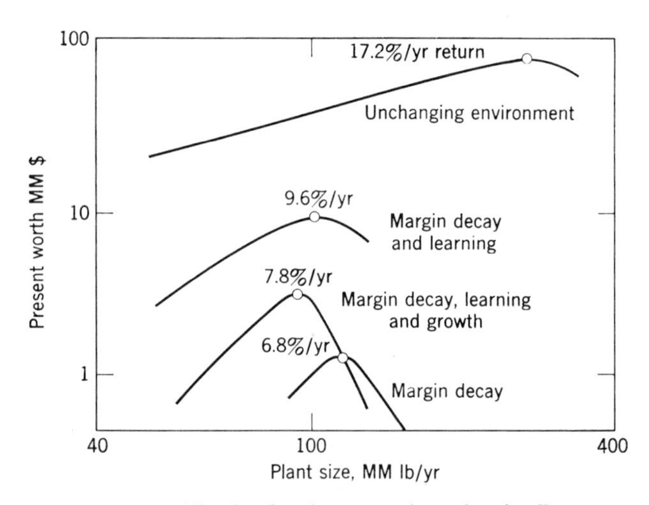

Fig. 11.5-1 Allowing for change can have drastic effects.

Investment costs. The investment costs are based on a \$40 MM cost for a 200 MM lb/yr plant, to be scaled up by the power law with exponent $M = 0.65$.

$$I = \$40\text{ MM}\left(\frac{Q}{200\text{ MM}}\right)^{0.65} \tag{11.5-6}$$

Increases in capacity. The capacity of a plant is expected to double ultimately with a time factor $K = 0.1$/yr. The cost of this increase in capacity is assumed to be one half the average cost per pound of initial annual capacity.

$$\frac{Q_a}{Q_d} = 1 + 2[1 - \exp(-0.1\theta)] \tag{11.5-7}$$

This illustrates part of the detail which enters into this realistic assessment of future developments.

Results. Figure 11.5-1 shows the present worth of processes of different sizes computed by neglecting a number of these factors. This shows how sensitive the recommended process size is to the several important factors. In summary:

If no allowance is made for change, the project appears very attractive, earning 17.2%/yr on capital for a process of size 220 MM lb/yr.

If price decay is included in the analysis, the project looks unpromising, earning only 6.8%/yr on capital for a process of size 120 MM lb/yr.

However, if price decay and the increase in capacity due to learning are accounted for, the process earns 9.6%/yr at a size of 100 MM lb/yr.

Including all factors, which results in the most realistic analysis, we would plan a process of capacity 97 MM lb/yr earning 7.8%/yr on invested capital.

This shows the vital importance of a complete and detailed analysis of the environment into which a process must integrate. Had the 220 MM lb/yr process been constructed, based on an unchanging environment, in the actual changing environment the process would have been a waste of capital which might have been used better in other engineering projects.

11.6 PARAMETRIC SENSITIVITY

A happy situation exists when the recommended initial decisions are insensitive to the possible sources of uncertainty. For example, consider the situation in Fig. 11.6-1 in which the recommended initial plant capacity is

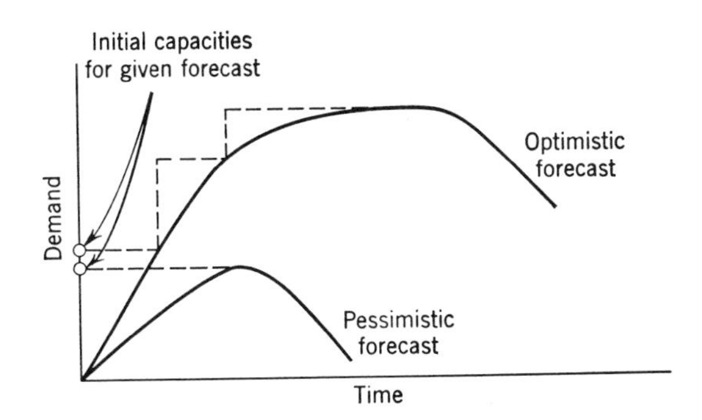

Fig. 11.6-1 A happy situation—initial decision insensitive to forecast.

nearly identical for two different forecasts. We would be quite confident in recommending that initial capacity if the forecasts happened to cover the range of possible future events, from optimistic to pessimistic. Regardless of what actually occurred in the future, a suitable first step would have been made. The initial plant capacity is insensitive to forecast errors.

However, suppose that the unhappy situation in Fig. 11.6-2 existed, in which the recommended initial capacity is extremely sensitive to the forecasts. What is the best initial plant capacity? This situation necessarily involves gambling. We must estimate the probability that a given forecast will obtain and estimate the penalty to be assessed an inaccurate decision. By playing the odds, the engineer can select among parametrically sensitive decisions. This section serves only as a warning so that a potentially dangerous situation may be recognized. The rational analysis of parametrically sensitive situations is discussed in Chapter 12 and again in Chapter 15.

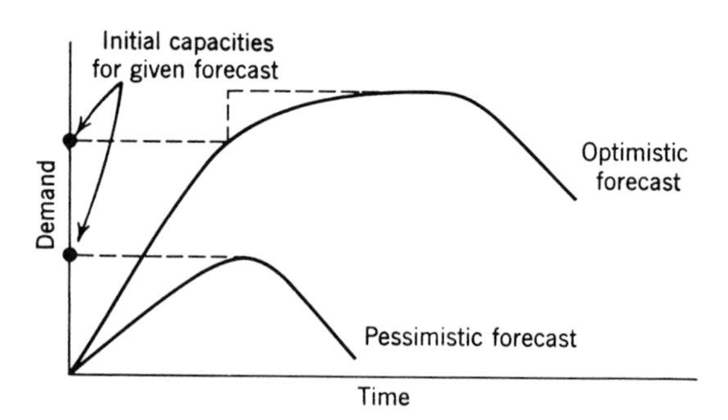

Fig. 11.6-2 An unhappy situation—parametric sensitivity.

11.7 A CURIOUS EFFECT OF HUMAN NATURE

The simplifying assumption made in parts of this chapter that the succeeding plant designs are identical except for size would appear rather unrealistic. Surely, we would think, the operating data gained from the initial plant would lead to an improved and different design for the second unit, and so on. This is indeed a possibility, but in a rapidly expanding market it is, in fact, only too likely that the human desire for certainty will lead to exact reproduction of the design on which the existing unit was successfully operated.

The logical procedure (if time permits) is to reexamine the entire process design, in the light of the actual operating experience and data, and to evolve a new and more nearly optimum design. If it can realistically be assumed that this is going to take place, then the analysis for optimal timing of the various units in the overall campaign should reflect this fact. Each unit will then serve as the full scale pilot unit for those to follow.

On the surface, it would appear that a treatment which takes account of the advances in technology which ought to result from the logical study of past experience is much more realistic than the simplified analysis. However, limitations of time and human nature unfortunately tend to make the simplest analysis occasionally the most realistic, after all. With an understandable desire for certainty in an inherently uncertain field, the designer will tend to read too much into the observation that his initial design actually has been shown to operate. He will feel that this design should now be "frozen" in every detail and copied, except for scale changes, for each succeeding unit. But all that it is logical to infer about the first design is that it is one of a set of designs which physically are able to operate but are not necessarily optimal. However, familiarity with successful systems tends to blind us to the logical evolution of even more successful systems.

Witness the following situation. An engineering firm is committed to meet a rising demand for the next five years, and their first system in this campaign is successful. At the time for considering the first expansion of the system, a new technology appears which *promises* to perform the desired function at less cost. Will the engineer incorporate this new technology into his plan or leave well enough alone and continue with the certain, still profitable, but perhaps not optimal technology?

We can make valid arguments for both sides of the question, and the designer of the initial process must assess the chances of one or the other course of action obtaining. In practice the true situation will fall between that of a completely static technology and that of the theoretically most advanced technology.

11.8 CONCLUDING REMARKS

A saving grace in all decisions involving the distant future is twofold. First, the financial effects of uncertainty about the future are mitigated by the exponentially decreasing function $e^{-i\theta}$ used to compute the present value. The distant future weighs less heavily in decision making; hence, the inherent uncertainties of the future are not fed back to strongly bias initial decisions. Second, while in our analysis, plans for expansion are made into the distant future, a commitment is made to only the first step in the form of a process design. Continual reevaluations occur over the life history of the campaign which may embrace the design of a number of processes. Errors in the original decisions are subject later to partial correction.

Accommodating to the future is but one of the considerations which tends to shape the initial design of a process. Allowances for a variable environment, for the failure of parts of the system, and for the inherent ignorance of basic data also enter. These topics are investigated in the next few chapters.

References

The use of the learning curve is discussed in:

W. B. Hirschmann, "Profit from the Learning Curve," *Harvard Business Review*, January to February 1964.

W. H. Hirschmann, "The Learning Curve," *Chem. Eng.*, March 30, 1964.
F. J. Andress, "The Learning Curve as a Production Tool," *Harvard Business Review*, August 1960.

See the following for methods for accommodating to forecasts.

W. W. Twaddle and J. B. Malloy, "Evaluating and Sizing New Chemical Plants in a Dynamic Economy," *Chem. Eng. Progr.*, **62** (1966).
R. M. Lawless and P. R. Haas, "How to Determine the Right Size Plant," *Harvard Business Review*, **40**, 3 (1962).
J. R. Coleman and R. York, "Optimum Plant Design for a Growing Market," *Ind. Eng. Chem.*, **56**, 1 (1964).
D. I. Satelan and A. V. Caselli, "Optimum Design Capacity of New Plants," *Chem. Eng. Progr.*, **59** (1963).
H. Hinomoto, "Capacity Expansion and Probabilistic Growth," *Management Science*, **11**, 5 (1965).
A. S. Manne, "Capacity Expansion with Facilities under Technological Improvement," *Econometrica*, **29**, 4 (1961).
H. A. Quigley, "Economics of Multiple Units," *Chem. Eng.*, August 29, 1966.

PROBLEMS

11.A. Suppose that the expected changes in technology and in the economic environment can be expressed by the following weighting factors which multiply the investment cost for a process.

$$(\text{Cost of a plant at time } \theta) = e^{-r\theta}e^{+s\theta}(\text{Cost at time zero})$$

Where r is the fractional reduction per year in the required investment in the process due to improvements in technology and s is the inflation rate of the dollar causing an increase in the investment.

Show that any of the simple analyses in Sections 11.2 and 11.3 can be corrected for these effects merely by correcting the expected earning rate for capital invested in the firm from i to $i - s + r$.

11.B. Determine the recommended size of the initial ammonium nitrate plant and the size and time of the first expansion for the situation described in Section 11.4 with the additional information that inflation is increasing the cost of chemical processes by 5 per cent per year in a static technology and increases in the competence of engineering ammonium nitrate processes are expected to reduce the required investment by 10 per cent per year in a static economy.

11.C. Suppose that there exists an immediate need for 1,000 square feet of heat-exchange surface for a preheater in an oil processing plant. Three years hence, when a planned expansion in the oil processing plant is realized, 4,000 square feet more will be required. The installed cost of the heat exchangers is approximated by the cost correlation $I = \$5{,}000(Q/100)^{0.8}$ where Q is the heat-exchange surface area. Should a heat exchanger be replaced three years hence, a salvage will be realized equal to one third the installed cost of the original equipment. To meet the forecast demand for heat-exchange capability, should we install an exchanger with 5,000 square feet of heat-exchange surface initially, install one with 1,000 square feet initially and replace it with one with 5,000 square feet in three years, or install one with 1,000 square feet initially and install 4,000 square feet of additional capacity three years hence? Solve this problem for industrial situations in which the firm is earning 5 per cent per year, 10 per cent per year and 30 per cent per year on invested capital. This problem illustrates the fact that questions of initial design alternative in a changing environment can be handled by the present value concept introduced in Chapter 4, and that the more extensive analyses of this chapter are also merely extensions of that concept.

11.D. E. Calanog gathered some interesting data on technological improvements in the blast furnace, the basic system in iron making. The blast furnace was first used in the fourteenth century, and in the years that followed considerable improvements were experienced and are even now being experienced. Oxygen enrichment, the use of auxiliary fuels, and so forth, contributed to the increase in iron production per volume of furnace since 1953 shown below.

Per cent increase above 1953 $\frac{\text{(production)}}{\text{(Volume of furnace)}}$	-6	12	27	30	28	32	30	49	62
Year	1954	55	56	57	58	59	60	61	62

How might we correct blast furnace investment data to accommodate to these changes? Would you expect this trend to continue?

11.E. Frequently, engineering an expansion is somewhat different from the original engineering job, the investment required per unit capacity change being greater. Suppose that this difference can be expressed in the factor M used in the investment power law, M_0 for the initial job, and M_e for any expansions. Develop a relationship for the best initial capacity to meet the increasing linear demand in an otherwise static environment.

$$D = a\theta$$

In the optimal plan, what plant additions would you expect to be identical?

11.F. Suppose that the differences between the first engineering job and subsequent expansions can be expressed as follows:

Type of job	*Investment*
Initial design	$I = I_0\left(\frac{Q}{Q_0}\right)^M$
All expansions	$I = I_e\left(\frac{Q}{Q_0}\right)^M$

Develop an expression for the best initial capacity to meet the linear demand

$$D = a\theta$$

11.G. The initial investment often carries with it terms which are not directly dependent on the capacity of the system to include such factors as site preparation, buildings, roadways and the like. Accommodate to the linear demand $D = a\theta$ in the following cost environment.

Type of job	*Investment*
Initial design	$I = I_0\left(\frac{Q}{Q_0}\right)^M + I_1$
All expansions	$I = I_2\left(\frac{Q}{Q_0}\right)^M + I_3$

11.H. Accommodate to the forecast demand

$$D = D_0 + a\theta$$

in the following cost environment.

Type of job	*Investment*
Initial design	$I = I_0\left(\frac{Q}{Q_0}\right)^M + I_1$
All expansions	$I = I_2\left(\frac{Q}{Q_0}\right)^M + I_3$

11.I. The assumption that a plant of any size can be designed economically is rather unrealistic; engineering efficiencies are often only realized when some reasonably large plant is being engineered. Incorporate this fact into the developments in this chapter by accommodating to the linear demand

$$D = a\theta$$

in the cost environment

$$I = I_0\left(\frac{Q}{Q_0}\right)^M$$

with the constraint $Q \geq Q_{\min}$ where $Q_{\min}$ is the minimum size plant that can be considered. Show that the initial plant size chart shown in Fig. 11.I-1 obtains.

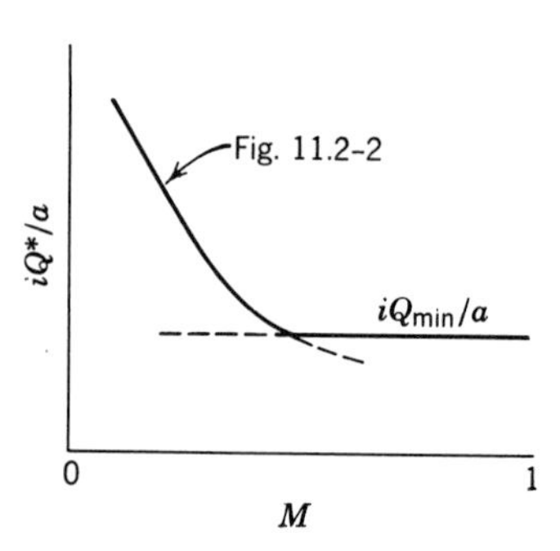

Fig. 11.I-1

11.J. Our company has signed a 10-year contract to sell a new product to a customer. The customer's demand will increase linearly until the end of the contract. Management knows that competitors will enter the market in a few years and that the market price will drop drastically. Accordingly, the decision has been made to build the new plant without provisions for later expansions. The new plant probably will not have the capacity to meet the peak demand. Production will be supplemented by purchases from competitors. Derive an expression for the capacity of the plant which minimizes the present value of the sum of plant investment and product purchases. Because of tax provisions, the money spent on product purchases costs the company treasury only two-thirds of the same amount of money spent on plant investment and operation.

11.K. Sanders has correlated the costs of electric arc furnaces as a function of capacity in pounds

$$I = I_0 \left(\frac{Q}{Q_0} \right)^{0.3}$$

and for a furnace of capacity of 10,000 lb, the cost is $140,000. The heat time is 2 hours.

If we assume that the melting rate, lb/hr, is independent of size, what would be the best size initial furnace to buy to meet an initial demand for metal of 440,000 lb/yr, if the forecast is for a linear increase in production at a rate of 44,000 lb/yr^2? What would be the size of future additions, and when would you make them?

12

ACCOUNTING FOR UNCERTAINTY IN DATA

No foolproof methods exist for forecasting future events be they a demand for the services of a system, a variation in environmental conditions, the rate of a reaction in a proposed reactor, or the efficiency of a separation process. Commonly, a cloud of uncertainty surrounds the data and equations upon which a system design rests, and much of this uncertainty usually is swept away only after the system is built and its operation observed, too late to correct design errors. In this chapter we describe a strategy for hedging against design errors caused by this persistent uncertainty.

Typical Problem

A recovery unit is to be sized to recover the vapor which will issue from a proposed storage tank for some 60,000 tons of liquid ammonia. We can only speculate about the actual vapor losses that will obtain, since the storage facility is still on the drawing board. How might we gamble on the size of the vapor recovery unit in the presence of this ignorance of the demands to which it must respond?

12.1 ENGINEERING ON THE SAFE SIDE

The process engineer protects his systems from the effects of uncertainty in design data by engineering on the safe side. In many situations the processing components are purposely designed to be more durable, more flexible, and of greater capacity than is demanded on the basis of the best information available, in an attempt to protect the system from unknown effects. On the other hand, when the uncertainty surrounds a critical feature of a novel system, the engineer may elect not to gamble on a full scale design and may recommend the construction of small scale pilot plant to test the processing concept, thereby putting less capital investment in jeopardy. Table 12.1-1 contains a summary of

Table 12.1-1 Survey of Industrial Practices for Accommodating to Uncertainty[a]

Name of Equipment	Is Pilot Plant Required	Environmental Variable	Design Variable	Scale-Up Ratio Based on		Over-design Range (Per cent)
				Design Variable	Flow Rate	
Plate and frame filters	Yes	Cake resistance or permeability	Filtration area	>100 : 1	>100 : 1	11–21
Rotary filters	Yes	Cake resistance or permeability	Filtration area	25 : 1	>100 : 1	14–20
Centrifugal pumps	No	Discharge head	Power input	>100 : 1	>100 : 1	
			Impeller diameter	10 : 1	>100 : 1	7–14
Reciprocating compressors	No	Compression ratio	Power input	>100 : 1	>100 : 1	
			Piston displacement	>100 : 1	>100 : 1	7–14
Screw conveyors	No	Bulk density	Diameter	8 : 1	90 : 1	8–21
Hammer mills	Yes	Size reduction	Power input	60 : 1	60 : 1	15–21
Liquid heat exchangers	No	Temperatures	Transfer area	>100 : 1	>100 : 1	11–18
Spray condensers	No	Latent heat vaporization temperatures	Height to diameter ratio	12 : 1	70 : 1	18–24
Plate columns	No	Equilibrium data	Diameter	10 : 1	>100 : 1	10–16
Packed columns	No	Equilibrium data	Diameter	10 : 1	>100 : 1	11–18
Cooling towers	No	Air humidity	Volume	10 : 1	>100 : 1	12–20
Cyclones	No	Particle size	Diameter of body	3 : 1	10 : 1	7–11

[a] L. Michel, R. D. Beattie, and T. H. Goodgame, *Chem. Eng. Progr.*, No. 7, **50** (1954).

a survey of industrial practices in this area indicating, when pilot plant work is used, how large a scale-up is commonly justified, and what per cent overdesign commonly has been added *empirically* to the scaled-up design to account for uncertainty in performance characteristics. For example, if we are designing a plate and frame filter attempting to estimate the proper filtration area, a pilot plant test is required to estimate the cake permeability. Existing theories of filtration are sufficiently reliable to enable a scale up of flow rate through the filter by a factor of 100 from the pilot experiment, with an overdesign of from 11 to 21 per cent commonly used to hedge against errors in the basic data and equations.

The evidence upon which such general overdesign factors are based is generally quite tenuous, and such factors must be considered more as articles of faith than reason.

A frequent danger is that a specific overdesign factor[1] for a given processing component might become a standard of practice and be applied in situations which are somewhat different from the original successful case which led to its adoption, thus becoming a source of design error in itself. For example, a reactor which operated successfully when designed to have a capacity 50 per cent greater than that recommended on the basis of the best available reaction kinetics information might lead to the adoption of an overdesign factor of 1.5 for all future reactors. This factor might be completely wrong for the design of a reactor to accomplish some other reaction in some other economic situation.

Moreover, a given processing component cannot be designed without regard for the system into which it must integrate, and, as Table 12.1-2 indicates,

Table 12.1-2 Uncertainty which Affects the Economic Definition of a Process Engineering Problem[a]

	Probable Variation from Forecasts over 10-Year Plant Life (%)
Cost of fixed capital investment	−10 to +25
Construction time	−5 to +50
Start-up costs and time	−10 to +100
Sales volume	−50 to +150
Price of product	−50 to +20
Depreciation method	None
Plant replacement and maintenance costs	−10 to +100
Obsolescence of process or equipment	Indeterminate
Income tax rate	−5 to +15
Inflation rates	−10 to +100
Interest rates	−50 to +50
Working capital	−20 to +50
Legislation affecting product	Indeterminate
Raw material availability and price	−25 to +50
Competition	Indeterminate
Salvage value	−100 to +10
Profit	−100 to +10

[a] From Bauman, *Fundamentals of Cost Engineering in the Chemical Industry*.

[1] The overdesign factor f might be defined as

$$f = \frac{\text{Recommended capacity}}{\text{Base calculated capacity}}$$

considerable uncertainty may exist even in the definition of the processing problem. For example, we would expect that a separation unit would be designed with a greater overcapacity when its effluent is the feed to a multi-million dollar process, than when its effluent enters a storage area. Undercapacity carries a severe penalty when the operation of a large system is interfered with, and it carries perhaps only a small penalty in the storage area case.

The engineer attempts to play the odds between erring on the low side and erring on the high side, striving to make decisions which have the best *chance* of eventually resulting in the most useful system. Thus, the elements of probability enter. Such corrections were made, and are still being made, intuitively. However, a theory of decision-making in the presence of uncertainty is evolving, and the application of that theory constitutes the bulk of this chapter. We expect that the existing empirical corrections for uncertainty and the developing theory will work hand-in-hand supporting and testing each other, as a more rational approach to this vital area of engineering evolves.

12.2 THE PROPAGATION OF UNCERTAINTY THROUGH DESIGNS

In nearly all processing problems certain bits of information are critical in the sense that small changes in the values assumed by those critical parameters greatly influence the solution to the problem. If uncertainty resides in the numerical values that should be assigned to such critical parameters, the optimal solution to the processing problem is uncertain. On the other hand, the optimal solution to a processing problem may be insensitive to wide variations in the values assigned to some of the other variables which enter into the design calculations, and great uncertainty can be tolerated there.

The sensitivity of the design equations to changes in parameters can be tested by incrementing each parameter in turn and computing the effect on the system of this change. For example, let the design objective function U depend in some complex way on the parameters $E_1, E_2, E_3 \ldots E_l$ which describe the environment into which the process is to integrate (for example, the rate of a chemical reaction, the anticipated load on a refrigeration unit, the tray efficiency of a distillation tower, the forecast rate of drop of price margin and so forth).

$$U = U(E_1, E_2, E_3 \ldots E_l) \qquad (12.2\text{-}1)$$

Also, the objective function will depend upon the design variables $d_1, d_2 \ldots d_F$ over which the engineer has control, such as vessel sizes, heat-exchange areas, and number of trays in the distillation tower. For the moment, we shall fix the values of the design variables and focus attention on the environmental parameters in which uncertainty resides.

Once a base design has been computed, assuming the values $\bar{E}_1, \bar{E}_2, \bar{E}_3 \ldots \bar{E}_l$ for the parameters and yielding a base value of the objective function $\bar{U}$, each parameter is incremented by a small fraction, say 5 to 10 per cent of its anticipated range $\Delta E_1, \Delta E_2 \ldots \Delta E_l$. The change in the objective function is computed for each change in the parameters, and the sensitivity coefficients $S_1, S_2 \ldots S_l$ are defined as the change in objective function caused by each change in the parameter divided by the change in the parameter.

$$S_1 = \frac{U(\bar{E}_1 + \Delta\bar{E}_1, \bar{E}_2, \bar{E}_3 \ldots \bar{E}_l) - \bar{U}}{\Delta E_1}$$

$$S_2 = \frac{U(\bar{E}_1, \bar{E}_2 + \Delta E_2, \bar{E}_3 \ldots \bar{E}_l) - \bar{U}}{\Delta E_2} \qquad (12.2\text{-}2)$$

$$\vdots$$

In the limit of small changes ΔE, the sensitivity coefficients become the derivative of the objective function with respect to the parameters

$$S_1 = \frac{\partial U}{\partial E_1}$$

$$S_2 = \frac{\partial U}{\partial E_2} \qquad (12.2\text{-}3)$$

However, in practice the finite form in Eq. 12.2-2 is the most convenient since it is generally impractical to attempt to differentiate the complex equations which arise in process engineering.

The parameters to which the objective function exhibits a high sensitivity, either positive or negative, are the critical parameters through which uncertainty can creep into a design.

A linear approximation to the detailed design equations is often useful in *estimating* the effects of uncertainty on the design. The sensitivity parameters are the coefficients in the linearization. Eq. 12.2-4, an equation which is a valid approximation only near the base designated by the overbars.

$$U - \bar{U} = \sum_{i=1}^{l} S_i(E_i - \bar{E}_i) \qquad (12.2\text{-}4)$$

It is convenient to think of the uncertainty in engineering and economic data in terms of a probability distribution function $p(E)$, where $p(E)dE$ is defined as the probability that E, the environmental variable in question, will take on a value in the range E to $E + dE$. Thus, with this interpretation of uncertainty, the engineer might assign a distribution function which rapidly drops to zero as E moves away from the assumed value $\bar{E}$ for a parameter in which uncertainty does not enter to any great extent. On the other hand, if the value of a parameter is in doubt, this doubt might be expressed by the assignment of a distribution function which has a greater spread. Figure 12.2-1 shows this concept.

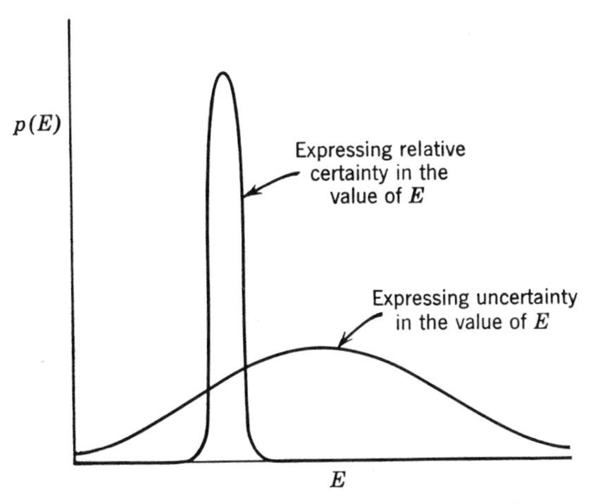

Fig. 12.2.1 The distribution function as a measure of uncertainty.

The engineer will generally be in doubt concerning the proper distribution function to assign each variable but will most likely have a feel for the degree of reliability of the data to be used in design. While there is considerable theoretical background abuilding on methods for encoding uncertainty such as that in Table 12.1-2 into distribution functions,[2] we shall employ in this text only the simplest distribution functions shown in Fig. 12.2-2; namely,

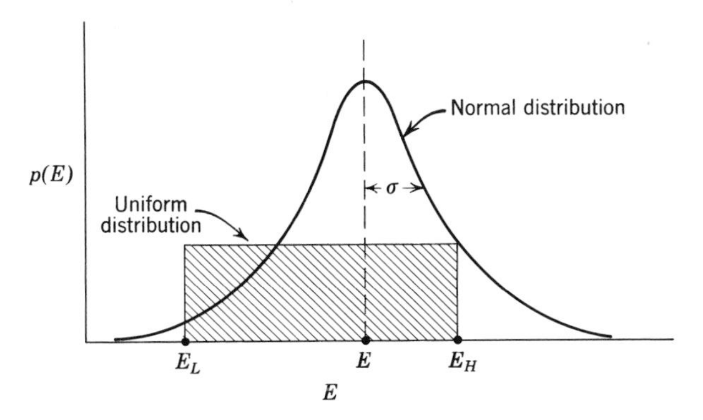

Fig. 12.2-2 Two common distribution functions for describing uncertainty.

[2] See, for example, *Jaynes, Physical Reviews*, **108**, 185 (1957).

The uniform distribution

$$p(E) = \begin{cases} \dfrac{1}{E_H - E_L} & \text{when} \quad E_L \leq E \leq E_H \\ 0 & \text{otherwise} \end{cases} \tag{12.2-5}$$

where $E_H - E_L$ is the range into which the uncertain variable E must fall.

The normal distribution

$$p(E) = \frac{1}{2\pi\sigma} e^{-(E-\bar{E})^2/2\sigma^2} \tag{12.2-6}$$

where $\bar{E}$ is the most probable value of E and σ is the variance, a measure of the spread about that value. In a normal distribution the probability that E will fall with $\bar{E} \pm \sigma$ is 0.68, and within $\bar{E} \pm 2\sigma$ is 0.98.

We shall now illustrate how the uncertainty in design data propagates through the design equations into the objective function, thereby clouding the view of the optimal process. Fig. 12.2-3 illustrates a solution to the design equations for a reactor which is to be cooled by a pumparound cooler using cold water as the cooling agent. The temperature of the cooling water at the process site is uncertain. Assuming that all other variables in the design are fixed at reasonable values, changing the assumed cooling water temperature changes the predicted yield from the system as shown in Fig. 12.2-3.

Consider two situations: situation A, in which the assumed uncertainty in the temperature of the available cooling water falls within a range of temperatures over which the yield is not greatly influenced by the cooling water temperature, and situation B, in which the uncertainty falls in a range in which the yield is sensitive to the temperature. In situation A, a uniform distribution function describing an uncertainty in the cooling water temperature is mapped into a nearly uniform distribution function describing the resulting uncertainty in product yield, with the range of uncertainty greatly suppressed. In situation B, a uniform distribution of uncertainty is mapped into a distorted distribution in the uncertainty of the yield in which the range is greatly expanded. In one situation the effect of the uncertainty in the cooling water temperature on the system product yield has been decreased, and in the other situation the effect of the uncertainty has been greatly expanded.

This example points to two phenomena which are common in process design; the form of the distribution function describing the uncertainty in the design data is distorted as the influence of the uncertainty propagates through a design, and the range of uncertainty expands and contracts according to the sensitivity of the design to the variables in which the initial uncertainty resides.

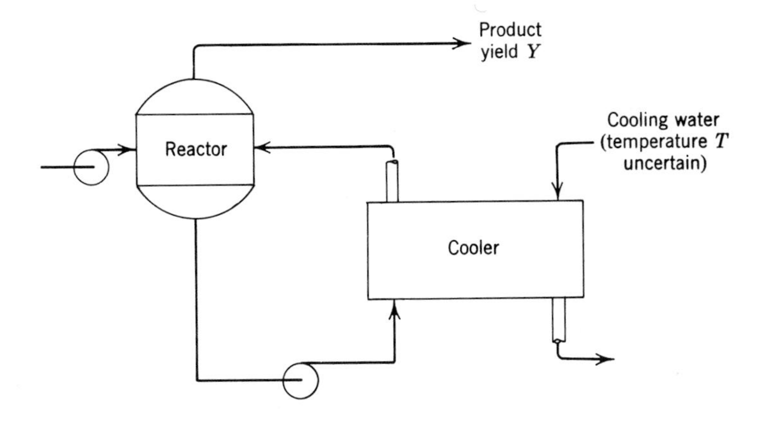

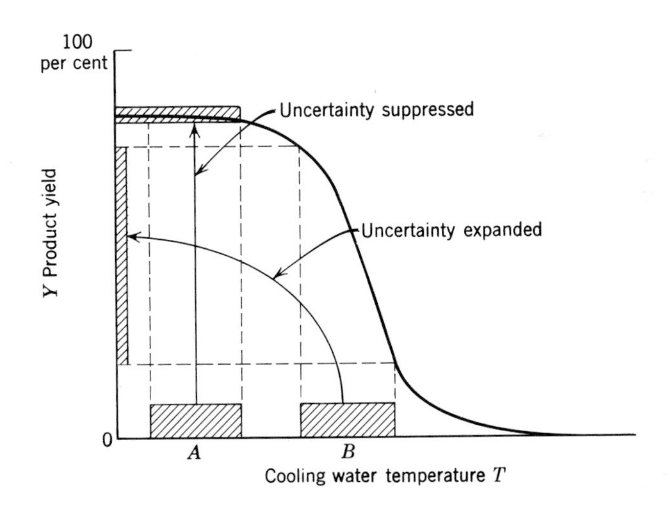

Fig. 12.2-3 Propagation of uncertainty through a design.

In the special case when the uncertain environmental variables $E_1, E_2, \ldots$ are related to the objective function U by a linear relationship, the analysis of the propagation of uncertainty is particularly simple. It can be shown that if the uncertainty about the environmental variable E_i is described by a normal distribution with mean $\bar{E}_i$ and variance σ_i^2, the resulting uncertainty in U is described by a normal distribution with mean

$$\bar{U} = U(\bar{E}_1, \bar{E}_2, \ldots) \tag{12.2-7}$$

and with variance

$$\sigma_U^2 = \sum_i S_i^2 \sigma_{E_i}^2 \tag{12.2-8}$$

In the more common case when the relationship between the uncertain parameters and the objective function is nonlinear, the sensitivity test recommended above leads to a useful linear approximation.

Example:
Table 12.2-1 contains cost estimates for a proposed process, reflecting uncertainty in the sales price, annual sales volume, and plant investment. How does this uncertainty propagate into the computed net profit?

Table 12.2-1 Economic Data

	Optimistic Estimate	Best Guess	Pessimistic Estimate
Price—p	40¢/lb	35¢/lb	30¢/lb
Annual sales volume—Q	12,000,000 lb	9,000,000 lb	6,000,000 lb
Total Investment—I	$1,700,000	$2,200,000	$2,700,000

Manufacturing costs $C = aI + bQ + cL$ \$/yr

Where	$a = 0.10$ \$/\$-yr	For insurance, maintenance, taxes, etc.
	$b = 0.10$ \$/lb	For raw materials
	$cL = \$250{,}000$/yr	For operating, labor, quality control, etc.

Tax computations

$d = 0.10$ \$/\$-yr	Depreciation of equipment for taxes
$t = 0.50$ \$/\$	Income taxes

Recovery of investment

$e = 0.10$ \$/\$-yr	To recover investment in process

The net profit, P \$/yr, is given by Eq. 12.2-9.

$$\begin{aligned} P &= (S - C) - (S - C - dI)t - eI \\ &= pQ(1 - t) - I[e - dt + a(1 - t)] - Qb(1 - t) - cL(1 - t) \\ &= pQ(0.50) - I(0.10) - Q(0.05) - 125{,}000 \end{aligned} \tag{12.2-9}$$

The best guess of the net profit is

$$\bar{P} = (0.35)(9{,}000{,}000)(0.50) - 2{,}200{,}000(0.10) - (9{,}000{,}000)(0.05) - 125{,}000$$
$$= \$780{,}000 \text{ per year}$$

We now estimate the probable uncertainty surrounding this estimate. Assuming that the uncertainties in the estimates of the sales priçe, annual sales volume, and plant investment are encoded in normal distribution functions with means at the best guess and variances of σ_p^2, σ_Q^2, and σ_I^2 respectively, we can take advantage of the linearized form of the estimate by using Eq. 12.2-8 to estimate the variance in the net profit, σ_P^2.

$$\sigma_P^2 = S_p^2\sigma_p^2 + S_Q^2\sigma_Q^2 + S_I^2\sigma_I^2 \qquad (12.2\text{-}10)$$

where

$$S_p = \frac{\partial P}{\partial p} = Q_{\text{best guess}}\ 0.50 = 4.5 \times 10^6$$

$$S_Q = \frac{\partial P}{\partial Q} = p_{\text{best guess}}\ 0.50 - 0.05 = 0.125$$

$$S_I = \frac{\partial P}{\partial I} = -0.10$$

What remains is the estimation of the variances of the sales price, sales volume, and investment from the data in Table 12.2-1. Clearly, the terms pessimistic and optimistic must be interpreted in terms of the variance in the normal distribution. If the pessimistic and optimistic limits have been solicited on the understanding that there is a 95.4 per cent chance that the uncertain variables will fall with these limits, then they represent the 2σ values in the normal distribution, Eq. 12.2-6.

$$\sigma = \frac{\text{Optimistic estimate} - \text{pessimistic estimate}}{4} \qquad (12.2\text{-}11)$$

Thus

$$\sigma_p = \frac{0.40 - 0.30}{4} = 0.025$$

$$\sigma_Q = \frac{12{,}000{,}000 - 6{,}000{,}000}{4} = 1{,}500{,}000$$

$$\sigma_I = \frac{2{,}700{,}000 - 1{,}700{,}000}{4} = 250{,}000$$

The variance on the computed net profit of $780,000 per year is

$$\begin{aligned}\sigma_P^{\,2} &= (4.5 \times 10^6)^2(0.025)^2 + (0.125)^2(1.5 \times 10^6)^2 + (-0.10)^2(2.5 \times 10^5)^2 \\ &= 1.27 \times 10^{10} + 3.5 \times 10^{10} + 6.25 \times 10^8 \\ &= 4.83 \times 10^{10}\end{aligned}$$

Thus

$$\sigma_P = \sqrt{4.83 \times 10^{10}} = \$220{,}000 \text{ per year}$$

The uncertainty in the net profit can be approximated by a normal distribution with mean $780,000 per year and $\sigma_P = \$220{,}000$ per year. Since in a normal distribution the probability that the variable will fall within σ_P of the mean is 0.68 and within $2\sigma_P$ of the mean is 0.954, we may state that the annual profit is in the range

$$P = 560{,}000 \text{ to } 1{,}000{,}000 \text{ \$ per year with 68 per cent confidence}$$

and

$$P = 340{,}000 \text{ to } 1{,}220{,}000 \text{ \$ per year with 95.4 per cent confidence}$$

The basic question we address in the next section is that of making rational decisions in the face of such a wide uncertainty.

12.3 THE PROBLEM OF DECISION-MAKING, A HOMELY EXAMPLE

Let us reflect for the moment on how a rational being might make decisions in the presence of uncertainty. To be specific, let us put ourselves in the shoes of an entrepreneur who has contracted to paint a set of buildings and has offered a one-year guarantee that the paint will not fade. Should the paint fade, the entrepreneur must repaint the buildings at his own expense.

The entrepeneur is to receive $500 for the contract and has available paints A, B, and C at a cost of $200, $100, and $5, respectively. Paint A will never fade, paint B will fade after 250 days of sunshine, and paint C will fade after 50 days of direct sun. The cost of labor is $200 regardless of the paint used.

The decision-making problem is summarized in the *payoff matrix*, Table 12.3-1. The entries in the table are the profits earned using the various paints for the several environmental conditions which might be experienced during the period of time when the guarantee is valid. For example, should the entrepreneur elect to use paint B and should the sun shine for more than 250 days the paint will fade. The net profit would then be

Original job	\$ 500	the original contract
	−100	paint B
	−200	labor
Repainting	\$− 5	paint C
	−200	labor
Net profit	−5	

The contract did not say when the repainting job must be done, so he can use paint C to repaint any time within 50 days of the expiration date of the contract and fulfill the guarantee according to the letter of the law, if not the spirit.

Table 12.3-1 Payoff Matrix

	n Number of Days of Sunshine Next Year		
	$0 \leq n < 50$	$50 \leq n < 250$	$250 \leq n \leq 365$
Paint A	\$100	\$100	\$100
Paint B	200	200	−5
Paint C	295	90	90

Now, how might the entrepreneur decide on the paint in the face of the uncertainty in next year's weather? He needs a *decision-making criterion*. Let us consider a few.

Maximin profit. To use the *maximin criterion* we search each row in the payoff matrix for the worst that could possibly happen (minimum profit) and then make the decision that results in the maximum-minimum profit.

This criterion is pessimistic and protects the decision maker against the worst that could possibly happen.

Payoff Matrix

Decision	*Environment* 0–50	50–250	250–365	*Minimum Profit*	
A	100	100	100	100	←Maximin profit (paint A)
B	200	200	−5	−5	
C	295	90	90	90	

Minimax regret. Should the entrepreneur focus attention on what he might have done, the regret in using a given paint could be expressed as the difference between the profit that might have been made and the profit that was made in given environment. This is expressed in the regret matrix below. For example, had he used paint A in the first environment he would have a regret of \$195, the additional profit he could have made had he used paint C.

Regret Matrix

		Environment 0–50	50–250	250–365	*Maximum Regret*	
	A	195	100	0	195	
Decision	B	95	0	105	105	← Minimax regret (paint B)
	C	0	110	10	110	

So far the entrepreneur has ignored a source of information that should be of considerable help, i.e., the meteorological records for the county. Contained therein is a record of the number of days of sunshine for each year in the past 50 years. How might he build such data as this into his decision making criterion?

Meteorological Records
Number of years in the past 50 years with given numbers of days of sunshine

	0–50	50–250	250–365
Number of years	26	14	10
Fraction	0.52	0.28	0.20

Expected value. The past records of sunshine give an estimate of the probability of a given environment in the year to follow. For example, we might utilize the records to estimate the probability of having 250 or more days of sunshine next year as 0.20 since this is the fraction of past years with that range of days of sun. The expected value criterion builds such information into the payoff matrix, by weighting each entry by the probability of the corresponding environment and defining the expected profit for each decision as the sum of the weighted profits. The decision that maximizes the expected profit is then optimal according to this criterion. The expected profit corresponds to the average profit to be realized over a large number of trials.

Payoff Matrix and Expected Value

	Environment			Expected Profit	
Decision A	100	100	100	100	
Decision B	200	200	−5	159	
Decision C	295	90	90	198	←Maximum expected profit (paint C)
Probability of environment	0.52	0.28	0.20		

Example:

$$\text{Expected profit for paint } C = 295\,(0.52) + 90(0.28) + 90(0.20) = 198$$

The financial success of the entrepreneur clearly depends on three factors:

The validity of his decision-making criterion.
The accuracy of the payoff matrix.
The ability to project past experience into the future.

This is generally true for any decision-maker, the engineer included.

We have seen that three seemingly reasonable decision-making criteria give three different answers. Any number of other criteria could be imagined. What is an appropriate criterion? The usual procedure has been to see whether a given criterion satisfies a number of common sense conditions, such as:

1. *Transitivity*. If D_1 is preferred to D_2, and D_2 is preferred to D_3, then D_1 must be preferred to D_3.
2. *Strong domination*. If for all environmental conditions the payoff from D_1 is greater than the payoff from D_2, and D_1 should always be preferred to D_2.

A dozen or so common sense conditions exist which have been used to test the internal consistency of a decision criterion.[3] While the details of the analysis can be quite tedious and will not be entered into here, the results are of interest in that of the several criteria discussed here only the expected value criterion passes all of the tests. Thus, in addition to the intuitive appeal of the expected value criterion, support is offered from purely mathematical conditions of internal consistency.

[3] R. D. Luce and H. Raiffa, *Games and Decisions*, Wiley, New York, 1958.

A few comments on the accuracy of the payoff matrix are in order. Suppose that the entrepreneur wishes to establish a reputation as a reliable contractor. A poor job, with the paint fading after a year or so, could penalize him even though he satisfied the letter of law in his contract, since his reputation might be ruined and future business lost. He must modify the payoff matrix to account for any changes in his business goals and methods of operation.

Finally, the concept of *gambler's ruin* must be considered, for there is the possibility that an unfortunate series of improbable failures may wipe out a firm's capital reserves before the expected long-range profits can be realized. For example, suppose an investor plans to risk a fraction f of his total capital reserves of I dollars each year on a project that has a probability of success, $p = 2/3$. If that year the project succeeds, the return on investment will be the amount invested, If dollars, and if it fails, this investment will be lost. The expected profits from this game will amount to

$$\text{Expected profit} = pfI - (1 - p)fI = \left(\frac{1}{3}\right) fI \text{ \$/yr}$$

Clearly, if the maximum expected profit criterion is to be followed, all capital should be risked, $f = 1$, for a maximum expected profit of $(1/3)I$ \$/yr.

However, it can be shown that the probability of ruin in this game is[4]

$$\text{Probability of ruin} = \left(\frac{1 - p}{p}\right)^{1/f}$$

and therefore to risk all the capital, striving for maximum expected profit, predisposes the firm to gambler's ruin, as Table 12.3-2 shows. Only when, say, one tenth of the assets are risked, is gambler's ruin improbable.

Table 12.3-2 The Effects of Limited Capital on Gambler's Ruin

Fraction of Assets Risked	Probability of Ruin for $p=2/3$
1.0	0.50
0.5	0.25
0.1	0.001

With this in mind, we might expect the character of engineering decisions made in the design of a steel or paper mill (which commonly requires the commitment of \$100,000,000 or more in equipment) to differ from the character of

[4] See, for example, W. Feller, *An Introduction to Probability and Its Application*, Vol. I, Wiley, New York, 1957, Chap. XIV, Sect. 2, "The Classical Ruin Problem."

those made in the design of a smaller system, which commits only a small fraction of the company's assets.

In Section 12.4 we present a more formal description of the expected value criterion and, in the remaining sections, we show how these concepts might apply to actual engineering problems. Such theories of engineering in the presence of uncertainty expose the structure of the overdesign factor, and should be used in conjunction with established standards of practice which might have evolved over the years in any successful engineering firm.

12.4 THE EXPECTED VALUE CRITERION

In this section we present the general expected value criterion for decision-making in the presence of uncertainty with the understanding that this is a reasonable criterion for most engineering design problems. Suppose that the variable E denotes the uncertain environment. This could represent a forecast of future demands for the services of a system, the value of a vital parameter such as a reaction velocity constant, or any other critical parameter which affects the design of the system.

Now let $U(D, E)$ be the objective function to be realized by the application of decision D in environment E. This corresponds to the entries in the payoff matrix in Section 12.3. The expected value of the objective function is then defined.

$$\overline{U}(D) = \int U(D, E)\, p(E)\, dE \qquad (12.4\text{-}1)$$

The proper decision in the presence of uncertainty according to the expected value criterion is then the D which maximizes the expected value.

$$\max_{\{D\}} \left[\overline{U}(D) = \int U(D, E)\, p(E)\, dE\right] \qquad (12.4\text{-}2)$$

In summary, the initial method recommended in this text for making design decisions in the face of uncertain data is outlined below.

1. Compute the value of the objective function for each design option for every set of environmental conditions which might obtain.
2. Estimate a probability of occurrence for each set of environmental conditions.
3. Multiply the objective functions computed in Step 1 by the probability estimates obtained in Step 2, and sum these weighted objective functions over all possible environmental conditions for each design option. The sums are the expected objective function for each design option.
4. Select the design option with the maximum expected objective function.

12.5 THE SIZING OF A CATALYTIC BED

We now apply a minimum expected cost criterion to a specific and simplified situation, that of sizing the catalyst bed in a reactor in the presence of inexact knowledge of the activity of the catalyst when used in this reactor environment. Our hope is to detect an underlying form for the overdesign factor by analyzing a number of such simplified situations.

Suppose that the reactor must produce Q pounds of product per day to integrate into a larger system, that excess production capacity goes unrewarded, and that underproduction requires the costly modification of the catalyst bed. The only reliable information on the catalyst activity is that it lies somewhere between A_L and A_H pounds of product per day per pound of catalyst. The cost of the catalyst is C_c dollars per pound, and the cost of a shutdown to rebuild the reactor, should the production be below the demanded Q, is C_s dollars. In the light of this information, what is the optimal recommended size D, pounds of catalyst, of the catalyst bed?

The uncertainty in the catalyst activity can be encoded as the uniform distribution function, Eq. 12.5-1.

$$p(A) = \begin{cases} \dfrac{1}{A_H - A_L} & \text{for } A_L \le A \le A_H \\ 0 & \text{otherwise} \end{cases} \tag{12.5-1}$$

and the costs are

$$C(D, A) = \begin{cases} \text{success on the first attempt} \quad C_c D & \text{if } DA > Q \\ \quad \text{or} & \\ \text{redesign needed} \quad C_c D + C_c\left(\dfrac{Q}{A} - D\right) + C_s & \text{if } DA < Q \end{cases} \tag{12.5-2}$$

The first term in Eq. 12.5-2 is the cost of the original design and the remaining terms include the penalty C_s for an inadequate initial design plus the cost of supplying just the right amount of catalyst.

The expected cost is then

$$\bar{C}(D) = \underbrace{\int_{A_L}^{Q/D} \left(\frac{C_c Q}{A} + C_s\right) \frac{dA}{(A_H - A_L)}}_{\text{(Redesign required)}} + \underbrace{\int_{Q/D}^{A_H} (C_c D) \frac{dA}{(A_H - A_L)}}_{\text{(Success in first design)}}$$

$$\bar{C}(D) = \frac{1}{(A_H - A_L)}\left[C_c Q \ln\frac{Q}{DA_L} + C_s\left(\frac{Q}{D} - A_L\right) + C_c D\left(A_H - \frac{Q}{D}\right)\right]$$

for

$$\frac{Q}{A_H} \leq D \leq \frac{Q}{A_L} \qquad (12.5\text{-}3)$$

The minimum expected cost is reached when

$$D^* = \frac{1}{2}\left(\frac{Q}{A_H} + \sqrt{\left(\frac{Q}{A_H}\right)^2 + \frac{4C_s Q}{C_c A_H}}\right) \qquad (12.5\text{-}4)$$

for

$$\frac{Q}{A_H} \leq D \leq \frac{Q}{A_L}$$

Equation 12.5-4 is obtained by differentiating Eq. 12.5-3 with respect to D, and setting that derivative to zero. Taking the upper catalyst activity A_H as an *arbitrary base*, an overdesign factor can be defined as

$$f = \frac{D^*}{(Q/A_H)} \qquad (12.5\text{-}5)$$

where f is the ratio of the recommended design D^*, pounds of catalyst to the optimistic design Q/A_H pounds of catalyst. Substituting this in Eq. 12.5-4 gives the optimal overdesign factor f as

$$f = \frac{1}{2}\left(1 + \sqrt{1 + \frac{4C_s}{(C_c Q/A_H)}}\right) \qquad (12.5\text{-}6)$$

for

$$f < \frac{1}{1 - \Delta}$$

$$f = \frac{1}{1 - \Delta} \qquad \text{otherwise}$$

where

$$\Delta = \frac{A_H - A_L}{A_H}$$

is the fractional uncertainty in the catalyst activity.

Equation 12.5-6 is plotted in Fig. 12.5-1 and has some interesting implications. The optimal factor for low uncertainty is $f = 1/(1 - \Delta)$, which corresponds to an overdesign sufficient to cover the worst possible situation (a minimax cost criterion). However, when the uncertainty becomes higher it is then wise to gamble on possible underdesign, and the possibility of paying the redesign cost C_s rather than be saddled with a system with excessive catalyst. The optimal gamble is determined completely by the parameter $C_s/(QC_c/A_H)$ which

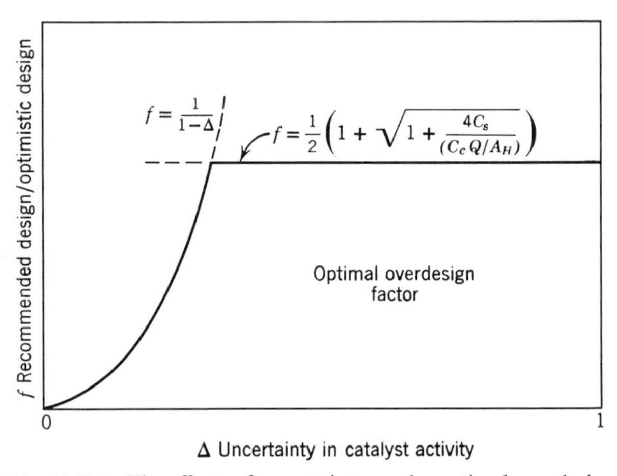

Fig. 12.5-1 The effects of uncertainty on the optimal overdesign.

is the ratio of the penalty for underdesign to the cost of the catalyst assuming the upper limit on activity. As this ratio increases the gamble is deferred to higher uncertainties. Observe that the overdesign factor f is *not a constant* for a given physical situation, but rather it is quite dependent on the uncertainty and the economics. This being the case, even in such simple situations, it is apparent that the time honored practice of using a constant universal overdesign factor is most likely in error.

12.6 OVERDESIGN OF A DISTILLATION TOWER

John Villadsen[5] worked out a practical example of statistical overdesign. Combining industrial experience in the design of distillation towers with the theory of overdesign presented in the previous sections, Villadsen investigated the effects of uncertainty on the recommended number of separation plates in distillation towers. In this section we review his work.

[5] J. Villadsen, 750, Report Chemical Engineering Dept., Univ. of Wisc., 1966 and *Transactions Udgivet af Dansk Ingeniorforening, Ingenior*, Nr. 8, 1967.

The uncertainties which attend the design of a distillation tower, one of the oldest and most extensively studied of process components, still persists to a significant degree. The advent of the digital computer widened the scope of design calculations to include multicomponent distillation tower design problems which were generally beyond routine computation only a few years ago. However, the removal of this computational obstacle has drawn attention more than ever to the unreliability of the basic data fed to the computer. In the past the accuracy of the basic data was on the same order as the accuracy of the methods of computation. Now, the numerical accuracy of the mathematics often far exceeds the accuracy of the basic data upon which the mathematical equations are based.

For example, correlations of multicomponent vapor-liquid equilibrium data required for distillation tower design are frequently in error by as much as 5 to 10 per cent for well studied chemicals. Correlations of plate efficiencies (which are related to the mass transfer rate between the liquid and gas phases on the plates) have been shown to deviate by 20 per cent from the data used to make the correlations! Considering these two factors alone, imagine the uncertainty attending the design of some novel plate construction or some unusual systems[6].

A common industrial design practice involves the computation of the required number of distillation trays to achieve the desired performance assuming complete knowledge, followed by the application of an overdesign factor to account for the expected errors. The overdesign factor employed frequently was not anchored to any basic theory.

Once a tower is built and put into operation, the uncertainties are swept away. Should the tower be inadequate to its task under normal operating conditions, the reflux ratio is increased and the operation of the tower is strained to meet the demands. This results in an increase in the operating costs (steam and cooling water). However, should the tower have an excess of trays, the operating costs might be somewhat lower but the investment too high. An economic balance is required.

One of the major uncertainties resides in the estimate of the *efficiency* of the distillation trays or plates. The correlations of plate efficiency are notoriously poor. This then is taken as the dominant source of uncertainty in the following application of the theory of overdesign.

Villadsen's analysis follows this line of thinking. In a given situation we can compute the yearly cost of separation by distillation (including the amortization of the investment) as a function of the number of trays in the tower, the reflux ratio for the overhead vapors, and an assumed tray efficiency.

$$U(N, R, E) \qquad (12.6\text{-}1)$$

[6] See, for example, J. Villadsen, *Brit. Chem. Eng.*, June 1966.

For a given number of trays, the reflux ratio R would always be adjusted to achieve minimum cost; but in the design calculations the two are constrained by the separation that is required of the tower. This constraint can be expressed

$$f(N, R, E) = 0 \tag{12.6-2}$$

Thus, for a given number of trays and tray efficiency, the required reflux ratio can be computed by the solution of Eq. 12.6-2, to give

$$R = F(N, E) \tag{12.6-3}$$

With the uncertainty residing in the tray efficiency E expressed by the distribution function $p(E)$, the expected cost is

$$\overline{U(N)} = \int_E U(N, F(N, E), E)\, p(E)\, dE \tag{12.6-4}$$

The recommended number of trays accounting for uncertainty is that number N which minimizes the expected total cost, Eq. 12.6-4.

The optimum overdesign factor f can be defined as the number of plates recommended above, divided by the number recommended assuming certainty in the tray efficiency E.

We now get into the details of implementing this line of thinking.

Cost data. The factors which enter into the estimate of the total cost, Eq. 12.6-1, include the fixed costs for the column, the reboiler, and the condenser; and operating costs for supplying steam and cooling water. The following correlations were used:

Fixed cost of column: C_1 (number of trays)(diameter of column)^{C_2}
Fixed cost of reboiler: C_3 (heat exhange area)^{C_4}
Fixed cost of condenser: C_5 (heat exchange area)^{C_6}

The parameters C in these correlations depend upon the materials used and upon the particular type of equipment employed and can be estimated from the literature as in Chapter 5, *Cost Estimation.*

Cost of water: C_7 (condenser duty in Btu per year)
Cost of steam: C_8 (reboiler duty in Btu per year)

In the numerical studies the cost of steam was taken as 1.4 dollars per ton and water as 0.02 dollars per ton.

Annual total cost: (operating costs) + C_9 (fixed costs)

The parameter C_9 takes on the values 0.30, 0.20, and 0.13 in the numerical studies.

Design Equations. The conventions employed in estimating the tower design follows.

The procedure for calculating the tower follow the principles set down by Smoker[7] and illustrated in Peters' text.[8]

The required heat-exchange areas can be computed from the heat loads required to achieve the separation. These areas depend on the prevailing temperatures of steam and cooling water, as well as the heat-transfer coefficients. Such rules of thumb as the recommended 10°C temperature difference in the condenser and a maximum vapor rate of 1.5 feet per second were used.

The basic calculations were performed for a binary mixture with the degree of split as one of the parameters to be varied.

Uncertainty. The uncertainty in the tray efficiency E was taken as uniformly distributed about the value given by the standard correlations E_0. The range of uncertainty is one of the parameters in this study.

The range of the study. The overdesign factor was then computed for a variety of conditions in the hope of detecting some general rules for the application of overdesign. The study included the variations in the several factors shown below.

1. The assumed value of plate efficiency E_0 (0.50, 0.60, 0.70).
2. The relative volatility of the two components (1.5, 2.0, 2.5).
3. The fraction of the more volatile component desired in the overhead (0.99, 0.97).
4. The amortization parameter C_9 (0.30, 0.20, 0.13).

This includes a rather wide range of industrially important distillation problems.

The results. The recommended overdesign factor was computed for this variety of conditions under various degrees of uncertainty of the plate efficiency. The results are plotted in Fig. 12.6-1. Notice that for all practical purposes the uncertainty in the plate efficiency is the dominant variable, and that the same overdesign is recommended for the variety of required splits of the feed, the relative volatilities of the components, and economic situations.

For example, should the design calculations for a certain distillation tower result in a recommendation of 50 trays, assuming a certain value of the plate efficiency, the engineer might account for the uncertainty in the assumed value of the efficiency as follows. If the source of the tray efficiency is felt to be in error by 20 per cent, Fig. 12.6-1 suggests an overdesign of $f = 1.06$. A design of 53 trays then properly accounts for uncertainty in the value of tray efficiency and its economic consequences.

[7] E. H. Smoker, *Trans. A.I.Ch.E.*, **34**, 165 (1938).

[8] M. S. Peters, *Plant Design and Economics for Chemical Engineers*, McGraw-Hill, New York, 1958.

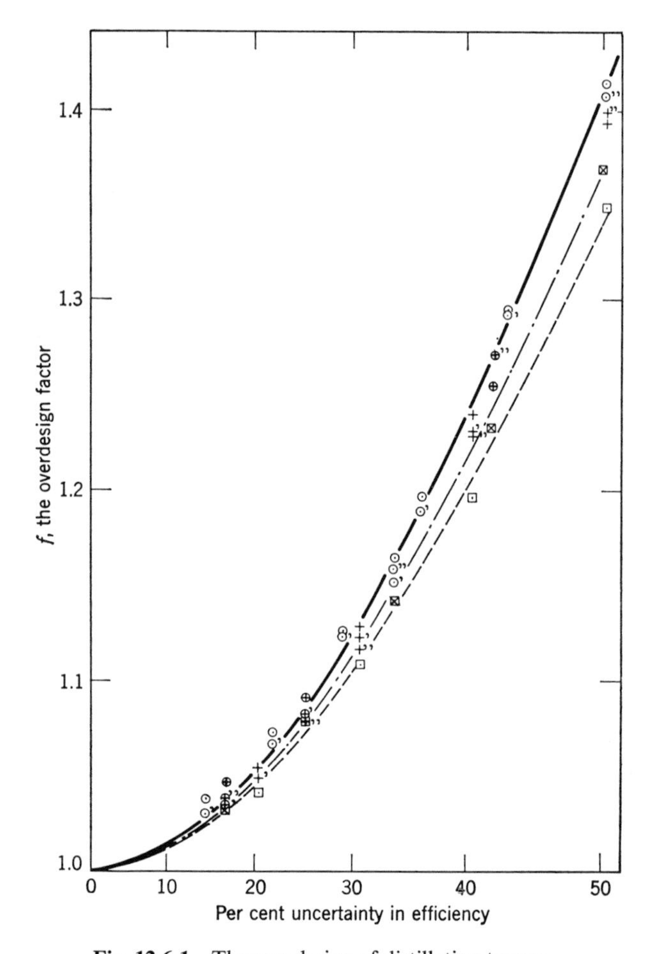

Fig. 12.6-1 The overdesign of distillation towers.

12.7 DESIGN OF A BACKMIX REACTOR[9]

Kittrell and Watson considered the design of a backmix reactor, Fig. 12.7-1, in the presence of uncertainty of the rate of the chemical reaction. We now further illustrate the engineering applications of the expected value criterion by presenting the results of their analysis.

One hundred gram moles of R are to produced hourly from a feed consisting

[9] J. Kittrell and C. C. Watson, *Chem. Eng. Progr.*, April, 1965.

of a saturated solution of A ($A_0 = 0.1$ gm moles/liter). The reaction is approximated by

$$A \rightarrow R$$

with the rate described by

$$r = KA \tag{12.7-1}$$

where $K = 0.2$ hr^{-1}, and A is the concentration of reactant in the reaction environment.

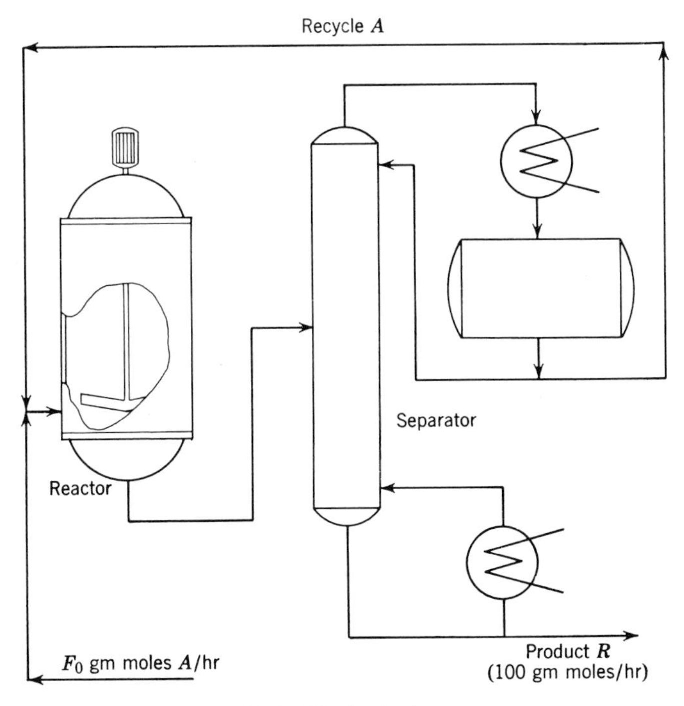

Fig. 12.7-1 The backmix reactor.

The cost of the reactant in the saturated solution is

$$C_A = \$0.50/\text{gm mole } A$$

The cost of the backmix reactor including installation, auxiliary equipment, instrumentation, overhead, labor, depreciation, etc., is

$$C_b = \$0.01/(\text{hr})(\text{liter of reactor volume})$$

The cost of the separation step is approximated by taking it as the value of the unreacted A recovered for recycle to the reactor.

What size of the reactor (V liters), the feed rate (F_0 gm moles A/hr), and the conversion (X_A, fraction A converted) should be used for minimum cost operation? What effect does uncertainty in the reaction velocity constant K have on the design?

The hourly cost of operation is

$$C_t = (\text{Volume of reactor})\left(\frac{\text{cost}}{(\text{Hr})(\text{Volume of reactor})}\right)$$

$$+ \left(\begin{matrix}\text{Feed rate of}\\ \text{reactant to reactor}\end{matrix}\right)(\text{Unit cost of reactant})$$

$$C_t = VC_b + F_0 C_A \tag{12.7-2}$$

A mass balance over the reactor gives

$$V = \frac{F_0 X_A}{KA(1 - X_A)} \tag{12.7-3}$$

In the case of a known reaction velocity constant $K = 0.2\ \text{hr}^{-1}$, we obtain the minimum total cost design by differentiating the total cost with respect to V and setting the derivative to zero, yielding

$$\begin{aligned} X_A^* &= 0.5 \\ F_0^* &= 200 \text{ moles } A/\text{hr} \\ V^* &= 10{,}000 \text{ liters} \\ C_t &= \$200/\text{hr} \end{aligned}$$

Now, the above design is based on the assumption of complete knowledge of the reaction velocity constant K, a situation which rarely exists in practice. Let us suppose that there exists an uncertainty in that vital parameter and calculate the design which best accounts for such ignorance.

A production rate of 100 gram moles of product per hour has been contracted for, so the reactor must be operated at a conversion X to fill that contract regardless of the designed volume of the reactor. The minimum expected cost is

$$\bar{C}_t(V) = \int C_t(V, K)\, p(K)\, dK \tag{12.7-4}$$

The first term in the integrand of Eq. 12.7-4 is the total cost, for given volume reactor and reaction velocity constant K, to achieve a production rate of 100 moles of product per hour ($F_0 X_A = 100$). The second term is the probability distribution expressing the uncertainty in the reaction velocity constant. The reactor volume V which minimizes this expected cost constitutes the optimal reactor design to account for the uncertainty.

Numerical studies were performed in the cases of a uniform distribution of uncertainty and a truncated normal distribution of uncertainty. Figure 12.7-2 shows the expected cost as a function of design volume for the case of a uniform distribution of uncertainty with $\Delta K = 0.075\ \text{hr}^{-1}$. Notice that the minimum expected cost is achieved when $V = 11{,}140$ liters, compared to $V = 10{,}000$ liters when no uncertainty was assumed.

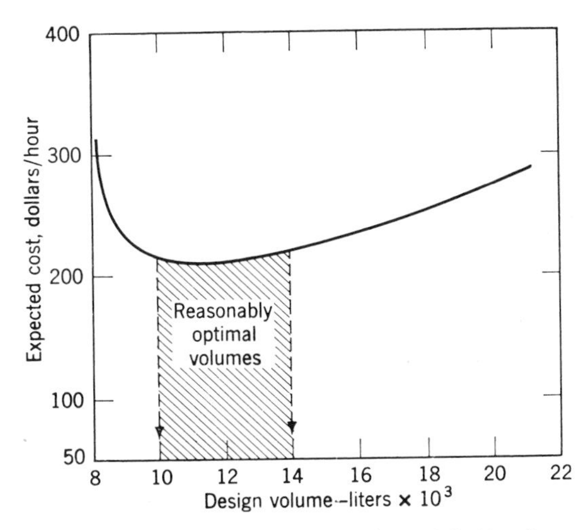

Fig. 12.7-2 The expected cost ($\Delta K = 0.075\ \text{hr}^{-1}$).

Defining the overdesign factor of

$$f = \frac{V}{V^*} = \frac{V}{10{,}000} \tag{12.7-5}$$

we see that an overdesign factor of

$$f = \frac{11{,}140}{10{,}000} = 1.114$$

is recommended. But even more important, notice how flat the expected cost is near the minimum cost. Any overdesign factor between $f = 1.0$ and $f = 1.4$ would not give an unreasonably large operating cost in this specific case For-tunately, here the efficiency of operation is insensitive to the specific value of the hedge against uncertainty. We would not expect this to be the rule.

Figure 12.7-3 is a plot of the recommended reactor overdesign as a function of the uncertainty in the reaction velocity constant. The use of a truncated normal distribution reflects a reduction in the ignorance of K with the values of K near 0.2 hr^{-1} being more probable than those nearer the outer reaches of the range of uncertainty. This additional information allows the reduction of the overdesign factor.

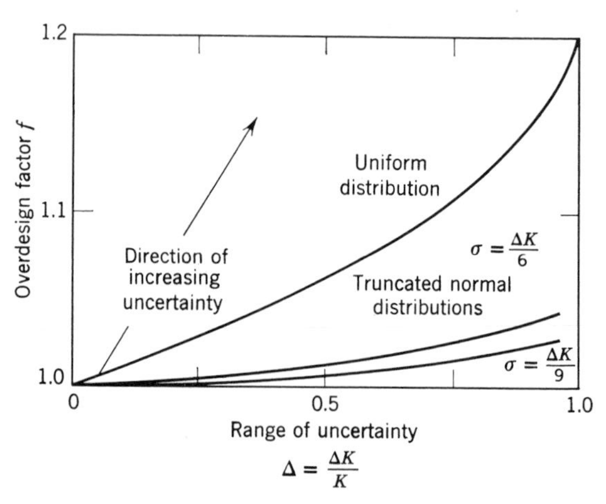

Fig. 12.7-3 The effects of uncertainty on the overdesign factor for the backmix reactor.

It can be shown that the uniform distribution of uncertainty constitutes a maximum degree of ignorance for a given range of uncertainty, and that the use of any other reasonable distribution function assumes a greater knowledge of the uncertain parameter. Thus, the overdesign factor obtained assuming a uniform distribution gives an upper limit on the effects of ignorance.

12.8 AN ANALYSIS OF THE RISK FACTOR

The several examples in the preceding sections dealt with correctives for uncertainty in the data for the design of processing components. The risk factor h introduced in Chapter 4 is an empirical corrective for the larger uncertainties surrounding the profitability of an entire system. The principles of statistical decision theory also apply to such economic uncertainties. This section summarizes a recent article by Rudd and Watson.[10]

[10] D. F. Rudd and C. C. Watson, "On Correctives for Project Uncertainty," *Can. J. Chem. Eng.*, December, 1965.

The risk factor h \$/\$-yr is a corrective on the rate of return on invested capital i \$/\$-yr, designed to account for possible errors in estimates of sales income, manufacturing costs, investment, project life, and the like. Uncertain and risky projects are assigned a higher minimum acceptable rate of return on invested capital, $i_m = i + h$, to reduce the chances of the investor making unwise commitments of capital on the basis of erroneous data.

The assignment of a numerical value to the risk factor was an empirical art not anchored to any rational analysis—a state identical to that of the over-design factor before the introduction of the expected value criterion.

It is far easier to estimate the uncertainty in the basic data upon which a profitability study is based than it is to estimate the proper risk rate to use in compensating for that uncertainty. By equating the expected value criterion of profitability to the risk factor corrected criterion of profitability there results

1. A simple relation between the risk factor corrective and the degree of uncertainty in the economic estimates.
2. A confirmation that the form of the well established minimum acceptable rate of return criterion and newer expected value criterion are in essential agreement.

We now outline these developments.

The venture profit is defined in Chapter 4 as

$$V = S - C - (S - C - dI)t - eI - i_m I$$

where the corrective factor i_m may be broken down into

$$i_m = i + h = i + h_S + h_C + h_I + h_n + h_f$$

where the h's are additional return on investment required to correct for uncertain estimates of sales income, S, manufacturing costs, C, required investment, I, project life n and probability of complete failure, p_f.

The expected value criterion which purports to make the same correction is

$$\overline{V} = \int V(E, h = 0) p(E)\, dE$$

where E denotes the uncertain parameters in the venture profit $V(E, h = 0)$ computed assuming no uncertainty $h = 0$. For example, considering only an uncertainty of from $S_0 - \Delta S$ to S_0 in the sales income, the expected venture profit is

$$\overline{V} = \int_{S_0 - \Delta S}^{S_0} [S - C - (S - C - dI)t - eI - iI] \frac{ds}{\Delta S}$$

$$= \frac{S_0^2 - (S_0 - \Delta S)^2}{2\Delta S}(1 - t) - C - (-C - dI)t - eI - iI$$

Now, these two criteria must both make the same correction for uncertainty and, hence, can be equated yielding a relationship between the risk factor h and the degree of uncertainty.

$$V(i_m) = \bar{V}$$

In the case of uncertainty in sales income

$$V = S_0 - C - (S_0 - C - dI)t - eI - (i + h_S)I$$

$$= \bar{V} = \frac{S_0^{\,2} - (S_0 - \Delta S)^2}{2\Delta S}(1 - t) - C - (-C - dI)t - eI - iI$$

or

$$h_S = \frac{\Delta S(1 - t)}{2I_0}$$

Thus, a \$50,000/yr uncertainty in the sales income for a million dollar project might properly be corrected for in profitability studies by assigning

$$h_S = \frac{(50{,}000)(1 - 0.5)}{2 \times 1{,}000{,}000} = 0.0125 \text{ \$/yr-\$}$$

Table 12.8-1 summarizes the contribution from other sources of uncertainty, and Table 12.8-2 illustrates three different situations, a cost saving project, an expansion, and a new project. The minimum acceptable rates of return compare favorably with the empirical values presented in Chapter 4.

Table 12.8-1 Contributions of Various Types of Uncertainty to h

Source	Range of Uncertainty	Contribution to h
Sales income	$S_0 - \Delta S$ to S_0	$h_S = \dfrac{\Delta S(1 - t)}{2I_0}$
Manufacturing costs	C_0 to $C_0 + \Delta C$	$h_C = \dfrac{\Delta C(1 - t)}{2I_0}$
Investment	I_0 to $I_0 + \Delta I$	$h_I = \dfrac{\Delta I(e + i - dt)}{2I_0}$
Project life	$n_0 - \Delta n_0$ to n_0	$h_n = \dfrac{\ln}{\Delta n}\left(\dfrac{1 - \exp(-in_0)}{1 - \exp(-in_0 + i\Delta n)}\right) - e$
Probability of failure	p_f	$h_f = p_f \dfrac{(S_0 - C_0)(1 - t)}{I_0}$

Table 12.8-2 Several Typical Situations

Type of Project	Cost-Saving	Expansion	New
I, 10^6 \$/yr	1.0	1.0	1.0
S, 10^6 \$/yr	1.5	1.5	1.5
d, fraction/yr	0.1	0.1	0.1
i, fraction/yr	0.1	0.1	0.1
n_0, yr	4	4	4
C, 10^6 \$/yr	0.5	0.5	0.5
t, fraction	0.5	0.5	0.5
ΔI, 10^6 \$/yr	0.1	0.2	0.4
ΔS, 10^6 \$/yr	0.05	0.15	0.60
Δn, yr	2	2	2
ΔC, 10^6 \$/yr	0.05	0.05	0.05
p_f fraction	0	0.05	0.30
h_I	0.015	0.030	0.061
h_S	0.013	0.038	0.150
h_C	0.013	0.013	0.013
h_n	0.094	0.094	0.094
h_f	0	0.025	0.150
$h = \Sigma h$	0.135	0.200	0.468
$i_m = \Sigma h + i$	0.235	0.300	0.568

12.9 CONCLUDING REMARKS

There is a natural human tendency to ignore factors which are not well understood and to attempt to avoid basing decisions on partial or conflicting information. Unfortunately, in process engineering uncertainties crop up in the most critical portions of our problems, and frequently there is not enough time to undertake an elaborate scientific study to reduce the uncertainty.

The empirical art of overdesign was evolved to provide some measure of security in this naturally insecure area. Now statistical decision theory appears to be removing some of the empiricism from overdesign, and the hope is entertained that further research will provide even more practical corrections for uncertainty.

The expected value criterion seems to be a fairly accurate measure of desirability in an uncertain environment, although in the case of limited capital the so-called gambler's ruin phenomenon must be considered. We suggest further reading, beyond this introduction to uncertainty, in the literature cited.

There is rapidly evolving, through the interaction of statisticians, chemists and chemical engineers, an area of research which goes under the names of experimental design, model building, and model discrimination. This offers hope in the troublesome problems of uncertainty and in the improvement of the quality of design data. The general line of development involves the postulation of a number of rival models to describe the nature of the process, using the limited and uncertain data which are available. And then, through this developing theory, it is possible to determine which experiments would be most likely to reduce uncertainty and to discriminate among the models. We expect within a few years that there will evolve guidelines which the engineer can use to gather best data about a design through further computation or by means of experimental pilot plant studies. We refer to some of the work in this area at the end of the chapter.

References

There are a number of textbooks available on decision making in the face of uncertainty:

R. D. Luce and H. Raiffa, *Games and Decisions*, Wiley, New York, 1957.

R. Schlaifer, *Probability and Statistics for Business Decisions*," McGraw-Hill, New York, 1959.

H. Chernoff and L. E. Moses, *Elementary Decision Theory*, Wiley, New York, 1959.

See also:

D. B. Hertz, "Risk Analysis in Capital Investment," *Harvard Business Review*, No. 1, **42**, 1964.

P. E. Green, "Risk Attitudes and Chemical Investment Decisions," *Chem. Eng. Progr.*, No. 1, **59** (1963).

P. E. Green, "Decision Making in Chemical Marketing," *Ind. Eng. Chem.*, No. 9, **54** (1962).

H. M. Hawkins and O. E. Martin, "How to Evaluate Projects," *Chem. Eng. Progr.*, No. 12, **60** (1964).

H. H. Isaacs, "Sensitivity of Decisions to Probability Estimation Errors," *Operations Res.*, **11** (1963).

P. S. Buckley, "Sizing Process Equipment by Statistical Methods," *Chem. Eng.*, **57**, 9 (1950).

For information on the developing area of experimental design and model building see:

G. E. P. Box and W. G. Hunter, "The Experimental Study of Physical Mechanisms," *Technometrics*, No. 1, **7** (1965).

W. G. Hunter and A. M. Reiner, "Designs for Discriminating between Two Rival Models," *Technometrics*, No. 3, **7**, (1965).

W. G. Hunter and M. L. Hoff, "Planning Experiments to Increase Research Efficiency," *Ind. Eng. Chem.*, **59**, 3 (1967).

J. R. Kittrell, R. Mezaki, and C. C. Watson, "Model-Building Methodology for Heterogeneous Kinetics," *Brit. Chem. Eng.*, *II*, No. 1, **2** (1966).
J. R. Kittrell, R. Mezaki, and C. C. Watson, "Reaction Rate Modelling in Heterogeneous Catalysis," *Ind. Eng. Chem.*, March 1967.
G. E. P. Box and W. J. Hill, "Discrimination among Mechanistic Models," *Technometrics*, No. 1, **9** (1967).
W. G. Hunter, J. R. Kittrell, and R. Mezaki, "Experimental Strategies in Mathematical Modelling," *Trans. Brit. Chem. Eng.*, May 1967.

PROBLEMS

12.A. You have been given the responsibility of ordering a large amount of fertilizer for resale to farmers during the next growing season. The farmer's demand can only be estimated to be between $D_L = 50{,}000$ tons and $D_H = 150{,}000$ tons. You can purchase fertilizer in the bulk at $C_p = \$80/\text{ton}$, to be sold later to the farmers at $C_s = \$110/\text{ton}$. However, should the farmers' demand exceed the original order, the demand must be satisfied by purchase from competitors at $C_D = \$120/\text{ton}$ and sale to the farmers at \$110/ton; you take a loss on those sales. Further, should there be any unsold fertilizer left at the end of the season, it must be disposed of by sale to your competitors at $C_c = \$30/\text{ton}$; no over-winter storage is allowed.

How much fertilizer should you purchase in the face of this uncertain demand?

Answer:

$$D^* = \frac{D_L(C_p - C_c) + D_H(C_D - C_p)}{C_D - C_c} \text{ tons}$$

12.B. A plant is to be built to recover a valuable material A from a waste product. The amount of feed available to the plant is uncertain and may be anywhere from zero to 20,000 tons per year of material A.

The investment required for the plant will be \$10 per ton of A per year recovered, and the total charges associated with investment will amount to 40 per cent per year of the total investment. The net sales income will be \$12 per ton of A recovered and the operating costs \$2 per ton recovered. What plant capacity do you recommend?

Answer: 12,000 tons per year

12.C. A waste treatment plant is to be designed with the hope of obtaining \$1.00 per ton of waste treated by the sale of the recovered material. However, the amount of waste available is uncertain, falling somewhere in the range of 100 to 300 tons per day. If a plant is constructed with insufficient capacity the excess waste must be disposed of at a cost \$0.10 per ton. The operating cost is \$0.20 per ton treated. The investment cost of the waste treatment plant is \$1.00 per ton year of capacity and the total cost per year associated with the fixed investment is \$0.30 per dollar invested. What size plant is recommended?

12.D. A seasonal product is to be manufactured for sale during a short period of annual demand. It is to be purchased and stored at a cost of C dollars per ton. The product can be sold for S dollars per ton and unsold material has no value after the season. Both the selling price S and the demand D tons are uncertain with rectangular distributions. S may range between S_l and S_h dollars per ton, while D may range between D_l and D_h tons. How many tons of this material should be ordered for maximum expected profit?

$$\text{Answer: } Q^* = D_h - \frac{2C(D_h - D_l)}{S_h + S_l}$$

12.E. The possibility exists that a new project may fail completely, and this is empirically accounted for in Table 4.6-1 by charging new and risky projects a higher minimum acceptable rate of return on invested capital. Let p_f be the estimated probability of project failure. If the project succeeds it will earn at a rate of $S - C - (S - C - dI)t - eI - iI$ dollars per year, and if it fails, $S - C = 0$ in the previous expression for there will be no products made. Derive the risk factor in Table 12.8-1 which compensates for the failure possibility.

12.F. Derive the proper risk factors to account for uncertainty in manufacturing costs, investment, and project life as discussed in Section 12.8.

12.G. We are about to design a vapor recovery unit to condense ammonia vapors that will be boiled off a proposed cryogenic storage tank for 60,000 tons of liquid ammonia. Since the storage tank has not yet been constructed, we can only speculate on the load on the vapor recovery unit. The ammonia is to be stored in an out-of-the-way area where any unrecovered ammonia vapor can be dissipated without danger of air or water pollution.

The vapor boil off rate is estimated to be within the range of 20 to 80 tons of ammonia per day, with ammonia worth \$80/ton. The investment in a vapor recovery unit depends on its capacity and is to be amortized at 10 per cent per year.

$$I = I_B\left(\frac{Q}{Q_B}\right)^{0.6}$$

$$I_B = \$50{,}000$$

$$Q_B = 20 \text{ tons/day}$$

The operating cost of the recovery unit is estimated to be \$5/ton of ammonia condensed.

What size recovery unit do you recommend, in tons of ammonia condensable per hour?

12.H. A catalytic reactor must be designed to produce feed for a large process at a rate $Q = 300$ tons per day. Only a rough estimate is available on the actual activity of the catalyst in the reactor, the uncertainty residing in the effects of channeling, heat transfer, and residence time distributions within the reactor bed. The activity limits are $A_L = 10$ tons product/day-ton of catalyst and $A_H = 30$ tons product/day-ton of catalyst. The cost of the reactor has been estimated as

$$C_R = C_B\left(\frac{D}{D_B}\right)^{0.6} \text{ \$/day}$$

where C_R includes operating costs and amortization of investment.

$$C_B = \$2{,}000/\text{day}$$
$$D_B = 10 \text{ tons catalyst}$$
$$D = \text{tons of catalyst to be used}$$

Any excess production, above 300 tons per day, can be sold at $C_e = \$20$/ton, and extra feed must be purchased at $C_p = \$150$/ton should the reactor be inadequate.

How much catalyst should be put in the reactor?

12.I. You are in charge of a process. Marketing informs you that they can sell product A at a good price for a short period of time. You have equipment available, and the only expense would be in catalyst which can selectively produce product A. Research has developed three catalysts; one will produce only A, and the other two will produce B by a side reaction. Which catalyst should you invest in, in view of the fact that marketing is not sure how long they can sell product A?

The following data are available.

Catalyst	Cost \$/pound	Pounds catalyst required for 834 lb/hr of feed
1	50	500
2	25	700
3	10	1,000

The existing reactor will hold 2,000 pounds of solids and must be full to operate. A special inert material is available at \$2.00 per pound to fill the voids not occupied by catalyst. The feed rate to the reactor is fixed at 834 pounds per hour. The value of the feed and products, which includes operating costs, follow.

	\$/lb
Feed	0.04
A	0.10
B	0.03 (fuel value)

The yields of the several catalysts are:

Catalyst	Per cent yield at 500°F	
	A	*B*
1	100	0
2	75	25
3	60	40

Marketing also tells you that the longest period they can foresee for selling product *A* is 60 days, with a 30 per cent chance it will sell for 60 days, a 50 per cent chance it will sell for 40 days, and a 20 per cent chance it will sell for 20 days. You cannot recover your investment in catalyst at the end.

12.J. A 13,500 barrel per day hydrocracker has been built for processing crude oil. Expansion studies are under way to expand the process to 25,000 barrels per day. The forecast of the available inplant hydrogen to meet this expansion is uncertain: the lower limit is 3×10^7 SCF per day and the upper limit is 4×10^7 SCF per day. The hydrogen consumption in the hydrocracker is 2,000 SCF per barrel cracked of crude oil processed. The cost of building a new hydrogen generating plant with a capacity of 10^6 SCF per day is \$1,500,000, and the value of the hydrocracker product is \$1.00 per barrel above feed and manufacturing costs. If the hydrogen plant is underdesigned, it can be expanded in one year at a cost of \$2,000,000 for a 10^6 SCF per day expansion. What size hydrogen plant do you recommend to supplement the uncertain inplant hydrogen supply?

12.K. Figure 11.2-2 is useful in determining the initial plant capacity to accommodate optimally to an expected forecast increase in demand. Suppose that the forecast demand is uncertain and that this uncertainty resides in the factor a, which falls somewhere in the range $\bar{a} - \Delta \leq a \leq \bar{a} + \Delta$. Determine the overdesign factor $f = Q'/Q$, where Q' is the initial capacity recommended in the face of this uncertainty, and Q is the capacity recommended assuming the demand slope $\bar{a}$. Assume that after the first plant is built all uncertainty is swept away. Can we assume that the initial capacity and the capacity of all expansions will be equal?

12.L. Determine the optimal overdesign factor to apply to Fig. 11.3-1 in the case where initial demand falls somewhere in the range $\bar{D}_0 - \Delta \leq D_0 \leq \bar{D}_0 + \Delta$. No other uncertainty is assumed to exist.

13

FAILURE TOLERANCE

The life of a processing system will not be trouble free, and the occasional failure of a component is to be expected. The process engineer has the responsibility of designing systems in which local failure is unlikely to trigger chains of events leading to system failure and even to disaster. In this chapter, we are concerned with the detection of failure prone systems and with the strategy for engineering around such difficulties. A tolerance to failure must be built into the design of every system.

Typical Problem

A tentative design of an acid purification plant has been completed, optimistic in the sense that complete reliability of all components has been assumed. Can the very real possibility of local misoperation lead to system failure? How might we engineer such characteristics out of the system?

13.1 INTRODUCTION

In the preceding chapter we discussed methods for overcoming the uncertainty in critical bits of information upon which a design must be based. A failure to account properly for such a lack in quality of information might result in the design of an economically inefficient process. In this chapter we investigate an even more critical kind of uncertainty which might lead to the complete destruction of an improperly designed system. These are the uncertainties which express themselves by the sudden and complete failure of parts of a system.

There exists a spectrum of modes of failure, ranging from failures which cause only minor reductions in efficiency, to those that lead to the violent destruction of the system. For example, the failure of a feed pump to a reactor may require the shutdown of the system while a new pump is put into operation, perhaps just a minor but costly inconvenience. On the other hand, a pump failure may be the start of a chain of events which leads directly to an explosion in the system and a major catastrophe.

It is not at all accidental that the majority of the processing systems created in industry eventually function. Systems for which the lack of reliability is too great to tolerate never reach the stages of construction and operation, since the successful engineer, as an experienced gambler, rarely enters into risks unless the odds are clearly favorable. However, in certain situations it will be economically feasible to engineer a design which contains a processing component which can fail, as long as the periods of successful operation are sufficiently profitable and the mode of failure does not constitute a safety hazard.

The strategy of engineering the optimal failure tolerance into a design involves a blend of the following concepts:

1. Each component in the system must be designed to be sufficiently reliable to reduce the chances of the initiation of a failure in the system.
2. Should a component be thought to be failure prone, the system must be designed to be tolerant to the failure and to reduce the chances of the failure propagating further.
3. The system should be designed so that the devastation resulting from unforeseeable and uncontainable failures is minimized.

13.2 CATASTROPHIC RESULTS FROM MINOR EVENTS

The instability of a house made from playing cards is obvious. The falling of one card initiates a chain of events which eventually can lead to complete collapse. A few well placed drops of glue or a different card orientation can isolate the weak areas and prevent the propagation of this failure through the cards, thereby imparting a degree of stability to the house.

This instability also occurs in process systems, and for the purposes of orientation we cite an actual example—the well documented failure of a hydroformer and subsequent fire in Whiting, Indiana, in August 1955.[1] Notice how the minor failure of a valve propagated into a major conflagration eventually encompassing a goodly portion of one of our major oil processing complexes.

At the time of the initial failure, operating personnel were starting the hydroformer following a shutdown for maintenance and repair. No catalyst was in the system, and the inert gas used to heat up the recycle system was in contact with naphtha in two vessels, the high pressure quench drum and the high pressure separator. The inert gas was contaminated with pentanes and hexanes

[1] From a report presented during the 36th Annual Meeting of the American Petroleum Institute, and published in J. C. Ducommun, *Process Safety*, Volume I, Copyright ©, American Oil Company, Chicago, 1966.

and was circulated through the recycle system. This in itself would not have been dangerous were it not for the prior failure of the closure of a valve on a recycle line, thereby allowing the leakage of air into the system.

We now trace the propagation of this minor failure of a valve to close.

At 6:12 A.M. there was a loud thud, followed by a sheet of flames which surrounded the reactor, the second step in the propagation of the disaster. Then, there were two loud sharp explosions which seemed to peel the reactor apart.

The 600-ton reactor shell broke into 13 major-sized pieces ranging from 3 to 136 tons, scattering debris over a quarter mile radius. A 60-ton fragment of the reactor was found 1,200 feet from the site of the initial explosion. The separator broke into 29 pieces.

The mechanical damage to the surrounding system from flying pieces of the hydroformer was not extensive and had the propagation of failure been stopped at this point relatively minor damage would have resulted.

The propagation did not terminate, unfortunately, the fragments having severed pipelines and damaged storage tanks. Gasoline, oil, and other dangerous materials gushed out. Workmen attempted to divert the flood into the sewers with sand dikes. Within a few minutes flash fires began and a perimeter of fire engulfed 17.5 acres of the processing complex.

After two hours, the fire area had increased to 20 acres; and to 42 acres by noon. Additional ignition and storage tank boilovers occurred, increasing the fire area to 47 acres by 4:00 P.M. It was impossible to extinguish the fire, it could only be contained.

A minor event, the air leakage, had triggered a major disaster. In theory, it is possible to trace through a definite chain of events which led directly to disaster.

What was wrong with the system design, for obviously a drastic and perhaps subtle error must have existed for such devastation to be initiated?

1. The valve arrangement was not sufficiently reliable nor were the operating instructions sufficiently foolproof to prevent the entrance of air into the vessels during start-up.
2. The system itself was not sufficiently tolerant to the fire and detonations which resulted from the failure of the valve. The scattering of debris propagated the failure far beyond the initial site of the failure. Tolerance to detonations is very difficult to provide in a system. Rupture disks and safety valves do not respond quickly enough and the directional nature of shock load is difficult to design for.
3. Certainly the way in which the debris was scattered could not have been predicted; however, the surrounding processing complex might have been designed so that the destruction caused by this unpredictable and then uncontrollable failure was not so complete.

Why then are systems designed and constructed to contain these chains of events which potentially lead to the catastrophic failure? The answer is that hindsight has twenty-twenty vision, and that the vision into the future is clouded by the complexity of the system being engineered. Many of the disasters are fully capable of being anticipated before the fact but are not foreseen, merely because the dangerous chains of events are deeply imbedded and hidden in the thousands of innocuous events which occur during the normal operation of the system.

Successful operation of a system over a period of time tends to lull us into a false sense of security. A near disaster in a system tends to point to weaknesses which must be remedied. But near disaster is a rather inefficient means for detecting the disastrous chains of events. A good example of an improbable but yet completely predictable chain of events is afforded by a disaster that occurred several years ago on the Gulf coast. An overhead crane was used in a plant to lift heat-exchanger tube bundles from their shells for cleaning. During a hurricane the winds caught the crane and carried it down its track where it jumped over an abutment. The crane landed on a pump and broke a line which was carrying butane. The butane was released and eventually exploded, leading to a further chain of events too complicated to follow. This disastrous result was probably considered a freak accident, yet might well have been predicted, and thus avoided. In fact, it *should* have been avoided.

These chains of events *must* be detected and broken. It is common practice, for example, to physically isolate the storage tanks for inflammable materials from the processing systems, surround the storage area with earthen dikes, and equip the lines that connect the storage tanks to the processing system with the proper valves and flame arresters to prevent the propagation of any fires. Another convention involves the practice of placing downwind from any source of ignition processing components which might issue an explosive gas. Pressure disks might be used to break a chain of events by allowing the contents of a vessel to spew out into an unoccupied area should a violent excursion in operating pressure occur.

Knowledge in this critical area of process engineering is essentially empirical, and the theory discussed near the end of the chapter is only now entering the outskirts of failure tolerant engineering.

13.3 PRELIMINARY FLOW SHEET REVIEW

In this section we illustrate the strategy of checking a proposed design for sensitivity to component or operator failure. This is the time when the process engineer should assume that nature and mankind operate in direct opposition to the success of the system. Each aspect of a system must be tested and strained mentally, to reveal weaknesses. The process must be started up, operated, and shut-down on paper under a variety of conditions.

Consider the design of a nitric acid purification plant in Fig. 13.3-1. Impure acid is fed from a storage tank (D-101) by an automatically controlled pump to a flash tower (V-101). The flash tower is heated with steam, and the vapors issuing therefrom are condensed (E-101) as the pure nitric acid product to be stored (D-103) for use in the next part of the processing system.

Table 13.3-1 Operating Procedure for Acid Purification System Start-Up

1. Check that all valves are closed with the exception of vent valves on D-101, D-102, and D-103.
2. Open the drain valve at V-101 and drain out all waste.
3. Close the valve opened in 2.
4. Open the following:
   (i) Valve downstream from D-101
   (ii) Valves immediately upstream and downstream from P-101
   (iii) Valve immediately upstream from V-101
   (iv) V-101 overhead valve
5. Set P-101 at 0.858 gallons per minute until normal liquid level is obtained in V-101, as noted on LG-103.
6. Shut down P-101.
7. Fully open cooling water valves upstream and downstream from E-101.
8. Open steam valve until flow rate on FR-102 is approximately 360 lb/hr. (This gives a boilup rate of 300 lb/hr.)
9. Vent noncondensables by opening vent valve upstream from steam trap.
10. Close the vent valve on the steam trap.
11. When TI-101 indicates a temperature of 265°F, open valves upstream and downstream from the level control valve.
12. Start P-101, and initiate automatic level controller, LC-101, and level alarm, HLA-102.
13. Manually control cooling water flow to E-101 by maintaining TI-103 less than 100°F and by keeping cooling water outlet temperature on TI-104 less than 130°F.

A portion of the proposed operating procedure is included in Table 13.3-1. We wish to detect any sensitivity of the system to equipment or operating failure and modify the design to break any chains of events which might lead to a system failure.

There exists an obvious weakness in the system which the reader can discover after an examination of the following description of an event which might occur during process start-up.

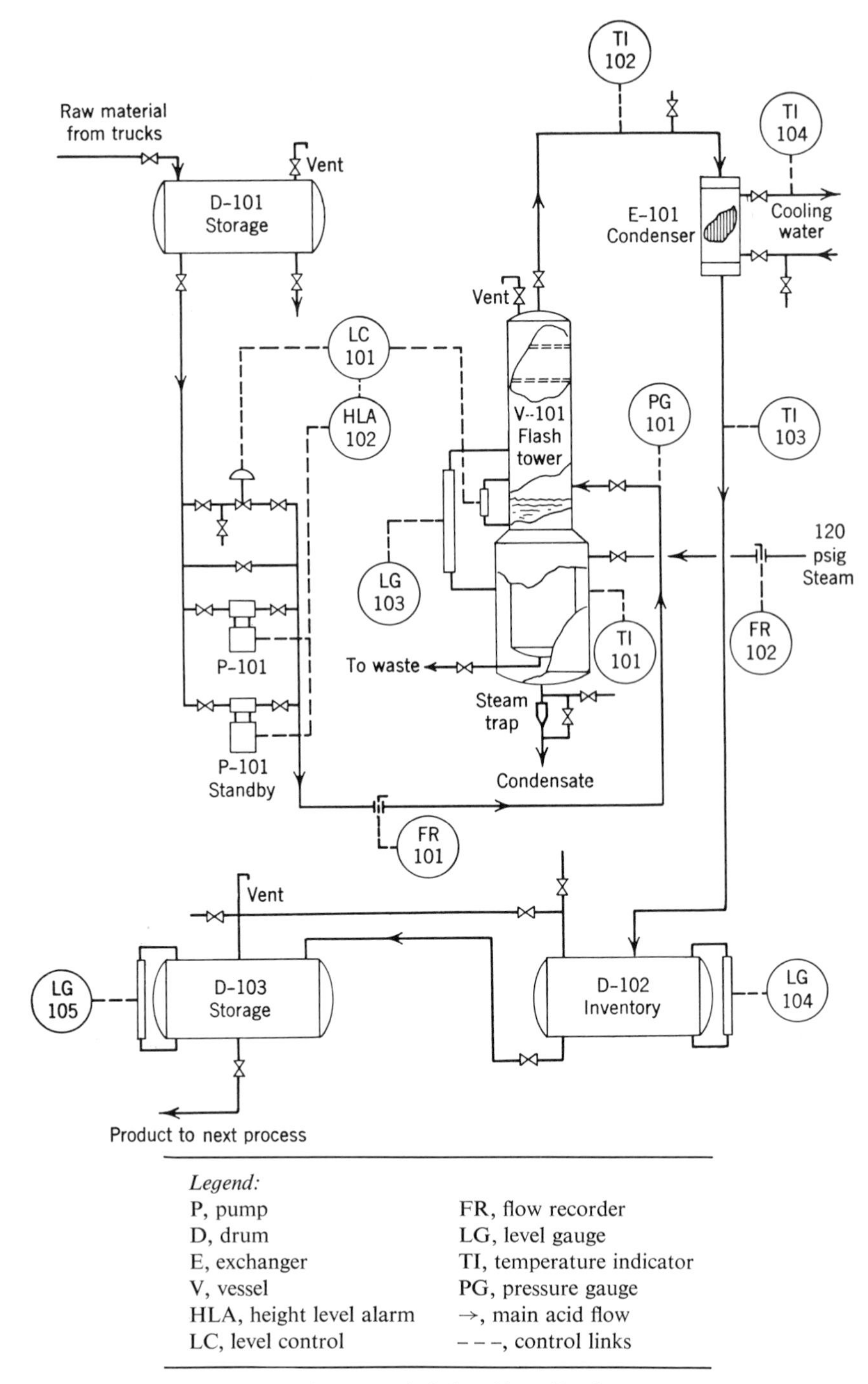

Fig. 13.3-1 A proposed nitric acid purification system.

Suppose that the processing system has been constructed and has been operating successfully and profitably for quite a while. Suddenly the engineer is called to the phone, and hears on the other end of the line:

There is trouble in the nitric acid purification system! We had shut the system down for some minor repairs and when we started it up again there was an explosion which filled the building with acrid orange-yellow gas. The emergency shutdown procedures were followed and the immediate area has been evacuated. What caused the explosion and what can we do to prevent its recurrence?

We ask the reader to discover the source of such an incident. The acrid orange-yellow gas is doubtless some mixture of nitrogen oxides formed by the overheating of the acid in the system. The explosion was probably caused by a build-up of pressure in the system. And the fact that the failure occurred during the start-up, after repairs, indicates that the system might not have reached steady operation. As a further hint we suggest that the failure might have been caused by an operator's error, a mispositioning of a valve.

How might this sensitivity to operator error be removed from the system by a simple design modification? The primary responsibility for this error in operation rests with the designer of the process; he failed to design a fail-safe system for the operators. Locate two other weaknesses of this design. A standard of practice in the petroleum industry is that a process must be tolerant to any single operation error but need not be designed to be fail-safe to two or more simultaneous operating errors. The chances of the latter occurring are considered low.

A considerable body of empirical knowledge has evolved which will be of use in the check out of the flow sheet of a proposed process, and in this text we allude to the existence of this information and offer references for further study at the end of the chapter.

Armistead[2] classifies the direct causes of destructive accidents in the petroleum industry, principally involving fire, in the order of severity of damage as follows.

1. *Misoperation or improper practices.* Mistakes in valve operation, opening or closing wrong valve, improper use of open flames, and so forth.
2. *Equipment failure.* Piping ruptures, vessel ruptures, pump and machinery failures resulting in ignition, and so forth.
3. *Repairing of equipment when operating.* Flow from disconnected piping, breaking of pipe or fittings, fracture of equipment under pressure, and so forth.

[2] G. Armistead, Jr., *Safety in Petroleum Refining and Related Industries*, 2nd ed., Simmonds, New York, 1959.

4. *Lightning, windstorm, and other effects of the elements.* Tank fires, flood, tank collapse, damage to structures and buildings, and so forth.
5. *Improper equipment and other.* Unforeseen action of new operations with equipment, electrical equipment igniting combustion, and so forth.

Having been alerted to the kinds of failures which have had disastrous effects in the past, we are in a better position to reduce the chances that these failures will appear to devastate a new design.

Hudson[3] suggests that the following might be among the important points to check in a design.

Reaction sections

Reactions must be in the stable range of composition, temperature and pressure.

Side reactions which might produce poisonous or explosive material or might lead to dangerous fouling must be controlled.

Poor distribution of reactants which might lead to hot spots resulting in equipment failure or to undesirable side reactions must be avoided.

Proper heat-transfer capability must be available for normal operation and to control any excursions caused by shutdown, start-up, or other variations in operation.

Emergency blowdown tanks, flare systems, and venting must be provided for.

Separation sections

Crushing, grinding, and size separation can lead to dust, with the possibility of explosive hazard, the acceleration of reactions, or the plugging of equipment.

Distillation systems may be endangered by the build-up of dangerous material during recycle, by polymerization in reboilers, or by the sudden and violent vaporization of water inadvertently left in the tower.

Extraction and absorption systems may contain large amounts of poisonous or flammable liquids.

Drying operations may involve fire and explosive hazard due to the production of dusts or vapors.

Materials handling

Proper piping and valving is of extreme importance. Relief valves must be provided on all positive acting pumps and compressors and wherever high excursions in pressure might be experienced. Equipment must be able to be isolated during start-up and shutdown or during emergency operation.

Mechanical conveyors may generate dust. Pneumatic conveyors handling flammable material should operate with inert gas under pressure to prevent the inclusion of air. Recovery of fines in the exit carrier gas must be considered.

[3] W. G. Hudson, "Process Design" in *Safety and Accident Protection in Chemical Operations*, H. H. Fawcett and W. S. Wood, eds., Wiley, New York, 1965.

Storage sections
Storage must be provided to smooth process upsets, contain dangerous products, and store "off-spec" materials, as well as for the normal operation of the process.

Purge systems of CO_2 or N_2 might be needed for the storage of flammable solids.

The storage of highly flammable liquids may require inert gas purging of vapor spaces, adequate venting, and provision for the disposal of spillage, for example, by diking.

Gas storage offers the same problem as liquid storage, with the added hazards resulting from the diffusion of gases to distant sources of ignition and so forth.

The activity discussed briefly in this section deals with methods for reducing the chances of a failure being initiated. In addition to the empirical guide lines alluded to here, a substantial body of design codes and standards have been established in certain areas of manufacture. For example, the American Society of Mechanical Engineers has laid out detailed specifications for the safe design of pressure vessels. The design of pressure vessels should be considered a specialty, which is best left to the qualified specialist. The process engineer should be familiar with the general provisions of the codes, however, so as to be warned against making alterations which would seriously weaken the vessels. We cannot undertake a survey of this body of knowledge here, but we include references to sources of more detailed information at the end of the chapter.

Unfortunately, failures will occur even in the most carefully designed systems. For this reason, the engineer must build into process designs a degree of immunity and tolerance to the propagation of failure. This will be examined in the next section.

13.4 RELIABILITY UNDER EXTREME CONDITIONS

Hopefully, a processing system will not be disaster prone under normal operating conditions. Of equal importance is an immunity to disaster under reasonably extreme conditions which are not part of the normal operating plan. This double level of security is necessary and often can be achieved by relatively simple changes in the design. A common strategy involves mentally testing the response of each component to harsh environmental conditions to make certain that the components cannot contribute to the propagation of a disaster.

To illustrate this kind of thinking, we shall examine the American Petroleum Institute Recommended Practice for the Design and Installation of

Table 13.4-1 Operational Difficulties and Required Relief Capacities

Item Number	Condition	Required Relief Capacity	
		Relief Valve for Liquid Relief	Safety Relief Valve for Vapor Relief[a]
1	Closed outlets on vessels	Maximum liquid pump-in rate	Total incoming steam and vapor, plus that generated therein under normal operation.
2	Cooling-water failure to condenser	—	Total incoming steam and vapor, plus that generated therein under normal operation, less vapor condensed by sidestream reflux
3	Top-tower reflux failure	—	Total vapor to condenser.
4	Sidestream reflux failure	—	Difference between vapor entering and leaving section.
5	Lean-oil failure to absorber	—	None.
6	Accumulation of noncondensables	—	Same effect in towers as for items 2 and 8 in other vessels.
7	Entrance of highly volatile material:		
	a. Water into hot oil	—	For towers—usually not predictable.
	b. Light hydrocarbons into hot oil	—	For heat exchangers—assume an area twice the internal cross-sectional area of one tube so as to provide for the vapor generated by the entrance of the volatile fluid.
8	Overfilling storage or surge vessel	Maximum liquid pump-in rate	
9	Failure of automatic controls:		
	a. Tower pressure controller, to closed position	—	Total normally uncondensed vapor.
	b. All valves, to closed position, except water and reflux valves	No operational requirement	No operational requirement.

10	Abnormal heat or vapor input:		
	a. Fired heaters or steam reboilers	—	Estimated maximum vapor generation including noncondensable from overheating.
	b. Split reboiler tube	—	Steam entering from twice the cross-sectional area of one tube.
11	Internal explosions	—	Not controlled by conventional relief devices but by avoidance of circumstances.
12	Chemical reaction	—	Estimated vapor generation from both normal and uncontrolled conditions.
13	Hydraulic expansion:		
	a. Cold fluid shut in	Nominal size	
	b. Lines outside process area shut in	Nominal size	
14	Exterior fire	—	Estimate by the method given in Sect. 6, API RP 520.
15	Power failure (steam, electric or other)	Study the installation to determine the effect of power failure, and size relief valve for the worst condition that can occur.	
	a. Fractionators	—	All pumps could be down with the result that reflux and cooling water would fail. Size valves as in Item 2.
	b. Reactors	—	The agitation or stirring would stop and the quench or retarding stream would fail. Size valves for product generation from a runaway reaction.
	c. Air-cooled exchangers	—	Fan failure. Size valves for the difference between normal and emergency duty.
	d. Surge vessels	Maximum liquid inlet rate	

[a] Consideration may be given to the suppression of vapor production as the result of the valve's relieving pressure being above operating pressure, assuming constant heat input.

Pressure-Relieving Systems in Refineries.[4] This document sets forth design practices for the localization of modes of refinery failures which result in the build-up of pressure in a vessel.

Table 13.4-1 outlines the extreme conditions that might cause the build-up of pressure in a process vessel and indicates the rates at which liquid and vapor might have to be released from the vessel to prevent bursting. For example, item 15 is the extreme condition of *power failure* within the refinery. If the vessel under consideration is a fractionator, all pumps would be down with the result that the reflux cooling water supply would fail, and the release device (say, a rupture diaphragm) should be sized as in item 2, *cooling-water failure to condenser*. The recommendation is then to provide for vapor release at a rate equal to the total incoming steam plus vapor, plus that generated during normal operation, less vapor condensed by side stream reflux.

Let us now examine in detail the abnormal condition described by item 14, *exterior fire*. It is necessary to determine the relieving capacity necessary to prevent excessive pressure build-up. It is recommended that the amount of heat absorbed by the vessel be estimated by Eq. 13.4-1 (Section 6, API RP 520) which is a summary of tests on vessels during experimental fire conditions.

$$Q = 21{,}000\ FA^{0.82} \tag{13.4-1}$$

Q = total heat input to vessel, Btu/hr
A = total wetted surface area of vessel, ft^2
F = environment factor for various installations, Table 13.4-2

It is recommended that the wetted surface area includes that wetted area within 25 ft above the base of the surrounding fire, taking into account the kinds of vessel as follows.

1. *Liquid-full vessels* (such as treaters) operate liquid full; therefore, the wetted surface would be the total vessel surface within the height assumed to be affected by a fire.
2. *Surge drums* (*vessels*) usually operate about half full, therefore the wetted surface would be calculated at 50 per cent of the total vessel surface.
3. *Knockout drums* (*vessels*) usually operate with only a small amount of liquid; therefore, the wetted surface would be in proportion.
4. *Fractionating columns* usually operate with a normal liquid level in the bottom of the column and a level of liquid on each tray. It is reasonable to assume that the wetted surface be based on the total liquid in the bottom and in the trays within the height assumed to be affected by a fire.
5. *Working storage tanks*' wetted surface is usually calculated on the aver-

[4] API RI 520, September 1960, American Petroleum Institute, Division of Refining, New York.

Table 13.4-2 Environment Factor

Type of Installation	Factor F[a]
1. Bare vessel	1.0
2. Insulated vessels[b]. (The following arbitrary conductance values are shown as examples and are in British thermal units per hour per square foot per degree Fahrenheit):	
a. 4.0	0.3
b. 2.0	0.15
c. 1.0	0.075
3. Water-application facilities, on bare vessel[c]	1.0
4. Depressuring and emptying facilities[d]	1.00
5. Underground storage	0.0
6. Earth-covered storage above grade	0.03

[a] These are suggested values for the conditions assumed in Par. 6.2. When these conditions do not exist, engineering judgment should be exercised either in selecting a higher factor or in providing means of protecting vessels from fire exposure.

[b] Insulation shall resist dislodgement by fire hose streams. For the examples a temperature difference of 1,600°F was used. In practice it is recommended that insulation be selected to provide a temperature difference of at least 1,000°F and that the thermal conductivity be based on a temperature which is at least the mean temperature.

[c] See recommendations regarding water application.

[d] Depressuring will provide a lower factor if done promptly, but no credit is to be taken when safety valves are being size for fire exposure.

age inventory. This should be satisfactory not only because it conforms with a probability, but also because it provides a factor of safety in the time needed to raise the usually large volume of the liquid's sensible heat to its boiling point.

The excess amount of vapor that must be released due to the fire is computed by Eq. 13.4-2.

$$V_{\text{excess}} = \frac{Q}{\lambda} \tag{13.4-2}$$

where λ is the latent heat of vaporization of the liquid in the vessel.

For example, consider the fire surrounding the fractionating tower in Fig. 13.4-1 which is bare of insulation but which is provided with a safety device which applies a spray of water onto the tower in case of emergency. The tower is 4 ft in diameter and will be filled with liquid butane to 20 ft.

$$A = 4\pi 20 \approx 250 \text{ ft}^2$$
$$F = 1.0$$
$$\lambda = 100 \quad \text{Btu/lb} \quad \text{at 20 atm}$$

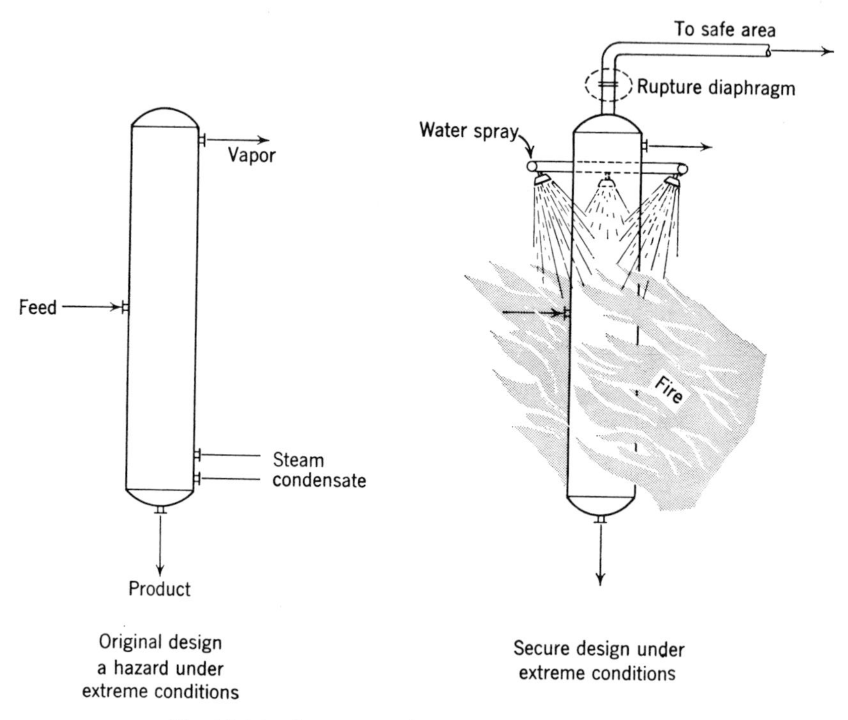

Fig. 13.4-1 A pressure release device for a butane still.

The comment on water deluge systems in the footnote to Table 13.4-2 is to the effect that the water deluge cannot be counted on to lessen the heat addition but only to prevent the high temperature weakening of the metal in the vessel.

$$V_{\text{excess}} = \frac{21{,}000(1.0)(250)^{0.82}}{100}$$
$$= 20{,}000 \text{ lb/hr}$$

Now in addition to this vapor excess, the excess vapor rates resulting from the other items in Table 13.4-1 might have to be included depending on the nature of the system. For example, a fire might cause the power failure in item 15 and hence the failure of cooling water.

The API recommendations then detail the designs of pressure relieving devices, such as rupture diaphragms, to accommodate to the vapor release rates. We shall not continue further in this text since the API recommendations are readily available.

It is interesting to note that the design of a pipeline system, to convey the vapor and liquids released during a disruption in a system to some safe area such as a flare tower, is in itself an extensive design problem. Such "safety design" problems are often as extensive and important as the design of the process.

13.5 SAFETY THROUGH PROPER LAYOUT

The experienced engineer knows full well that the best laid plans often go awry and that complete faith cannot be placed in any particular safety feature, such as, water flooding nozzles, rupture diaphragms, and quench pools. A final level of security is then built into the system by locating the components in the proper positions, relative to the other components in the system and to the surrounding community.

In this section we outline the practices of site selection, plant layout, and unit plot planning, as discussed by D. M. Liston[5] in an article of the same name.

The major hazards in chemical operations involve

Fire
Explosion
Toxic material

Any incidents involving these must be confined to the location in which they will do the least harm.

The processing units are the most dangerous parts of a chemical plant and should be removed from the plant boundaries and reasonably consolidated rather than scattered. The consolidation improves the identification of a hazardous area and allows a reduction in traffic passing through. These units should be downwind, both from major concentrations of people and from any sources of ignition, such as incinerators, open flames, or workshops. The actual winds may well vary in direction and we might use the direction of the prevailing winds to play the odds. This would minimize the adverse effects from the release of inflammable or toxic material.

Boiler plants and maintenance shops are major ignition sources, the latter a heavy concentration of personnel, and should be upwind of the processing area.

[5] D. M. Liston in *Safety and Accident Prevention in Chemical Operations*, H. H. Fawcett and W. S. Wood, eds., Wiley, New York, 1965, Chap. 5.

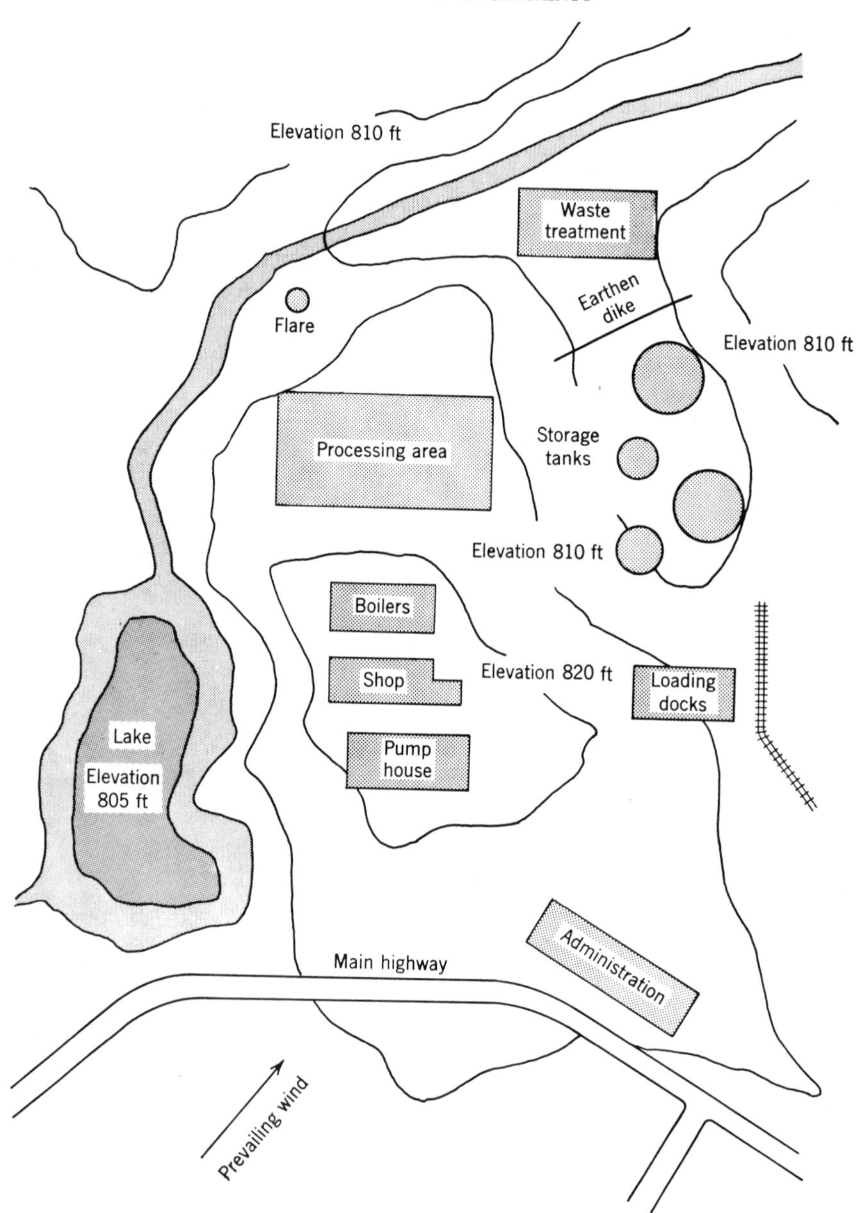

Fig. 13.5-1 A reasonable plant layout.

Loading docks should be downwind since spilling is likely to occur there. Waste water treatment systems are the ultimate destination of anything spilled in the plant and, hence, should be in a remote downwind location.

Flares and burning pots represent a source of ignition and should not be downwind. On the other hand, large amounts of combustibles might belch from a flare during an upset, and, hence, the flare should not be upwind of the plant. A side wind location seems the best compromise.

Storage tanks represent a tremendous storehouse of energy, and tankage should be relegated to its own area downwind from the plant. Guides exist for positioning tankage.[6]

Each processing unit should be surrounded by roadways and accessible from several directions. There should be several routes of egress for the operating personnel.

Often it is not possible to build the plant on a smooth flat location, and topographical considerations enter into the planning. Sources of flammable or toxic liquids should not be uphill from concentrations of people or ignition sources. The site may be subject to flooding, and, hence, boiler houses, electrical substations, and pumping stations should be on high ground to remain functional during an emergency. Perhaps we might consider a system of barriers and dikes to prevent the spread of fire from one part of the plant to another.

Figure 13.5-1 shows a reasonable plant layout. This is not the place to go into any more detail beyond the barest of introductions. The reader is referred to the literature at the end of this chapter for sources of further information and to the problems which draw upon this abundant literature. Keep in mind, however, that plant safety is a matter of common sense, good judgment, and experience, and that empirical rules-of-thumb do not apply in unusual situations, just when such guidance is needed most.

13.6 THE THEORY OF RELIABILITY

We now leave the real world with all its changing winds, sticky valves, corroding pipes, and frustrating exceptions to any rule and enter onto a brief excursion in the world of mathematics in which the rules to any game can be assumed to be followed with perfect accuracy. An understanding of an idealized world is often helpful in beginning to understand the real world, as long as we realize that essential differences exist between the two.

One goal, in this brief excursion, is to show how the mathematical theory of probability can be used to predict the optimal replication of failure-prone components.[7] While this *theory of reliability* is applicable only to certain classes of

[6] See G. Armistead, *op., cit.* and Section 13.9 of this text.

[7] See for further detail, D. F. Rudd, "Reliability Theory in Chemical System Design," *Ind. Eng. Chem., Fundamentals*, **1** (1963).

problems associated with failure, the mental processes involved are of sufficient generality to offer a valuable "way of thinking" even when numbers are not directly forthcoming. This theory has been developed quite extensively and has found application in the design of rocket and computer systems, as well as in the process industry.

We associate a probability of success, a reliability R, with each component in a batch or stage wise operation. This is a measure of the chance the component has of completing its task, perhaps estimated by the fraction of previous successes.

$$R = \frac{\text{number of past successes}}{\text{number of trials}} \tag{13.6-1}$$

Such a measure then describes the reliability characteristics of each component.

Now suppose that a system consists of N components connected in *series*, and that the failure of any one of the N components results in the failure of the system. The probability of the success of all the components is a measure of the system reliability. Assuming that the components fail independently of each other.

$$R_s = \prod_{i=1}^{N} R_i \tag{13.6-2}$$

For example, should the system consist of three operations, 1, 2, and 3, which have succeeded on 1/3, 1/2, and 3/4 of the past trials, the chances of the simultaneous success of all three is

$$R_s = R_1 R_2 R_3 = \left(\frac{1}{3}\right)\left(\frac{1}{2}\right)\left(\frac{3}{4}\right) = \frac{1}{8}$$

The reliability of a system with a series dependence in the operation of its components is less than or equal to the reliability of any of the components.

$$R_s \leq R_i \tag{13.6-3}$$

where i denotes any of the components in the system.

Now suppose that the reliability structure of the system is *parallel* rather than series, that is, only one of the N components need operate successfully to achieve the success of the system. All N components must fail to cause a system failure. Such a structure would arise when two pumps operate in parallel feeding a reactor. Both pumps must fail before the reactor has to be shut down.

The probability of failure of component i is

$$(1 - R_i)$$

The probability of the failure of all N components is

$$\prod_{i=1}^{N} (1 - R_i)$$

The probability of system success, the system reliability, is then

$$R_s = 1 - \prod_{i=1}^{N} (1 - R_i) \tag{13.6-4}$$

For example, should the three components above be placed in a parallel arrangement, rather than in a series arrangement, the system reliability would be

$$R_s = 1 - (1 - R_1)(1 - R_2)(1 - R_3) = 1 - \left(1 - \frac{1}{3}\right)\left(1 - \frac{1}{2}\right)\left(1 - \frac{3}{4}\right) = \frac{11}{12}$$

The reliability of a system with parallel structure is greater than or equal to the reliability of any of its components.

$$R_s \geq R_i \tag{13.6-5}$$

where i denotes any component in a parallel structure.

Therein lies the *strategy of redundant design*, the replication of failure-prone components to improve the reliability of a system. We nów show how these ideas could be used to estimate the economically optimal number of replicate batches to produce in order to improve the operation of a fine chemicals system.

13.7 THE OPTIMAL REPLICATION OF AN INTERMEDIATE REACTOR

Suppose now that a batch reactor used to produce a delicate chemical intermediate is known to fail frequently, producing off-quality material, thus disrupting the operation of a larger system which consumes the intermediate. Consider now the possibility of replicating this reactor so that several batches of the intermediate are started in the hope that at least one will reach completion and insure the smooth operation of the system. Clearly this is an economic problem in which we must balance the savings achieved by greater systems reliability against the cost of providing the replication of reactors.

Suppose that C_f \$/batch is the cost of a system's failure; i.e., the cost associated with not having a batch of the intermediate material available upon the demand of the system. Let C_R \$/batch be the cost of providing the reactor and reactants to attempt to generate a single batch of the intermediate. If N

reactors are used, where only one would be needed if the system were completely reliable, the cost of the reactor assembly is

$$NC_R$$

Now a systems failure occurs only when all of the N reactors fail, with a probability

$$(1 - R)^N$$

where R is the reliability of a single reactor. The *probable* cost of failure is then

$$(1 - R)^N C_f$$

and the total cost is

$$C_{\text{total}} = (1 - R)^N C_f + NC_R \tag{13.7-1}$$

The optimal replication can be estimated by differentiating the total cost with respect to N and setting the derivative to zero giving the following relationship

$$N^* = \frac{\ln\left\{\dfrac{C_R}{C_f \ln[1/(1 - R)]}\right\}}{\ln(1 - R)} \tag{13.7-2}$$

For example, suppose that the cost of a failure of the system is 100 times the cost of the reactor

$$C_f = 100C_R$$

and the reliability of the reactor is one half

$$R = \frac{1}{2}$$

The optimal number of reactors is then

$$N^* = \frac{\ln(1/100 \ln 2)}{\ln (1/2)} \cong 6$$

13.8 AN EXAMPLE OF STANDBY REDUNDANCY

We illustrate two points in this section. First, the concept of standby redundancy is introduced. Second, methods for analyzing reliability structures more elaborate than simple parallel and serial are presented.

Consider the five-step batch process shown in Fig. 13.8-1 which produces a very delicate fine chemical product. Three raw materials are reacted in separate vessels, producing unstable intermediate materials which then react in a prescribed manner in other vessels to form the final product. Due to difficulties in control, frequent batches of the intermediates are off quality, as summarized in Table 13.8-1. This disrupts the operation of the system.

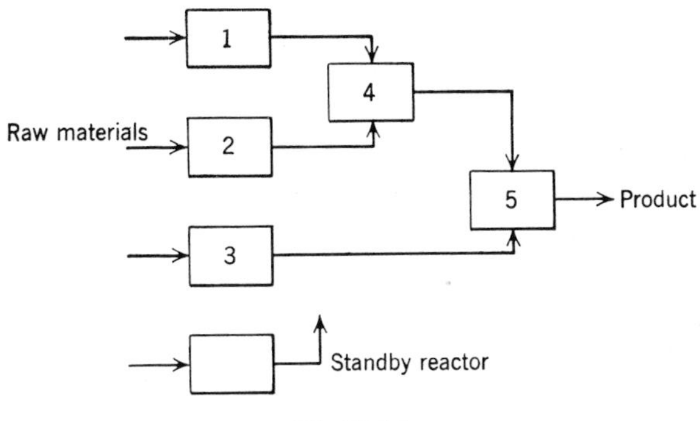

Fig. 13.8-1

Table 13.8-1

Reactor	Fraction of Failures	Reliability
1	1/2	1/2
2	2/3	1/3
3	1/5	4/5

This system has a serial reliability structure; the failure of any one of the reactors leads to a system failure. The reliability of the initial system is

$$R_s = R_1 R_2 R_3 = \left(\frac{1}{2}\right)\left(\frac{1}{3}\right)\left(\frac{4}{5}\right) = \frac{2}{15} = 0.134$$

We would expect that 13 out of every 15 trials would lead to failure.

Now suppose that it is possible to detect an incipient failure early enough to call on a standby reactor to take the place of the failing reactor. Should the standby succeed in producing the desired material on time, no system failure will occur. We wish to predict the increase in productivity which might accompany the use of such a standby reactor.

Suppose that the component failure is chemical rather than mechanical and that the standby reactor assumes the reliability of the reactor it replaces.

The reliability analysis proceeds as follows:

1. List all independent chains of events that can lead to a system success.
2. Determine the probability of each chain.
3. Sum these probabilities to obtain the probability that the system succeeds by any one of the ways.

The complete list of mutually exclusive events that can lead to a system success is shown in Table 13.8-2, with the corresponding probabilities. For

Table 13.8-2

Chain of events	Probability
1. No failures	$R_1R_2R_3$
2. Reactor 1 fails and standby succeeds	$(1 - R_1)R_2R_3R_1$
3. Reactor 2 fails and standby succeeds	$R_1(1 - R_2)R_3R_2$
4. Reactor 3 fails and standby succeeds	$R_1R_2(1 - R_3)R_3$

example, the sequence of events, reactor 1 fails and the standby reactor succeeds, leads to a system success, whereas any sequence involving the failure of two or more reactors cannot lead to success.

The reliability of the system with a standby reactor is the sum of the probabilities.

$$\begin{aligned} R_s &= R_1R_2R_3 + (1 - R_1)R_2R_3R_1 + R_1(1 - R_2)R_3R_2 + R_1R_2(1 - R_3)R_3 \\ &= R_1R_2R_3[4 - (R_1 + R_2 + R_3)] \end{aligned}$$

In the example problem

$$R_s = \left(\frac{1}{2}\right)\left(\frac{1}{3}\right)\left(\frac{4}{5}\right)\left[4 - \left(\frac{1}{2} + \frac{1}{3} + \frac{4}{5}\right)\right] = 0.316$$

The standby reactor will probably increase the productivity of the system by a factor of about two and one-half. An economic analysis can now be made to see if this increase justifies the investment in the additional reactor.

13.9 THEORETICAL STUDIES OF DISASTER PROPAGATION

In the preceding sections we have investigated means for determining process design characteristics which will result in an economically optimal tolerance to certain kinds of failures. We now show how a theoretical study of disaster propagation can be used to determine design characteristics which will result in a greater tolerance to the propagation of explosive-like violence.[8] The theories of disaster propagation are now being developed and have only led to a clearer understanding of certain limited but important aspects of disaster susceptability.

We begin by posing a very simple problem of storing Q tons of a material which can undergo explosive detonation in a narrow storage area of length L feet. How best can the material be stored?

Clearly, we would like to segregate the material so that should a detonation occur a minimum amount of material would participate in sympathetic deto-

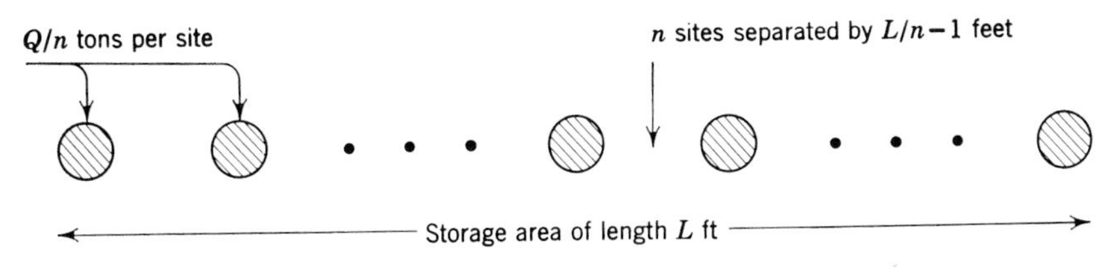

Fig. 13.9-1 Segregation of dangerous material.

nations. Such a segregation is illustrated in Fig. 13.9-1 in which the Q tons of material has been placed into n storage sites each containing Q/n tons and separated from each other by a distance of $L/n - 1$ feet. We might now restate our problem as that of finding the degree of segregation n which leads to the minimum expected participation in a disaster which is initiated, say, at the left-most site.

Before any progress can be made in such a disaster propagation study, information must be available on the mechanism by which explosive-like violence is propagated from site to site.

Given the detonation of material at site i, the probability $p_{i/j}$ of a sympathetic detonation at site j is expressed adequately by Eq. 13.9-1.

$$p_{i/j} = \begin{cases} \dfrac{K_i Q_i^{1/3}}{R_{ij}^{\ 2} B_j} & : \quad p < 1 \\ 1 & \quad \text{otherwise} \end{cases} \tag{13.9-1}$$

Where K_i describes the explosive energy released by a unit mass of the particular detonating material i, Q_i is the amount detonating, R_{ij} is the distance separating the detonation site i and the sympathizing material j, and B_j measures the sensitivity of the sympathizing material to detonation. For a given kind of material being stored, the parameters K and B will be known.

[8] A. H. Masso and D. F. Rudd, *Studies in Disaster Propagation*, Part I, "The Propagation of Explosive-Like Violence," Ind. Eng. Chem. Fundamentals, **7**, 131, 1968.

Notice in Eq. 13.9-1, that the propagation of a detonation will occur with certainty at a particular distance of separation; namely, the critical distance R_{ij}^* when

$$p_{i/j} = 1 = K_i Q_i^{1/3}/(R_{ij}^*)^2 B_j \tag{13.9-2}$$

Under no reasonable circumstances would material be stored that close.

Now we return to the problem posed in Fig. 13.9-1, and ask under what conditions will all of the sites participate in a detonation which initiates at the end site. Clearly, there are two limiting cases: $n = 1$, when all the material is contained in one site: and $n = n^*$, the segregation of material is sufficiently fine to produce a fuse effect in which the detonation travels from site to site through the storage area. The critical segregation n^* is obtained from Eq. 13.9-2 as

$$1 = \frac{K(Q/n^*)^{1/3}}{[L/(n^* - 1)]^2 B} \tag{13.9-3}$$

Thus two bounds have been established on the safest degree of segregation n_{safest}

$$1 < n_{\text{safest}} < n^* \tag{13.9-4}$$

We shall now investigate means for establishing the safest segregation between these bounds.

To further simplify the analysis we shall consider the situation in which the sites shadow each other and the explosive violence cannot jump over a site to leave it undamaged while detonating a site beyond. The effects of removing such simple restrictions have been studied in the work of Masso and Rudd and will not be included here.

The propagation probability of Eq. 13.9-1 is bounded between 0 and 1 for $n \leq n^*$; $0 \leq p \leq 1$. The possible chains of detonation propagation and their probabilities of occurrence are listed in Table 13.9-1.

The amount of material that is expected to participate in the disaster is thus

$$\bar{Q} = \frac{Q}{n}\left\{\sum_{k=1}^{n-1} [k(1 - p)p^{k-1}] + np^{n-1}\right\}$$

$$\bar{Q} = \frac{Q}{n}\left(\frac{1 - p^n}{1 - p}\right) \tag{13.9-5}$$

In Fig. 13.9-2, the relative expected loss in disaster $\bar{Q}/Q$ is plotted against the degree of segregation n, for the case when a segregation of ten sites will result in the fuse effect. Notice that for a relative segregation of the material into five sites, the theory predicts the least participation in disaster, namely an expected loss of only one-quarter of the material.

Table 13.9-1 Possible Chains of Detonation

Chain	Probability of Occurrence	Loss of Material
1 Site only	$(1-p)$ since an initiating detonation is assumed	$\frac{Q}{n}$ tons
2 Sites only	$p(1-p)$	$\frac{2Q}{n}$
3 Sites only	$p^2(1-p)$	$\frac{3Q}{n}$
k Sites only	$p^{k-1}(1-p)$	$\frac{kQ}{n}$
All n sites	p^{n-1}	Q

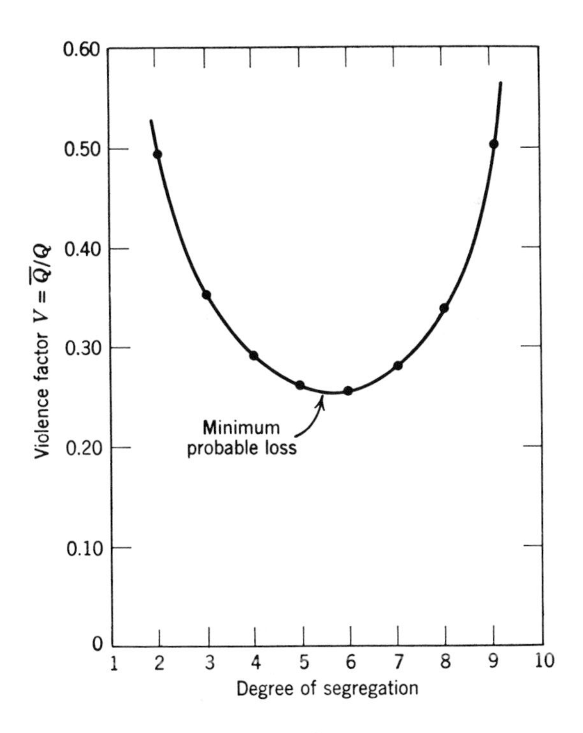

Fig. 13.9-2

A general rule of thumb has evolved from a number of theoretical studies such as that outlined here: a safe degree of segregation is one-half the critical segregation

$$n_{\text{safe}} = \frac{n^*}{2} \tag{13.9-6}$$

We have attempted to give a glimpse of the kind of thinking which will lay a theoretical foundation under the existing empirical studies in disaster propagation. Work in this area is only beginning; and the theory of probability and Monte Carlo simulation (Chapter 15) will form the basis of these theories. The reader is referred to the work of Masso[9] for further discussion of the simple theory of the propagation of explosive-like violence and to a study of the possible effects of debris scattering.

13.10 CONCLUDING REMARKS

Only the surface has been scratched to reveal some of the considerations that must be made to develop safe processing systems. The ideas to carry away from a study of the material in this chapter are:

1. No process design can be assumed to be free of possible failure; designs can only be failure tolerant at best.
2. There is an extensive body of empirical information which is useful in building failure tolerance into a design.
3. The theory of failure tolerance engineering is only in its infancy, and significant developments along these lines are needed.

We must remark that the theory of reliability of continuously operating systems is more complicated than that of batch systems, for the duration of downtime must be included in the analysis. In the several problems below you are to regard the continuous processes as batch processes, operating over small periods of time, and apply the simple theory presented here. See problem 14.I for an introduction to the more complete problem, analyzed by probability balance methods.

References

The following books offer an introduction to the variety of factors which enter into the design of a safe process.

H. H. Fawcett and W. S. Wood, *Safety and Accident Prevention in Chemical Operations*, Wiley, New York, 1965.

[9] A. H. Masso, Ph.D. Thesis, University of Wisconsin, Madison, 1968.

G. Armistead, Jr., *Safety in Petroleum Refining and Related Industries*, 2nd ed., Simmonds, New York, 1959.
J. C. Ducommun, *Process Safety*, American Oil Company, Chicago, 1966.

Certain nationally accepted codes, specifications and standards have been set forth, and are commonly employed in industry even when not legally enforced. A survey of these will be found in J. C. Ducommun, *Process Safety*, cited above; particular sources worth citing here include:

National Fire Codes: Vol. 1, *Flammable Liquids;* Vol. 2, *Gases;* Vol. 3, *Combustible Solids, Dusts and Explosives*, National Fire Protection Association.
American Standard Code for Pressure Piping, American Society of Mechanical Engineers.

For further study in the theory of reliability we recommend:

I. Bazovsky, *Reliability: Theory and Practice*, Prentice-Hall, Englewood Cliffs, N.J., 1961.

There is the risk of system instability, and it is only in the area of chemical reactor design that an adequate study has been made. See the scholarly work of Amundson:

N. R. Amundson, *et al.*, "An Analysis of Chemical Reactor Stability and Control," a continuing series of research publications in *Chem. Eng. Sci.*

See the following for testing of stability of chemicals.

J. S. Snyder, "Testing Reactions and Materials for Safety," Chap. 19 in *Safety and Accident Prevention in Chemical Operations*," Wiley, New York, 1965.

As an example of the kinds of information which appear in the literature see:

"Process Safety Manual," *Chem. Eng. Progr.*, No. 8, **62** (1966) and continuing.
A series of articles on safety in pilot plant procedures in *Chem. Eng. Progr.*, No. 11, **63**, 49–68 (1967).

For information on toxicology see:

"Threshold Limit Values for 1966," American Conference of Industrial Hygienists.

PROBLEMS

13.A. The accidental inclusion of water in processes which are to operate at high temperatures constitutes a severe hazard, as the water may violently vaporize, expanding in volume by factors of thousands. The location of the motor-driven block valve between the reactor and fractionator shown in Fig. 13.A-1 was a contributing factor in a violent accident.

Catalyst circulating through the reactor had been heated to 700°F, and hot oil in the fractionator had been heated to 470°F during process start-up with the block valve closed. The valve in the drain line was opened to remove any water and then closed. The block valve was then opened and there quickly occurred the complete disruption of the fractionator, with all the trays, supporting trusses, and internal piping damaged. The drain line had been plugged and the vertical section of the pipe above the block valve was filled with water condensate. How can the system be made fail-safe to this form of misoperation?

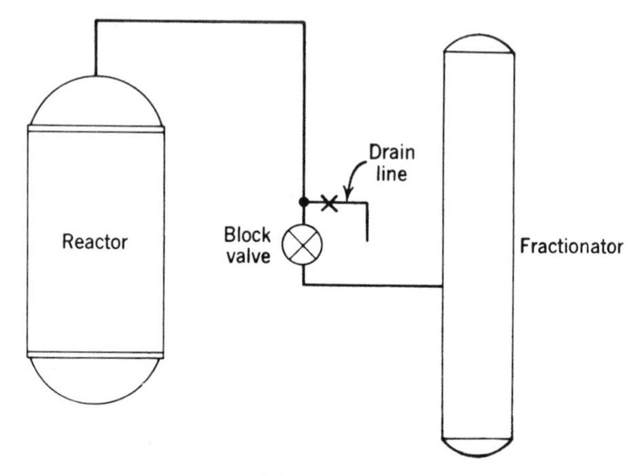

Fig. 13.A-1

13.B. Flammable hydrocarbons are often stored as liquids under pressure. The storage spheres may not be insulated but are equipped with pressure release devices. However, the supporting legs may be insulated. Why?

13.C. Suggest a safe three dimensional layout for the nitric acid purification process shown in Fig. 13.3-1. The process is to be constructed out of doors to service a high explosives manufacturing system. List the safety features which are important in your layout.

13.D. A three component batch processing sequence is to be designed in which secondary chemical species are to be prepared expressly for use at a given place in the processing sequence. (See Fig. 13.D-1.)

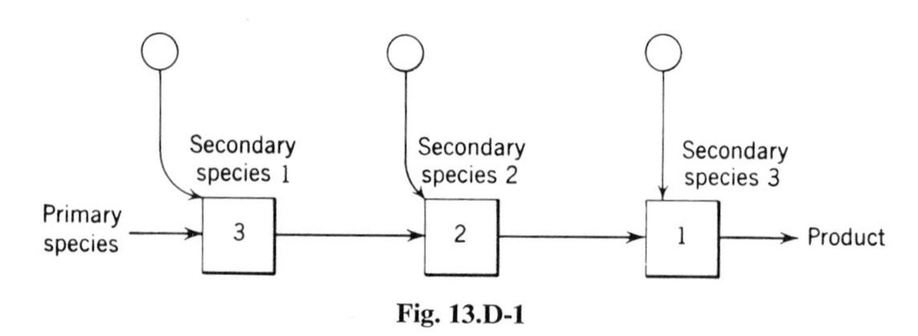

Fig. 13.D-1

The following data are available on the profit P obtained from the completion of a processing sequence, the amortization cost C_i of the reactors used to manufacture the secondary species, the operating costs O_i, and the reliability R_i.

$P = \$1{,}000$ per batch

	C_i	O_i	R_i
Stage 3	10	10	1/3
Stage 2	5	5	1/2
Stage 1	5	5	3/4

(a) Demonstrate that if only one reactor is employed per secondary species required, the expected profit for the system is

$$PR_1R_2R_3 - \sum_{i=1}^{3}(C_i + O_i) = \$85/\text{batch}$$

(b) Demonstrate that an expected profit of \$580 per batch is obtained if seven reactors producing species 3, three reactors producing species 2 and two reactors producing species 1 are employed.

(c) The reliability structure in this problem is acyclic and therefore the optimal redundancy can be obtained by dynamic programming. Taking the probability of feed availability as the state variable for suboptimization, solve for the optimal redundancy.

13.E. Suppose now that the reactors in Problem 13.D can be run on a standby basis and the operating costs are charged only when a reactor is brought into service. Compare the expected profit obtained in the problem stated in 13.D to that stated here for the special case when two reactors are used at each stage.

13.F. A pilot plant has been assigned to you to operate. (See Fig. 13.F-1.) The following information is available on the component *service factors*, the fraction of time available accounting for repair after failure, comparable to the reliability in batch systems.

Component	Service factor	Cost
Pumps	0.95	\$ 2,000
Compressors	0.80	\$12,000
Control valve	0.99	\$ 200

It must be on stream a total of 100 stream days, with a cost of \$500 per calendar day. A stream day is 24 hours of running time and a calendar day includes running and downtime. Each system failure results in a one day shut down.

(a) How many calendar days should you allow for 100 stream days?

(b) You have the option of installing one spare pump (for either the fresh feed or recycle), a compressor in parallel as a spare for the compressor, and spare control valves. How many calendar days might you save by these additions? Is it worthwhile?

(c) What if you have funds only to install spare control valves. Should you?

(d) You might feel that only the compressor should be spared, since it has the lowest reliability. Is that feeling justified?

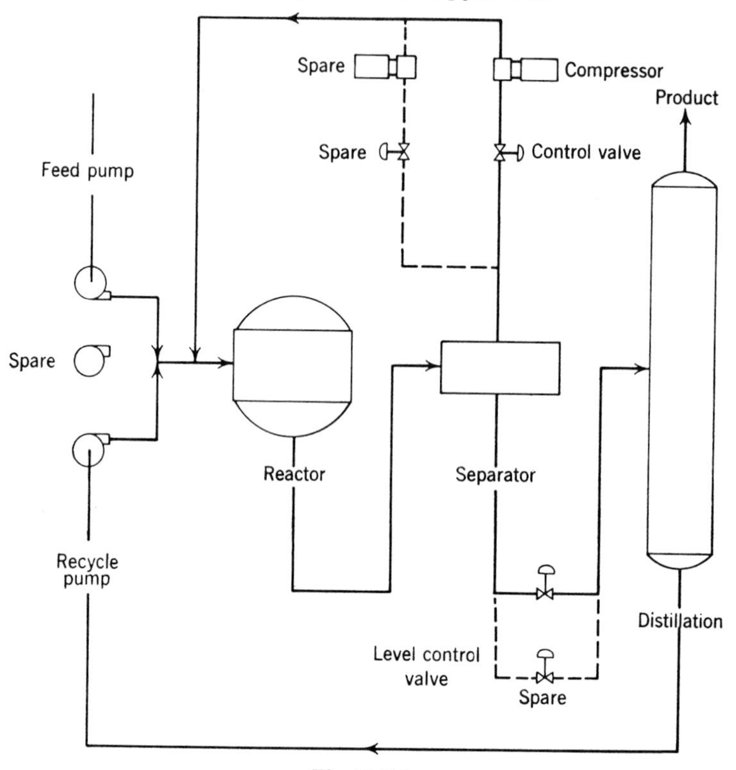

Fig. 13.F-1

13.G. We have one filter on each of three processing lines separating liquid and solid phases of a polymer slurry. The reliability factor for each filter is 0.9 (the fraction of time the filter operates). A standby filter is being considered, at a cost of \$60,000. The cost of downtime in slurry filtering is \$300/line-day, and downtime is one day. If the filter must pay for itself in 3 years what should be done?

$$\text{Answer: } \Delta R_S = (1 - R^3)R$$

13.H. A plastics plant has three separate production lines and each line consists of two reactors in series with a capacity of 5,000 pounds per hour. (See Fig. 13.H-1.) A seventh reactor is available to replace any of the six in use. Each reactor is available 95 per cent of the time. When the heat-transfer surfaces become fouled with polymer, the reactor must be shut down and washed with hot solvent three times. This, plus random mechanical failures, account for the 5 per cent unavailability. Two solutions have been proposed to improve the availability of the reactors, and both cost the same: increase the size of the hot wash facility so that a fouled reactor is only out of service 4 per cent of the time (96 per cent availability), or add another reactor to serve as an additional spare. Which solution will probably lead to a higher production rate from the three production lines?

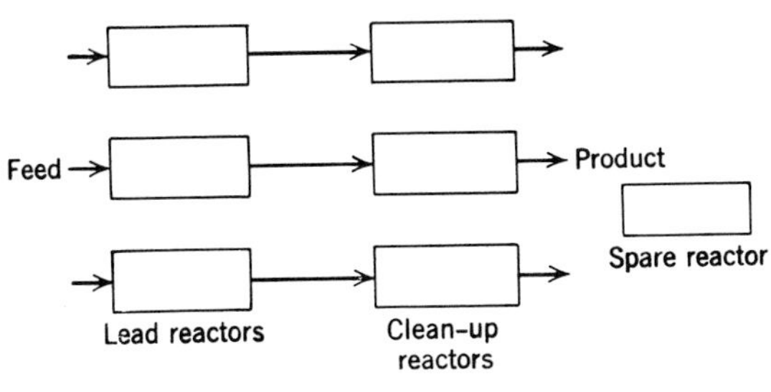

Fig. 13.H-1

13.I. A supply of highly explosive rocket fuel is to be stored in spherical tanks dispersed uniformly over a square tract of land. (See Fig. 13.I-1.) Two

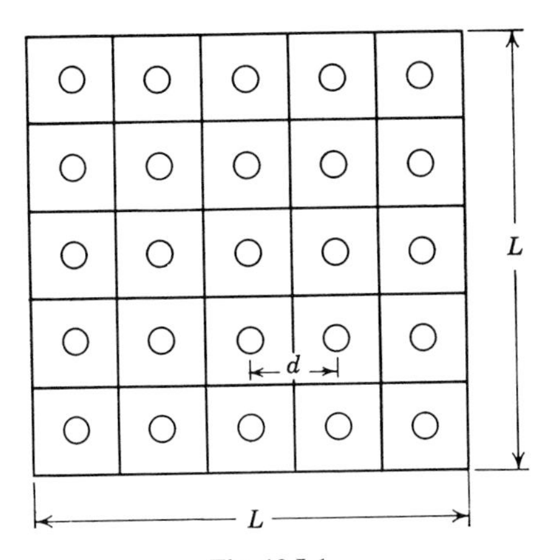

Fig. 13.I-1

sizes of tanks are being considered for the fuel depot; one size twice the diameter of the other. Should one tank detonate, a wave of overpressure will hit an adjacent tank, perhaps causing a second detonation; thus propagating a disaster through the storage depot. We wish to select the tank size that minimizes the overpressure at adjacent tanks in case of detonation.

Definition of Terms

L ft^2	length of tract
$A = L^2$ ft^2	area of tract
$V =$ ft^3	volume of fuel to be stored
n	number of storage tanks
$Q = r/n$ ft^3	volume of single tank
$d = \frac{L}{\sqrt{n}}$ ft	distance between tanks
F	explosive force of detonation
B	blast effect, overpressure of the detonation wave

The explosive force of a detonation is proportional to the volume of fuel detonating.

$$F \propto Q$$

The blast effect B (the overpressure resulting from the detonation) is proportional to the one-third power of the explosive force and drops off with distance squared as the pressure wave propagates out from the blast.

$$B \propto \frac{(F)^{1/3}}{(d)^2}$$

13.J. In the development of the expected loss during the propagation of explosive-like violence through a linear array, Section 13.9, it was assumed that the initiating detonation occurred at the end storage site. Derive the following expression for the expected loss when the detonation initiates at the kth site.

$$\bar{Q} = \frac{Q}{n}\left[p\left(\frac{1-p^{k-1}}{1-p}\right) + \left(\frac{1-p^{n-k+1}}{1-p}\right)\right]$$

What initiating site results in the largest expected loss?

13.K. Suppose that a single initiating detonation can occur at any site in Problem 13.J with equal probability. Derive the following expression for the expected loss.

$$\bar{Q} = \frac{Q}{n}\left[\left(\frac{1+p}{1-p}\right) - \frac{2}{n}\left(\frac{p}{1-p}\right)\left(\frac{1-p^n}{1-p}\right)\right]$$

14

ENGINEERING AROUND VARIATIONS

A process must respond to persistent and perhaps unpredictable deviations from average conditions, as it is buffeted by variations in power costs, weather conditions, the quality of its feed stock, steam pressure, and so forth. Should a system be designed with only the average environment in mind, such deviations might render the system inefficient and even inoperable. In this chapter we discuss the problems of engineering around variations.

Typical Problem

The off-peak power available to our electrolysis plant varies throughout the day and season as the dominant domestic power consumption varies. How might we adjust to this variable supply? This unique problem occurred during the design of an ammonium nitrate plant in Iceland.

14.1 VARIABILITY, A CONSTANT ANTAGONIST

It is obvious that a process must be able to respond to at least the average demands for its services. Imagine the mess that would ensue if the waste treatment plant for a paper mill could only process or store half of the waste continually belched from the mill.

However, engineering to handle only the average conditions is unacceptable, since any variations about the average may disrupt operations completely. The engineer has the responsibility for designing flexible systems which can ride out reasonable deviations from the average since deviations tend to be the rule rather than the exception.

We might list the origin of several variations to which process systems must respond. There appears to be a strong correlation between the natural cycles of nature and the periodic variations which buffet a system, even when the variation is not of direct natural origin. The most important cycles of nature are the yearly, monthly, and daily variations in weather. In addition to directly influencing the systems, these cycles alter the level of activity of the human community which the processing systems service and thus exert an indirect influence on the systems.

It is obvious that a manufacturer of antifreeze, fuel oil, or fertilizer would experience a yearly surge in the demand for his products. In addition to these obviously weather-dependent variations, other variations occur, such as seasonal changes in the cost of natural gas, coal, and electric power, reflecting more distant and obscure changes in the activity of the larger community.

Iceland's fertilizer plant offers a striking example of such an indirect variation. The production of hydrogen by electrolysis in the plant had to be programmed to respond to a variable off-peak electric power supply. Reykjavik, the neighboring city, had precedence in the demand for power and, as the level of activity in that city changed throughout the day and year altering the domestic power consumption, the operation of the fertilizer system had to respond. To have disregarded that factor in designing the process would have been inexcusable. This power availability variation was smoothed by providing large storage facilities for hydrogen, accumulating hydrogen during the upswings in power availability, and depleting the storage during the downswings when the hydrogen production had to be lowered. The bulk of the process was thus protected from the drastic power variations.

In addition to variations which tend to correlate with the cycles of nature and the concomitant cycles of man, other variations occur, the origin of which is either obscure or under the direct control of the engineer.

For example, we would expect that the successive loads of ore arriving at a beneficiation system would vary in quality should the deposit from which the ore is being recovered be extensive and varied. This variation might well disrupt the operation of the processing system, requiring the frequent adjustments of operating conditions.

Should the crude oil for a refinery arrive by tanker after a long sea voyage, variations in arrival time are to be expected. In fact, any transportation system —truck, barge, rail, or pipeline—can be a source of disrupting variations.

Internal variations within the system may also disrupt operations. For example, if an intermediate chemical is being manufactured within the system in a batch operation, an excessively long reaction time for a given batch might cause the shutdown of the system for want of available intermediate chemical.

The engineer is not at the mercy of his variable environment, however. By some very simple changes in design a system can be endowed with a certain degree of immunity to and flexibility in the presence of rather extreme excursions in environmental conditions. In fact, the engineer sometimes can take advantage of variations to pump more out of a situation than could be obtained by smooth operation.

One approach is to overdesign the system to the point that it can handle the variations within economic limits, wasting or rejecting the extreme peaks. A further adaptation is to provide storage capacity to smooth the peaks passed on to the processing system, to the extent justified economically. Even further, a costly multiproduct process may be designed to allow a rapid shift to new areas of manufacture as a varying demand dictates.

We attack this problem of engineering around variations by discussing a variety of typical industrial situations in which variability is a prominent feature. Some of our examples have been dignified by the mathematical community with extensive theoretical treatment, such as the theory of waiting lines. Others seem to yield best to a homemade analysis, and are not sufficiently general in nature to have attracted the attention of the mathematician, or are sufficiently messy to escape general mathematical solution. Here again, common sense appears to be the dominant feature, more important than pure theory. Hopefully, the mathematics will be common sense put to numbers. We shall not discuss the classical areas of process dynamics and control, since to do an adequate job would require a full text in itself.

14.2 THE EFFECTS OF STORAGE ON A PULSED SUPPLY

In this section we analyze the special situation, shown in Fig. 14.2-1, in which material arrives in discrete batches to be processed by a continuously operating system. A storage tank is provided for the arriving material, and we are concerned with the effect such storage has on the efficiency of the system.

Material arrives in pulses of duration A hours at a rate of H pounds per hour during the pulse. No material arrives during the next B hours, and then the cycle repeats itself. The process system can operate only at a rate of Q pounds per hour or less, and excess material that cannot be stored must be shunted for use in another facility, the overflow.

This type of pulsed supply is quite common in the processing industry, as material arrives via barge, truck, rail, or programmed pipeline. Any excess materials that cannot be stored might even be burned in the boilers, thus being put to a less valuable overflow use, but not wasted. While industrial variations are rarely as predictable, a careful analysis of this ideal situation will add considerable insight to the more messy real problems.

How much of this supply can be utilized by our system?
How does storage smooth the variations?
How frequently does the storage tank overflow?
How frequently does the system stop for want of feed?
In a given situation, what is the optimal storage design strategy?

When $H < Q$ no problem in analysis exists, since the system can handle the material as it comes. However, when $H > Q$ the process cannot handle the instantaneous arrival rate during the pulse, and with *no storage* $(H - Q)A$ pounds of material must be shunted to the overflow every cycle.

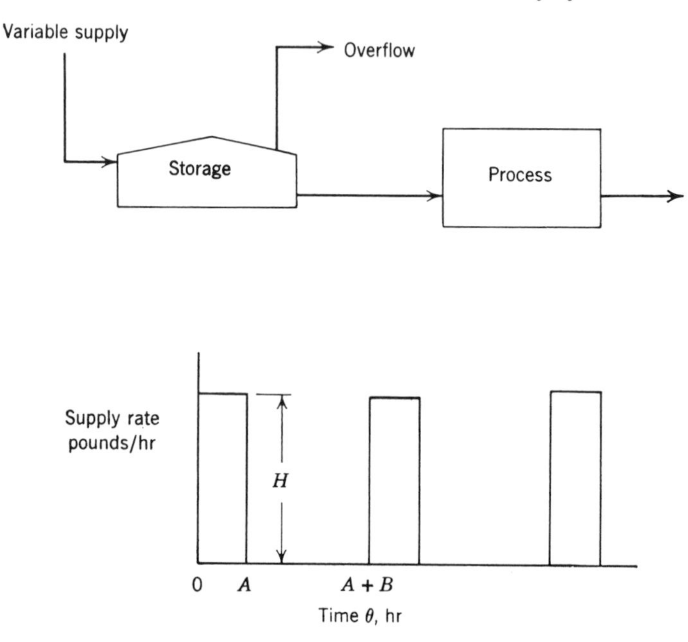

Fig. 14.2-1 Pulsed supply to a continuous process.

We might define the *supply utilization factor* Φ_s as the fraction of the supply that can be used by the system. In the case of no storage

$$\Phi_s = \frac{HA - (H - Q)A}{HA} = \frac{Q}{H} \leq 1 \tag{14.2-1}$$

That is, with no storage only a fraction Q/H of the arriving feed can be utilized by our system, and the remaining fraction $1 - Q/H$ must be routed to the overflow.

We now determine the effect that storage has on the utilization of the supply. Suppose that a storage tank is present with the capacity of C pounds of material. During an arrival pulse, $0 \leq \theta \leq A$ hours, the amount of the material stored in the tank, S pounds, will increase until the tank is full, at which time the tank will remain full as the excess arrivals are shunted to the overflow. Thus, in the period $0 \leq \theta < A$, a transient material balance over storage gives

$$\frac{dS}{d\theta} = \begin{cases} (H - Q) & \text{for} \quad S \leq C \\ 0 & \text{for} \quad S = C \end{cases} \tag{14.2-2}$$

Now, we might arbitrarily select the amount of the material in storage at the beginning of the pulse to be zero, $S(0) = 0$, since any material in the tank at that time can never be processed in this mode of operation and could be removed without changing the operation of the system. Such material would be *dead inventory* in our simple problem. Equation 14.2-2 can then be integrated to give

$$S(\theta) = \begin{cases} (H - Q)\theta & \text{for} \quad (H - Q)\theta \leq C \\ C & \quad\quad \text{otherwise} \end{cases} \tag{14.2-3}$$

valid when $0 \leq \theta \leq A$.

The time at which overflow first occurs, θ_0 is obtained thus

$$S(\theta_0) = (H - Q)\theta_0 = C$$

or

$$\theta_0 = \frac{C}{(H - Q)} \leq A \tag{14.2-4}$$

That is, overflow must occur before time A if it is to occur at all, since after time A the supply pulse stops.

Now, we might argue that storage capacity greater than $(H - Q)A$ pounds would never be used and, hence, that capacity can be thought of as the upper limit of optimal capacities, the supremum of the economically optimal capacities C^*.

$$C^* \leq (H - Q)A \tag{14.2-5}$$

The amount of material that must be routed to overflow while the storage facility is full is

$$(H - Q)(A - \theta_0)$$

Thus, the supply utilization factor in the presence of storage is

$$\Phi_s = \frac{HA - (H - Q)(A - \theta_0)}{HA} \tag{14.2-6}$$

$$= \frac{Q}{H} + \frac{C}{HA} \leq 1$$

where

$$C \leq (H - Q)A$$

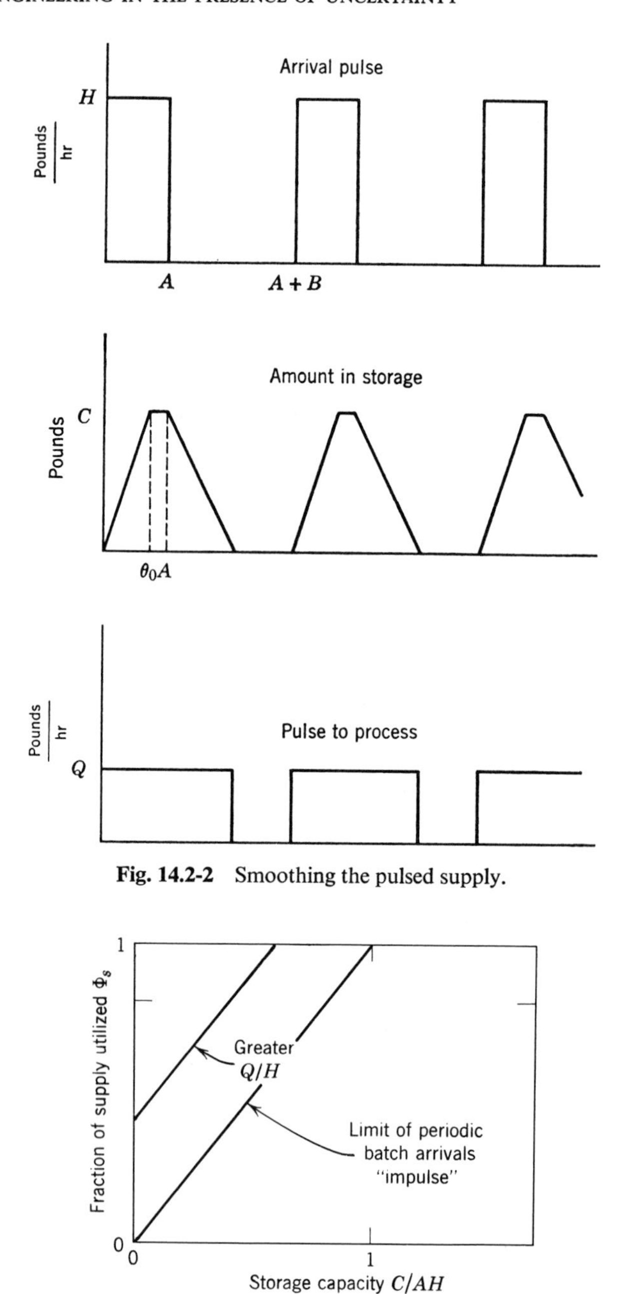

Fig. 14.2-2 Smoothing the pulsed supply.

Fig. 14.2-3 The supply utilization.

Figure 14.2-3 shows this supply utilization factor as a function of the amount of storage provided, in units of pounds of material arriving during one pulse. An impulse supply is one in which HA pounds of material arrive in single batches ($A = 0$) every B hours.

Now we might define the *process utilization factor* Φ_p as the fraction of the possible production capacity per cycle, $Q(A + B)$, that is actually utilized. This possible production capacity might not be realized for want of feed during the period when no supply of material is entering the system should the storage be empty.

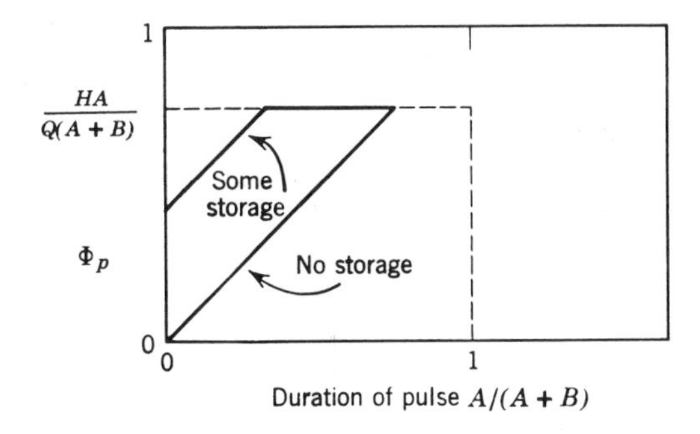

Fig. 14.2-4 The effects of pulse duration on process utilization.

The total amount of feed that appears for processing per cycle is HA pounds, and of that $(\Phi_s)HA$ pounds remain to be processed, the remainder overflows. Thus, the process utilization factor is

$$\Phi_p = \frac{(\Phi_s)HA}{Q(A + B)}$$

$$= \frac{A}{A + B} + \frac{C}{Q(A + B)} \leq 1 \qquad (14.2\text{-}7)$$

Notice how, for a given storage capacity, a semicontinuous supply results in a greater utilization of the process than a batch supply. Figure 14.2-4 shows this expected effect.

We defer our discussion of the economically optimal storage capacity to Section 14.5 since the ideas there are general to the analysis of any mode of supply and storage. We merely seek the storage which offers the most profitable situation, balancing the profits to be gained by increased utilization against the costs of providing the storage and holding the inventory. First we must gain more experience in estimating utilization factors.

14.3 ANALYSIS BY QUEUING THEORY

It is frequently enlightening to compare and contrast the opposite of extremes, especially when the analysis requires the use of new and novel ways of thinking. In the previous section, we analyzed one extreme in variability, a periodic pulsed supply to a continuous process. Now we attack an opposite extreme, a batch supply which arrives randomly at a batch system which itself has a random processing time. The new way of thinking is that of the mathematical theory of queuing, the stochastic theory of waiting lines.

A. E. Erlang, an engineer in Copenhagen, at the beginning of this century investigated the effects of variations in the demands for the services of a telephone exchange and the effects of variations in the duration of telephone conversations. He developed a theory which describes the effects of these random variations on the effectiveness of telephone systems. It was not until after World War II that Erlang's ideas received general attention, and there then evolved the theory of waiting lines—queuing theory.

Queuing theory has to do with the prediction of the effects of random variations in the arrival rate of items to a service facility and random variations in the time of service for a given item. We see that queuing theory deals with the random counterpart of the pulsed system discussed in Section 14.2.

We cannot attempt to discuss queuing theory in all its generality; such discussions have filled many volumes the size of this text. Instead, we focus attention first on the simple situation illustrated in Fig. 14.3-1, consisting of a

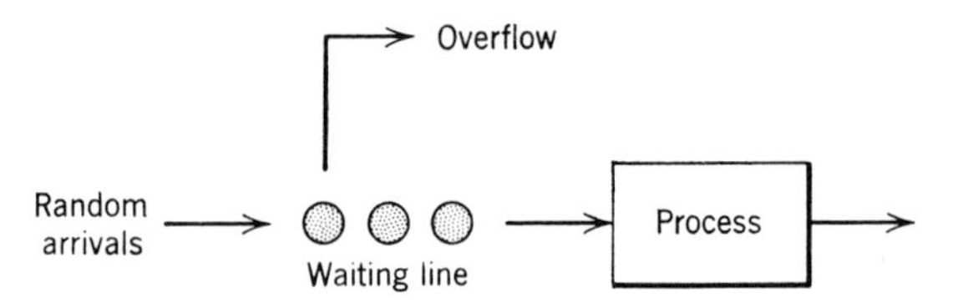

Fig. 14.3-1 A simple queuing problem.

service system, a storage area in which N items can wait for service, and a transportation facility through which items arrive. The processing or service time for a given item once it is able to enter the process is assumed to be completely random with average rate $\bar{Q}$ items per hour. The times between the arrivals of new material is also randomly distributed, with an average arrival rate of $\bar{Q}'$ items per hour. An item retains its identity during processing and storage; this is a batch system. It must be noted that the average processing rate $\bar{Q}$ describes the process performance assuming its feed is readily available, and that the nonready feed will result in a reduced actual processing rate. We wish to predict that actual rate.

We might describe the distribution of times between successive arrivals by $F(\theta)$, the probability that a given arrival comes later than θ hours after its predecessor. The probability that an arrival will occur in the period of θ to $\theta + d\theta$ hours after its predecessor is then, according to this definition

$$F(\theta) - F(\theta + d\theta) \tag{14.3-1}$$

Now, in the simplest case, the chance of an arrival in an interval of time $d\theta$ is *assumed* to be constant, that constant value being denoted by $\lambda\, d\theta$. This particular arrival can only occur in time θ to $\theta + d\theta$ if it has not occurred previously; thus the chances of an arrival in time period $d\theta$ is

$$F(\theta)\lambda\, d\theta \tag{14.3-2}$$

Equating this to the previous expression gives

$$F(\theta) - F(\theta + d\theta) = F(\theta)\lambda\, d\theta$$

and as $d\theta \to 0$ this becomes

$$\frac{dF}{d\theta} = -F\lambda \tag{14.3-3}$$

with $F(0) = 1$ and $F(\infty) = 0$. The solution for the distribution $F(\theta)$ for purely random arrivals is then

$$F(\theta) = e^{-\lambda\theta} \tag{14.3-4}$$

This is the *Poisson* or *exponential distribution function*, which describes this class of random events.

In Fig. 14.3-2 this exponential distribution of arrival times is compared to the distribution that would occur for purely periodic arrivals.

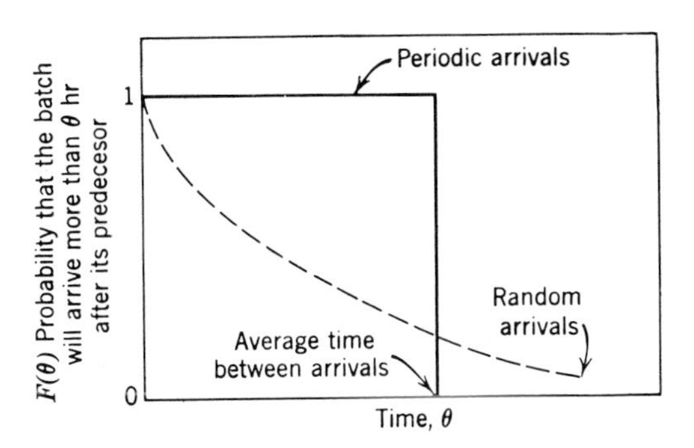

Fig. 14.3-2 Distribution of time between arrivals.

We now relate the parameter λ to some measurable quantity. The average time between arrivals is defined as

$$\bar{\theta} = \int_0^\infty \begin{pmatrix}\text{Probability that an arrival}\\ \text{occurs in the interval } \theta \text{ to } \theta + d\theta\end{pmatrix}(\theta)\, d\theta$$

$$= \int_0^\infty \left(-\frac{dF}{d\theta}\right)(\theta)\, d\theta = \int_0^\infty \lambda e^{-\lambda\theta}\,\theta\, d\theta = \frac{1}{\lambda} \qquad (14.3\text{-}5)$$

Thus, $\lambda = 1/\bar{\theta} \equiv \bar{Q}'$ the average arrival rate.

In summary, if arrivals occur purely randomly the chances of an arrival occurring in a given period of time $d\theta$ is $\bar{Q}'\, d\theta$ where $\bar{Q}'$ is the average arrival rate.

Applying the same reasoning to the random service times, we get the probability that a service will occur in the period of time $d\theta$ (assuming there is material in the system being processed) is $\bar{Q}\, d\theta$ where $\bar{Q}$ is the average service rate.

So far we have only defined what is meant by random arrivals and service, and we now begin the analysis of the effects of such variations. The line of thinking which follows is similar to that which led to the transient material balances on the storage tank in the previous section. The analysis now is a bit more slippery, however, since we necessarily must deal with probability rather than certainty. In the previous section we denoted the amount of material in storage by the variable S pounds. We now describe the number of items in the line by P_i, *the probability that i items are in the system.* Now if only N items are allowed to wait in line (one extra batch is in the process being processed) then

$$\sum_{i=0}^{N+1} P_i = 1 \qquad (14.3\text{-}6)$$

The system must exist in some state.

We now ask what single transitions can cause the line to build up or deplete.[1] These transitions are shown in Fig. 14.3-3. For example, state P_2 (two items in the system) can be formed by having one item in the system and having a new one arrive, or by having three items in the system and having one item be serviced. These are transitions $P_1 \curvearrowright P_2$ and $P_2 \curvearrowleft P_3$, respectively. The state P_2 can disappear either by an arrival or a service, the transitions $P_1 \curvearrowleft P_2 \curvearrowright P_3$. Notice that P_0 and P_{N+1} are special cases, the case when the system is empty and the case when the line is full.

[1] We can reduce the increment of time over which the system is under observation to so small a value that only one event can occur. The simultaneous occurrence of two events is a second order phenomenon which will drop out of the mathematics.

The probability that one of these transitions will occur in time $d\theta$ is

$$\begin{pmatrix}\text{Probability that}\\ \text{the state exists}\end{pmatrix} \times \begin{pmatrix}\text{Probability that the}\\ \text{particular event occurs}\end{pmatrix}$$

For example, the probability of the transition $P_1 \curvearrowright P_2$ is

$$(P_1)(\bar{Q}'d\theta)$$

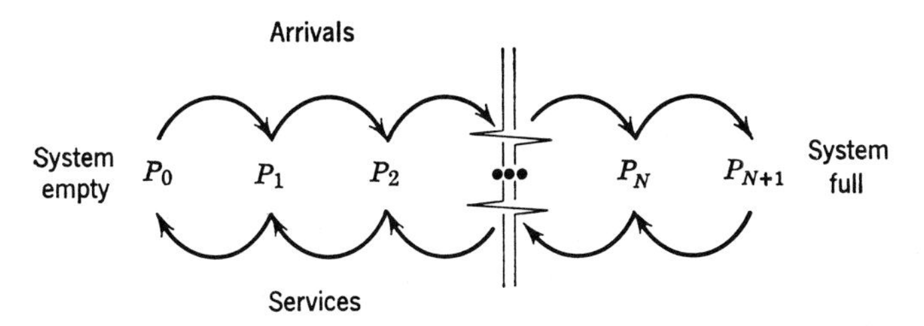

Fig. 14.3-3 All possible states and transitions.

We now require that the rate of formation of any state equal the rate of disappearance during steady state operation. Such a balance for state P_i gives

$$\text{Rate of formation} = \text{Rate of disappearance}$$

$$P_{i-1}\bar{Q}'\,d\theta + P_{i+1}\bar{Q}\,d\theta = P_i(\bar{Q}' + \bar{Q})\,d\theta$$

or

$$P_{i+1} = P_i\left(1 + \frac{\bar{Q}'}{\bar{Q}}\right) - P_{i-1}\left(\frac{\bar{Q}'}{\bar{Q}}\right) \tag{14.3-7}$$

For $i = 0$, the balance is

$$P_1\bar{Q}\,d\theta = P_0\bar{Q}'\,d\theta$$

or

$$P_1 = P_0\left(\frac{\bar{Q}'}{\bar{Q}}\right) \tag{14.3-8}$$

For $i = N + 1$, the balance is

$$P_N\bar{Q}'\,d\theta = P_{N+1}\bar{Q}\,d\theta$$

or

$$P_{N+1} = P_N\left(\frac{\bar{Q}'}{\bar{Q}}\right) \tag{14.3-9}$$

In general, by solving Eqs. 14.3-7, 14.3-8, and 14.3-9

$$P_i = P_0 \left(\frac{\bar{Q}'}{\bar{Q}}\right)^i \tag{14.3-10}$$

Employing Eq. 14.3-6 the general solution for the probability that i batches are in line is

$$P_i = \left(\frac{1 - \bar{Q}'/\bar{Q}}{1 - (\bar{Q}'/\bar{Q})^{N+2}}\right)\left(\frac{\bar{Q}'}{\bar{Q}}\right)^i \tag{14.3-11}$$

We are now in a position to compute the utilization factors for this system.

The supply utilization factor Φ_s is the fraction of the arriving batches that are free to enter the line and wait for service. A batch can only enter the line when there are less than $N + 1$ batches in the system, line plus process. That probability is $1 - P_{N+1}$. Thus, the supply utilization is the fraction of the time the system can accept the arriving material.

$$\begin{aligned} \Phi_s &= 1 - P_{N+1} \\ &= \frac{1 - (\bar{Q}'/\bar{Q})^{N+1}}{1 - (\bar{Q}'/\bar{Q})^{N+2}} \end{aligned} \tag{14.3-12}$$

The process can only function when there is at *least* one batch in the system. That probability is $1 - P_0$. Thus, the process utilization factor is the fraction of the time the process has material to process.

$$\begin{aligned} \Phi_p &= 1 - P_0 \\ &= \frac{(\bar{Q}'/\bar{Q}) - (\bar{Q}'/\bar{Q})^{N+2}}{1 - (\bar{Q}'/\bar{Q})^{N+2}} \end{aligned} \tag{14.3-13}$$

Notice that in the limit of infinite storage when $\bar{Q}' < \bar{Q}$, $(N \to \infty)$, $\Phi_s \to 1$, all the supply is utilized, and $\Phi_p \to \bar{Q}'/\bar{Q}$, the utilization of the process is limited by the available supply.

Notice also when no extra storage is provided $(N = 0)$

$$\Phi_s = \frac{\bar{Q}}{\bar{Q} + \bar{Q}'}$$

$$\Phi_p = \frac{\bar{Q}'}{\bar{Q} + \bar{Q}'}$$

In Fig. 14.3-4 we compare the two extremes of variability, the periodic batch arrival of material at a continuously operating process (Section 14.2) and the present case of the random arrival of batches to a process with random distributed service times. The batch system can hold one batch of material in the process during operation, an advantage that the continuous process does not have. We have corrected for that fact in the figure, so that the effects of the

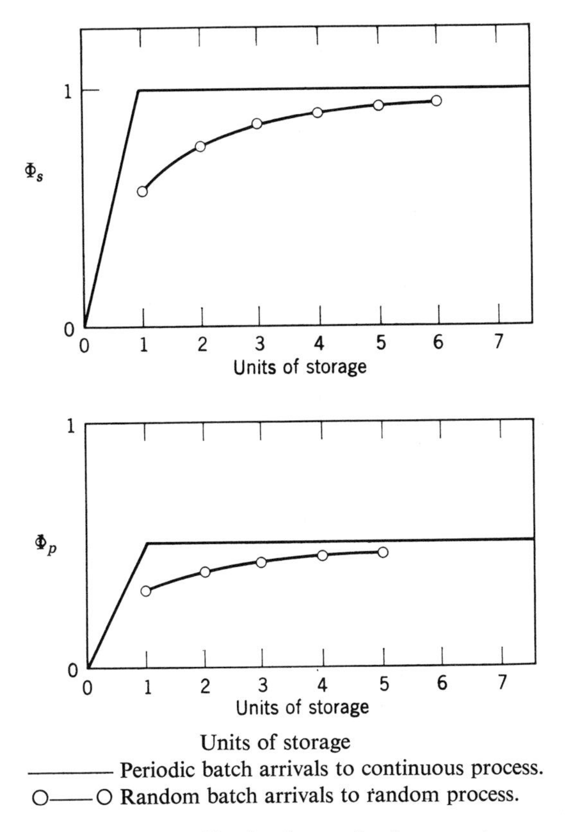

Fig. 14.3-4 The utilization factors for the two extremes.

variations would appear unbiased. As we might expect, the random situation is always more difficult to handle than the predictable periodic case and by sufficiently large storage the limits of complete utilization of the supply and of the process can be reached.

Enough for the analysis of the effects of variations in supply and demand, for the time being. We must now concern ourselves with internal variations which might disrupt a system.

14.4 INTERSYSTEM VARIATIONS[2]

In addition to the external variations which might disrupt the operation of a system, internal variations often appear and must be compensated for. This section has two purposes: One, to illustrate that basing a design on average operating conditions can be misleading; two, to illustrate further the use of probability balances in the preliminary analysis of processing systems. We wish to estimate the effects of variations in the batch times for two interacting processes.

Consider the two processes A and B (Fig. 14.4-1) which are to be connected to form a system, process B following process A. The operating data for the two processes operating separately indicates that on the average process A completes $\bar{Q}_A = 0.256$ batches per hour and process B completes $\bar{Q}_B = 0.312$ batches per hour. However, for some unknown reason the batch times vary, and in both cases the variation can be approximated by an exponential distribution function. The variations *appear* to be random.

In Fig. 14.4-2 we see how these variations disrupt the operation of the system when no provision is made for storage between the two processes. Process A is frequently *blocked*, being unable to get rid of a completed batch when process B is full. Process B is frequently *empty* for want of feed from process A. These interactions cause a lowering in the efficiency of the system. The utilization of process B during the period of this experiment is about 70 per cent, a rough estimate based on a ten batch trial.

It is erroneous to assume that the average productivity of the system is a simple and obvious function of the average productivities of the two processes. A thesis of *systems engineering* is that the *arrangement* of the units of a system plays just as important a role in determining the system performance as do the performance characteristics of the units. Without some general theory it is impossible to extrapolate to new situations, such as attempting to double the processing rate for process A. Such a theory is required to explore new designs for this system.

We wish to derive a relation between the average processing rate for the system and the average processing rates for the unit processes in the system, in order to enable the evaluation of design modifications.

The probability that process A will complete a batch in time $d\theta$ (assuming that it has a batch in operation) is $\bar{Q}_A \, d\theta$, for the case of randomly distributed

[2] N. H. Smith and D. F. Rudd, "On the Effects of Batch Time Variations on Process Performance," *Chem. Eng. Sci.*, **19** (1964).

Fraction of batches that take longer than θ hr

$F(\theta)$

$F(\theta) = e^{-0.256}$

1

0

0 2 4 6 8

θ hr

$\overline{Q}_A \cong 0.256$ Batches/hr

Process A—operating data

Batch Number	Batch Time
1	6 hr
2	2
3	7
4	3
5	1
6	4
7	7
8	1
9	2
10	6
Average 3.9 hr	

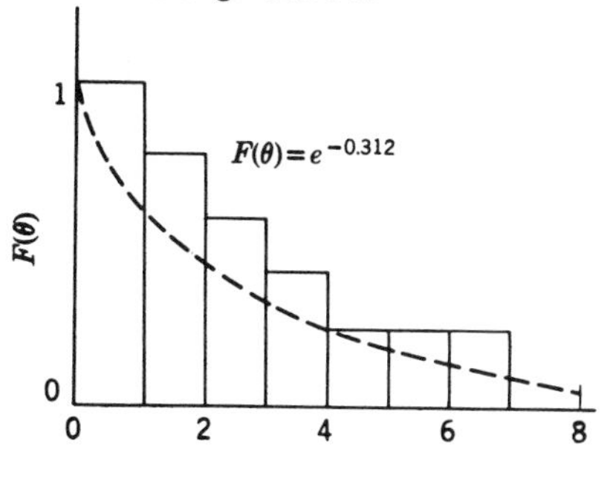

$\overline{Q}_B = 0.312$ Batches/hr

Process B—operating data

Batch Number	Batch Time
1	2 hr
2	7
3	4
4	3
5	1
6	2
7	3
8	4
9	1
10	5
Average 3.2 hr	

Fig. 14.4-1 Variability in batch times.

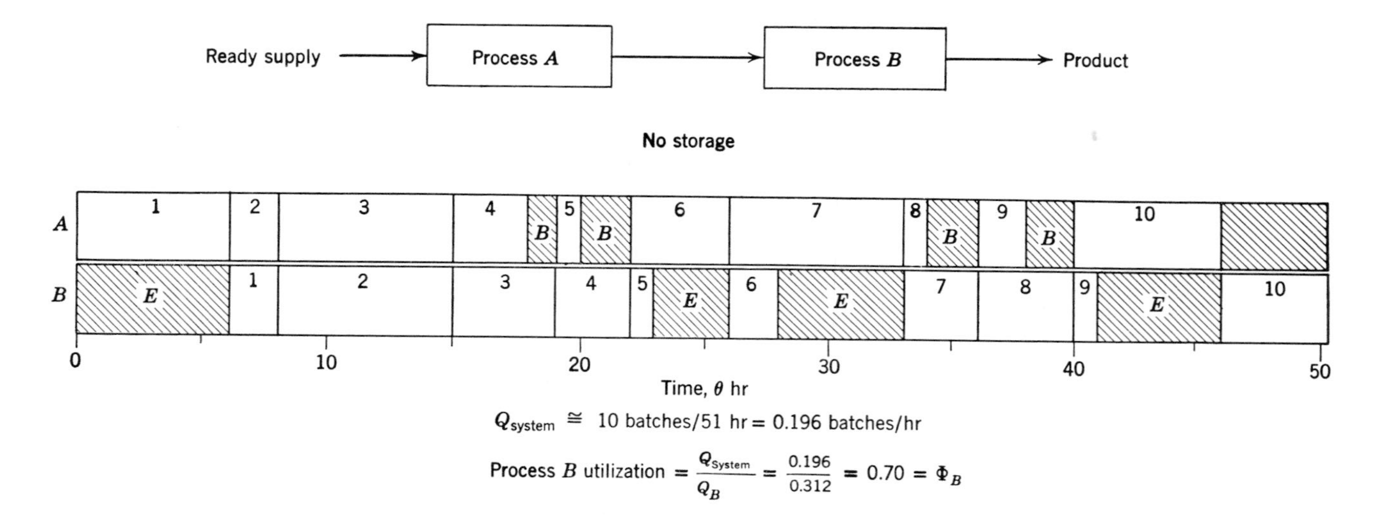

Fig. 14.4-2 Interaction between the two processes.

processing times. Process B has the probability of completion of $\overline{Q}_B\, d\theta$ if it is operating. Since the final product is produced by process B

$$\begin{pmatrix} \text{Probability of a} \\ \text{batch of product} \\ \text{in time } d\theta \end{pmatrix} = \begin{pmatrix} \text{Probability that} \\ \text{process } B \text{ is} \\ \text{operating} \end{pmatrix} \begin{pmatrix} \text{Probability that process} \\ B \text{ produces a batch once} \\ \text{operating in time } d\theta \end{pmatrix}$$

The rightmost term is $\overline{Q}_B\, d\theta$.

The first step in a probability analysis is to list all possible states in which the system can be found, and trace all single event transitions among them. The system in question can be found in one of three states, assuming negligible time to charge and discharge the processes.

Process A	*Process B*	*Probability*
Operating	Operating	P_{OO}
Operating	Empty	P_{OE}
Blocked	Operating	P_{BO}

For example, the state (process A-empty, process B-operating) cannot exist with a ready supply of raw materials, since process A would be filled with no delay.

The transitions among the states are shown in Fig. 14.4-3, For example, the

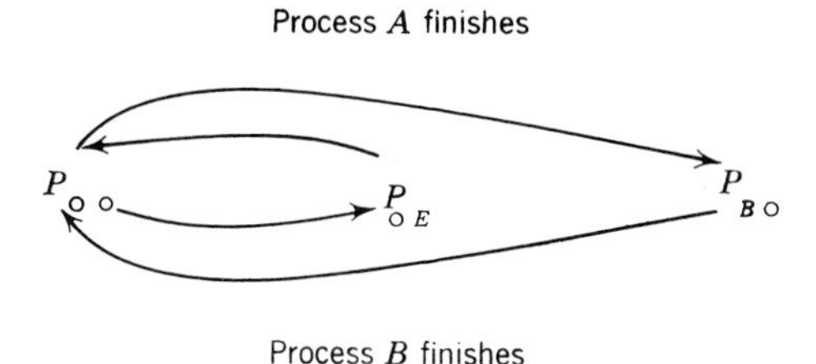

Fig. 14.4-3 Probability balance.

transition from P_{BO} to P_{OO} would follow the sequence—process B finishes, the blocked batch from A moves into process B, and a new batch enters process A.

The probability balances on all three states take the form:

State	*Probability of formation = Probability of disappearance*
P_{OO}	$P_{BO}\overline{Q}_B\, d\theta + P_{OE}\overline{Q}_A\, d\theta = P_{OO}(\overline{Q}_A + \overline{Q}_B)\, d\theta$
P_{OE}	$P_{OO}\overline{Q}_B\, d\theta = P_{OE}\overline{Q}_A\, d\theta$
P_{BO}	$P_{OO}\overline{Q}_A\, d\theta = P_{BO}\overline{Q}_B\, d\theta$

with the condition

$$P_{OO} + P_{OE} + P_{BO} = 1$$

Note that any one of the four equations above can be derived from the other three.

The probability of finding the system in any one of the three states can then be determined, by the solution of the probability balance.

$$P_{OO} = \frac{1}{1 + \bar{Q}_A/\bar{Q}_B + \bar{Q}_B/\bar{Q}_A}$$

$$P_{OE} = \frac{(\bar{Q}_B/\bar{Q}_A)}{1 + \bar{Q}_A/\bar{Q}_B + \bar{Q}_B/\bar{Q}_A}$$

$$P_{BO} = \frac{(\bar{Q}_A/\bar{Q}_B)}{1 + \bar{Q}_A/\bar{Q}_B + \bar{Q}_B/\bar{Q}_A}$$

$$\begin{pmatrix}\text{Probability that process } B\\ \text{is operating}\end{pmatrix} = \left(P_{OO} + P_{BO}\right) = \frac{1 + \bar{Q}_A/\bar{Q}_B}{1 + \bar{Q}_A/\bar{Q}_B + \bar{Q}_B/\bar{Q}_A}$$

Thus

$$\begin{pmatrix}\text{Probability of a}\\ \text{batch of product}\\ \text{in time } d\theta\end{pmatrix} = \frac{\bar{Q}_B\, d\theta(1 + \bar{Q}_A/\bar{Q}_B)}{1 + \bar{Q}_A/\bar{Q}_B + \bar{Q}_B/\bar{Q}_A}$$

Hence, the probable production rate for the system is

$$\bar{Q}_S = \frac{\bar{Q}_B(1 + \bar{Q}_A/\bar{Q}_B)}{1 + \bar{Q}_A/\bar{Q}_B + \bar{Q}_B/\bar{Q}_A} \tag{14.4-1}$$

In the special case shown in Fig. 14.4-2

$$\bar{Q}_S = \frac{(0.312)(1 + 0.256/0.312)}{1 + 0.256/0.312 + 0.312/0.256} = 0.153 \text{ batches/hr}$$

Any theory derives its real usefulness by enabling extrapolations into new and untested regions, and the reproduction of known data merely serves to test the theory. We observed that our limited trial with ten batches suggests a productivity of about 0.20 batches per hour, and that the queuing analysis suggests a productivity of about 0.15 batches per hour. What productivity increase might be achieved by a doubling in the processing rate of process A, assuming the randomness persists? For such extrapolations we must turn to the queuing theory.

Doubling the production rate of process A might yield a systems productivity of

$$\bar{Q}_S = \frac{0.312(1 + 0.256 \times 2/0.312)}{1 + 0.256 \times 2/0.312 + 0.312/0.256 \times 2} = 0.192$$

A theoretical increase of

$$\frac{100(0.192 - 0.153)}{0.153} = 25 \text{ per cent}$$

This estimate is completely inaccessible without a theory, and without such an estimate no rational economic analysis can be made to guide the design of the system. The change in the design of process A must be shown to more than pay its way with increased system productivity.

Other design modifications could be entertained.
Intermediate storage might be considered.
Two processes A might be used to feed into a process B.
Two processes B might receive product from one process A.
Two A-B systems might be placed side-by-side with cooperative usage of intermediate products.

Any of these modifications can be analyzed by the development of simple theories along the lines of this section. These form problems for the end of this chapter.

14.5 ECONOMICALLY OPTIMAL UTILIZATION

In the previous sections we have shown how we might estimate the supply and process utilization factors. In this section we assume that such an analysis has been completed for any particular variation at hand, i.e., the functions Φ_s and Φ_p have been estimated. What storage provision S leads to economically optimal operation?

We merely determine the storage capacity S which maximizes the economic objective function for our project.

For example, if the net profit to be earned by the processing of a pound of material by the processes is P_p and the profit to be earned by the use of a pound of material in the overflow facility is P_0, the total net profit for the system is

$$Q'[\Phi_s P_p + (1 - \Phi_s) P_0]$$

Now, if the investment in the system is estimated by the cost equation

$$I = I_0 \left(\frac{S}{S_0} \right)^M + I_1$$

where I_1 is the constant investment in all equipment except storage and the parameter α is the collection of interest terms and the like required to amortize this investment on a unit time basis, the venture profits to be maximized might be

$$\max_{\{S\}}\left\{Q'[\Phi_s P_p + (1 - \Phi_s)P_0] - \alpha\left[I_0\left(\frac{S}{S_0}\right)^M + I_1\right]\right\} \qquad (14.5\text{-}1)$$

14.6 ADAPTING TO A VARIABLE POWER SUPPLY

In the previous sections we concerned outselves with the provision for storage, assuming that the arrival and processing rates were known and fixed. Quite frequently we are free to adjust the maximum processing rate best to meet an incoming variation. In this section we illustrate that concept by considering a realistic example.

Power companies which contract to supply a community with electric power must respond to a variable demand for their product. During the hours of peak domestic activity the power generating facilities might be taxed to the limits, and during the night-time slack periods the demands fall well below the generating capacity. It is most efficient for the power company to run generators continuously, around the clock, even when there is little demand for the power.

Two methods for engineering around such variations have found common use. The first involves the creation of an artificial demand for power by offering the "off-peak" power to industry at a reduced cost. For example, the Iceland Fertilizer Company purchases off-peak power to generate hydrogen by water electrolysis from the Sog Power Company, which also supplies prime power to the city of Reykjavik. The fertilizer processing system then becomes an integral part of the community, forming what might be called a "macro system."

The second method involves the storage of the off-peak power for use later during periods of greater demand. For example, the Union Electric Company, serving Eastern Missouri, chopped off the top of Proffit Mountain to make a huge reservoir there. At the bottom of the mountain, pumping and power stations and a second reservoir were constructed. At night 5 million tons of water are pumped to the upper reservoir using low cost off-peak power available from other power stations. During the peak load hours the water flows from the upper to lower reservoirs generating electric power. The rated capacity of this enormous storage battery is 350,000 kw, charged during the night and discharged during the day.

In this section we show how we might make use of the variable off-peak power shown in Fig. 14.6-1 illustrating the kinds of thinking which lie behind the peak-smoothing systems alluded to above. Suppose a reservoir and power station exist in our plant, but it is troubled by a lack of water and cannot meet our demands for power. We have been forced to purchase power from the

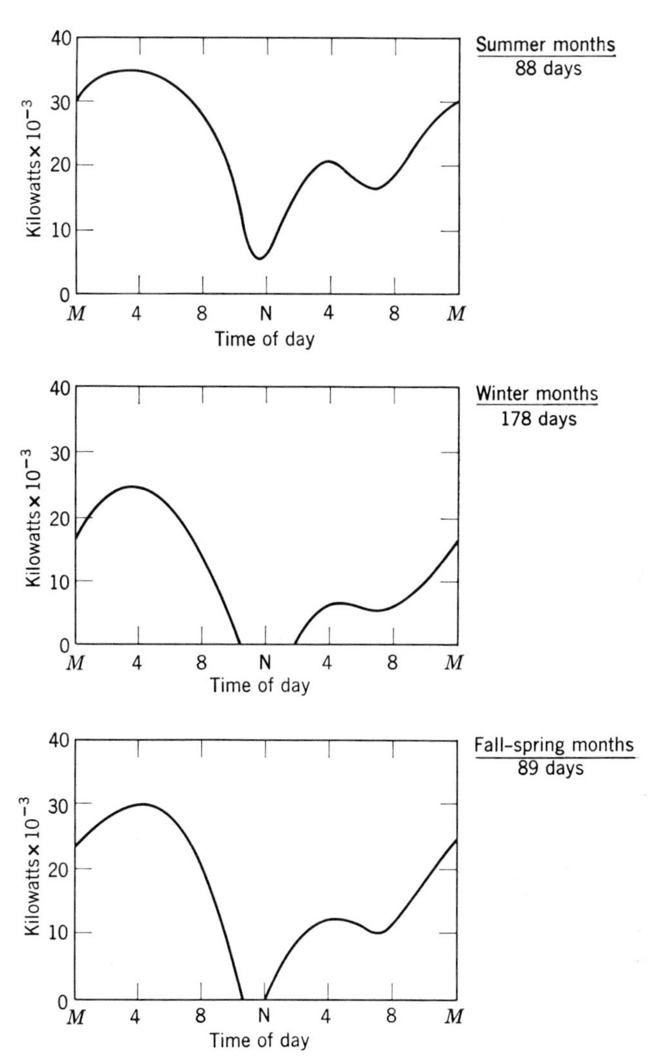

Fig. 14.6-1 Forecast of available off-peak electric power.

Table 14.6-1

Design Capacity Q_d kw	Average Power Used Q_a kw	Φ_s
10,000	7,700	0.51
20,000	12,400	0.82
30,000	14,400	0.95
36,000	15,100	1.0

Public Power Company to supplement this inadequate facility. The Public Power Company has offered its off-peak power for our use at a rate of $0.004 kwhr., the rate for peak power being $0.012 kwhr. We wish to consider the possibility of constructing a pumping station which could be used to fill our reservoir using off-peak power, thus lowering our demands for expensive peak power, allowing the expanded use of our own generating facilities. How large a pumping facility is justified?

In this simple analysis, we might think of the pumping station as a device which converts cheap electric power to more valuable peak power, the value of the power generated in our plant being equated to the cost of peak power. The venture profit might then be expressed as

$$V = Q_a \alpha \Phi_s (C_{pp} - C_{op}) - iI_0 \left(\frac{Q_d}{Q_0} \right)^M$$

where Q_a is the actual power available to the system, Q_d the design capacity of the pumping facility (both measured in kilowatts), C_{pp} and C_{op} are the costs of *p*eak *p*ower and *o*ff-*p*eak power, respectively, and $\alpha\Phi_s$ is fraction of the power entering our system as off-peak power that appears as more valuable peak power, the parameter α being an efficiency accounting for the consumption of power for driving pumps and other uses. Table 14.6-1 summarizes the calculations which might lead to the selection of a facility which can consume most effectively off-peak power. The design capacity of the pumping station Q_d is the power it consumes while pumping water in the reservoir at the maximum rate. Any off-peak power appearing at a greater rate cannot be utilized by this facility. The average available power for pumping $\Phi_s Q_a$ is then the average power over the seasons, which can be used, namely, that amount below Q_d. Notice how the 36,000 kw pumping station increases the available power by a relatively small amount over a 30,000 kw station, picking up only the brief increase in off-peak power available between midnight and 4.00 A.M. during the summer months.

14.7 THE PARAMETRIC PUMPING OF PROCESSES

There recently has evolved considerable interest in the possibility of pumping more production out of a process by operating it in a periodic fashion. The term *parametric pumping* has been coined to describe this mode of operation. In the references at the end of the chapter[3] we refer to specific instances in

[3] See Proceedings of the Symposium on Cyclic Processing Operations, *I.E.C. Prod. Design & Devel.*, No. 1, **6**, January 1967.

which this mode of operation has found use, and in this section we elect to describe a simple processing situation which might give rise to this phenomenon.

Consider the process in Fig. 14.7-1 in which material is continuously received by a process and continuously processed. We observe that the conversion of the material to product depends in a *nonlinear* fashion on the flow rate F lb/hr. A reduction in the flow rate of the material increases the conversion by an amount significantly greater than the reduction caused by a like increase in the flow rate.

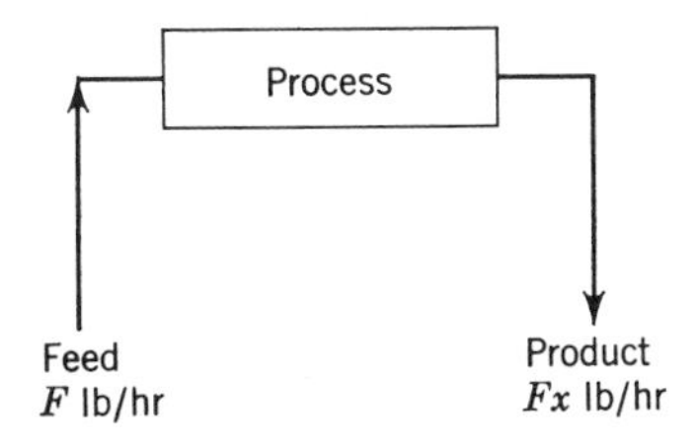

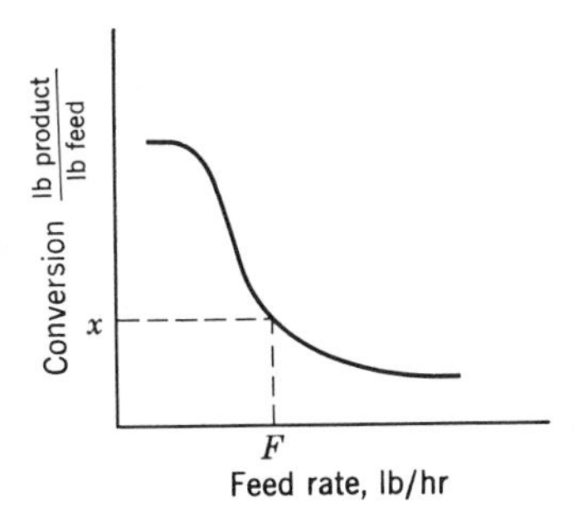

Fig. 14.7-1

This suggests the obvious: a storage tank should be installed and periodically filled and emptied in a way that will cause the process to experience pulses in supply alternately above and below the average. The net result is shown in Fig. 14.7-2, an increase in the average productivity of the system.

The effects of parametric pumping tend to be reduced by the dynamic lags of processes and a carelessly applied pumping generally reduces the efficiency of the process. Thus, a detailed study of process dynamics is necessary for the useful application of the principle presented here in a much too simplified form.

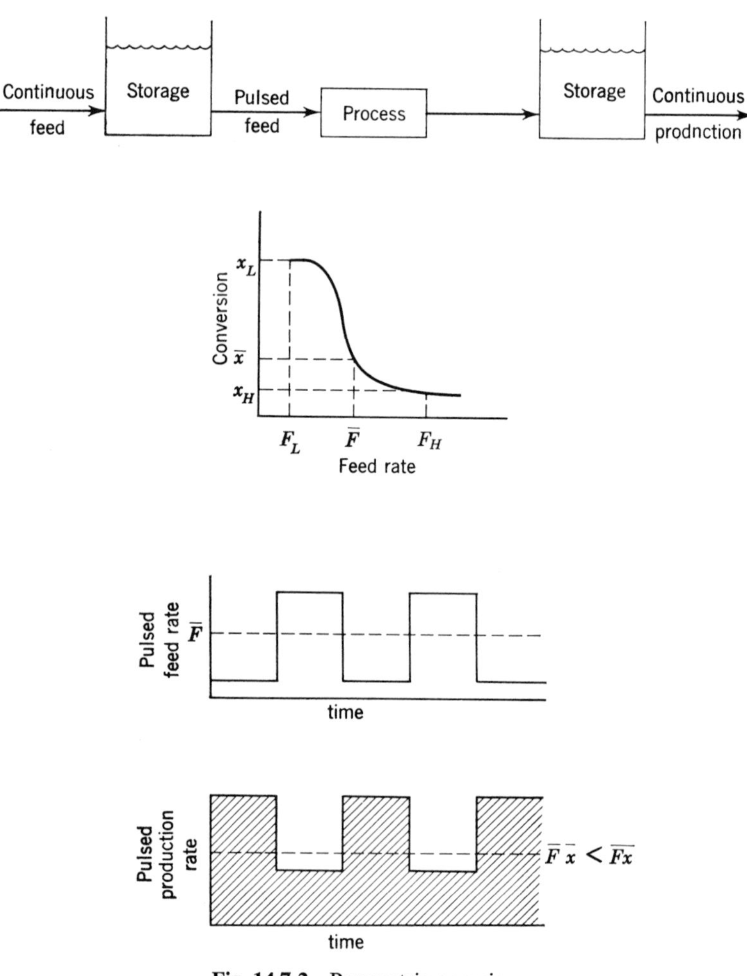

Fig. 14.7-2 Parametric pumping.

14.8 BLENDING RANDOM QUALITY VARIATIONS

We have seen how buffer storage smooths out troublesome variations in the flow rate of material, and how a system might be sized best to respond to a variable supply. We now consider the problem of smoothing *quality* variations rather than *quantity* variations. Such variations might arise as successive loads of ore arrive from different regions in a deposit to be processed. Rather wide and frequent variations in the mineral content of the ore might disrupt the operation of the ore processing system. Another frequent source of variations is the batch chemical reactor from which successive batches of product vary in quality for some unknown or unavoidable reasons.

Now, ideally, the system which is the recipient of these variations should be insensitive thereto, much as the submarine is insensitive to the waves on the surface of the ocean. However, should the variations be at a critical level, the system must be provided with a buffer to accommodate to this variable environment, much as the gyrostabilizers allow the ocean liner to accommodate to rough seas.

In some cases it may be practical to interpose a control system to protect the processing system from these environmental variations. Such a control system must be able quickly to analyze incoming feed streams or shipments and to compute an optimum and stable set of operating conditions for the unit, including the proper ratio of feed from the several feed storage areas in use. A case in point is the digitally controlled blending system used in smoothing the effects in moisture content of sand entering a cement plant. We shall not delve further into the theory of optimal process operation in the presence of variations.

Frequently a more practical means of protecting the processing unit from such input variations is a simple uncontrolled blender. This may be the only possible buffering action to use when rapid analysis of incoming materials is impracticable. The blender consists of a vessel in which material is well mixed with material that has been retained from previous batches and the quality of the blend is then a weighted time average of the quality of all the material which has arrived. This tends to smooth and integrate out any variations.

In this section we show how we might determine the proper size for a blender which must smooth uncorrelated random variations in batch quality. In Section 14.9 we analyze the effects of blending on a batch variation at the other extreme, i.e., a periodic step variation. In Chapter 15, *Simulation*, we discuss the problems of estimating the performance of blending systems being operated under more sophisticated policies.

We now confine attention to a situation similar to that illustrated in Fig. 14.8-1 in which successive batches of material arrive to be processed. The quality A of the batches has been observed to vary according to no well defined pattern, and is said to be a random uncorrelated variation.

$$A_i = \bar{A} + R_i \tag{14.8-1}$$

where A_i is the quality of the ith arrival, $\bar{A}$ is the long-range average quality of all arrivals, and R_i is the random variation. The random variation averages to zero.

$$\bar{R} = \lim_{N \to \infty} \frac{1}{N} \sum_{i=1}^{N} R_i = 0$$

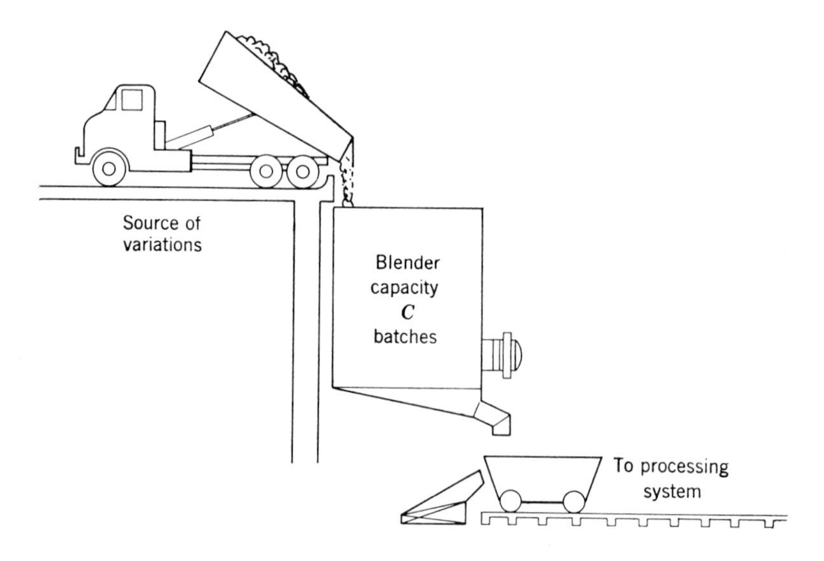

Two modes of batch quality variation

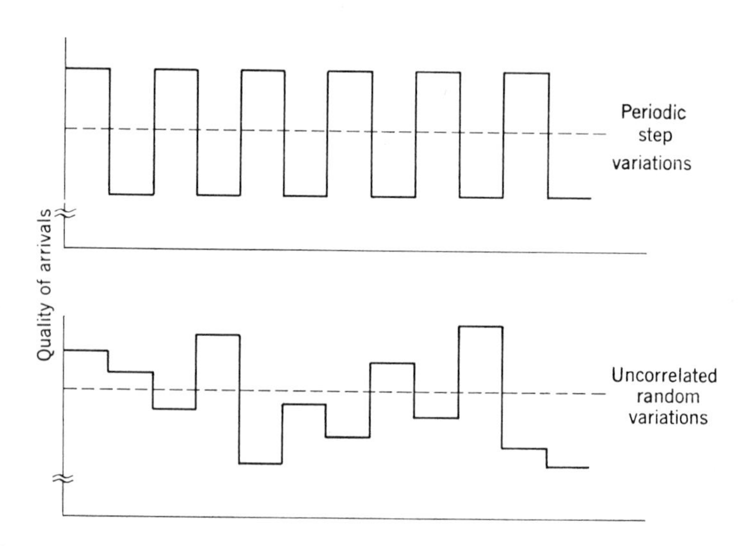

Fig. 14.8-1 Smoothing variations in the quality of successive batches.

It is common practice to measure the degree of variation by the average of the squares of the deviations from the mean, called the variance,[4] σ^2.

$$\overline{(A_i - \bar{A})^2} = \overline{(R_i^2)} = \lim_{N\to\infty} \frac{1}{N} \sum_{i=1}^{N} (R_i^2) \equiv \sigma_A^2 \qquad (14.8\text{-}2)$$

The smaller the variance the less the variation about the mean.

The blender consists of a vessel with a capacity of C batches. A total of $C - 1$ batches of the previous blend are retained for mixing with each new arrival. Once the C batches are blended, one batch leaves the blender for processing. Thus, the quality of the blend is given by

$$B_i = \frac{(C-1)B_{i-1} + A_i}{C} \qquad (14.8\text{-}3)$$

where B_i is the quality of the ith blend, A_i is the quality of the ith arrival, and B_{i-1} is the quality of the batches of the previous blend retained in the blender. We wish to determine the effect of the design parameter C on the reduction of quality variance achieved by the blending operation.

We take advantage of the fact that

$$\lim_{N\to\infty} \frac{1}{N} \sum_{i=1}^{N} B_i = \lim_{N\to\infty} \frac{1}{N} \sum_{i=1}^{N} B_{i-1}$$

and average Eq. 14.8-3 to get

$$\bar{B} = \frac{(C-1)\bar{B} + \bar{A}}{C}$$

or

$$\bar{B} = \bar{A} \qquad (14.8\text{-}4)$$

Thus, as to be expected, the average quality of the blended material is the same as the arriving material.

Now consider the variance of the blend about this average, using Eqs. 14.8-3 and 14.8-4, for a sufficiently large number of batches.

$$\sigma_B^2 = \overline{(B_i - \bar{B})^2} = \overline{\left[\frac{(C-1)(B_{i-1} - \bar{B}) + (A_i - \bar{A})}{C}\right]^2}$$

$$= \overline{(B_{i-1} - \bar{B})^2}\left(\frac{C-1}{C}\right)^2 + 2\overline{(B_{i-1} - \bar{B})(A_i - \bar{A})}\,\frac{C-1}{C^2}$$

$$+ \frac{\overline{(A_i - \bar{A})^2}}{C^2}$$

[4] For reasons of statistical significance in the definition of σ^2 the sum is divided by $N - 1$ rather than N. However, for large N, $\overline{R^2} \to \sigma_R^2$.

The first term on the left is, by definition

$$\sigma_B^{\,2}\left(\frac{C-1}{C}\right)^2$$

The second term is the average of the product of two uncorrelated random numbers each with mean zero; hence, it is zero. The third term is, by definition

$$\sigma_A^{\,2}\left(\frac{1}{C}\right)^2$$

This reduces to

$$\sigma_B^{\,2} = \sigma_B^{\,2}\left(\frac{C-1}{C}\right)^2 + \sigma_A^{\,2}\left(\frac{1}{C}\right)^2$$

or

$$\frac{\sigma_B^{\,2}}{\sigma_A^{\,2}} = \frac{1}{2C-1} \qquad (14.8\text{-}5)$$

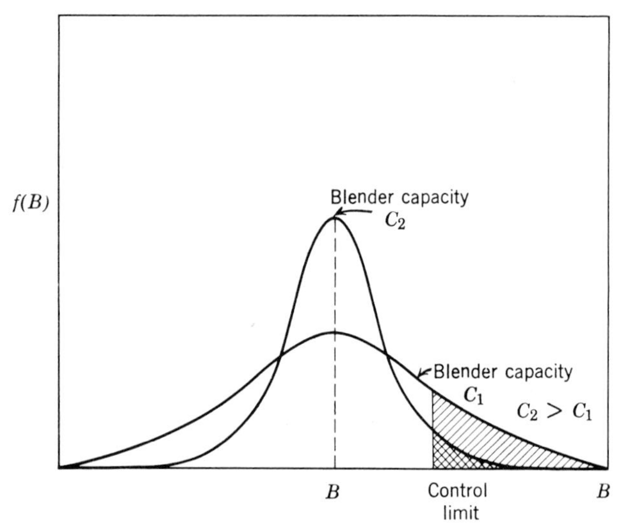

Fig. 14.8-2 The smoothing effect of blending.

Equation 14.8-5 describes the effect that the blender has in smoothing these random variations. Notice when $C = 1, \sigma_B^{\,2}/\sigma_A^{\,2} = 1$, the blender has no effect. Also as C becomes large $\sigma_B^{\,2}/\sigma_A^{\,2} \to 0$, a large enough blender can smooth the quality variations to zero. A blender of size $C = 5$ reduces the variance by a factor of nine.

A reasonable design criterion for such a blender does not involve smoothing the variations, rather it involves the *economic* smoothing of the variations. That is, to determine the optimal value of the blending parameter C we must balance the cost of the blender against the loss of profitability to the system caused by the variations.

For example, should the quality variations be distributed about the average according to the bell-shaped distribution shown in Fig. 14.8-2, and should a quality control limit be imposed beyond which batches require special and expensive treatment, the economic analysis would proceed as follows. We can compute the reduction in the variance of this distribution to be expected by any particular blender size and, hence, can compute the fractional reduction in the batches which require special treatment. The resulting saving in cost would be the gross profit associated with the blender design, and the equipment investment would be just that of the blender, a balance typical of those encountered again and again.

14.9 SMOOTHING PERIODIC STEP VARIATIONS

It is always of interest to look at opposite extremes, since, hopefully, reality might be bracketed between. An opposite extreme to the uncorrelated random variation in batch quality is a periodic step variation in which successive batches fall alternately above and below the average quality unbeknown to the operator of the process when rapid analysis cannot be made.

$$A_i = \bar{A} + (-1)^i D \qquad (14.9\text{-}1)$$

where D is the extent of the variation, which averages to zero. The variance for a large number of batches is

$$\sigma_A{}^2 = \overline{(A_i - \bar{A})^2} = \overline{(-1)^{2i} D^2} = D^2 \qquad (14.9\text{-}2)$$

The quality of the material issuing from the blender will also exhibit a periodic step variation, however, of a lesser variation.

$$B_i = B_{i+2} \qquad \text{for all } i \qquad (14.9\text{-}3)$$

We can easily solve for the variation in blend quality by considering the blend Eq. 14.8-3 for two successive blends, and invoking Eq. 14.9-3

$$B_i = \frac{(C-1)B_{i-1} + \bar{A} + (-1)^i D}{C}$$

$$B_{i-1} = \frac{(C-1)B_i + \bar{A} + (-1)^{i-1} D}{C}$$

Eliminating B_{i-1} gives

$$B_i = \bar{A} + \frac{(-1)^i D}{2C - 1} \tag{14.9-4}$$

Notice how $\bar{B} = \bar{A}$ and the extent of variation has been reduced from D to $D/(2C - 1)$.

The variance in the blend quality is

$$\sigma_B{}^2 = \overline{(B_i - \bar{B})^2} = \overline{\left[\frac{(-1)^i D}{2C - 1}\right]^2} = \frac{D^2}{(2C - 1)^2} \tag{14.9-5}$$

Thus, the reduction in variance is

$$\frac{\sigma_B{}^2}{\sigma_A{}^2} = \left(\frac{1}{2C - 1}\right)^2 \tag{14.9-6}$$

Compare the reduction in variance for the random variation, Eq. 14.8-5, to that for the periodic variation. We would expect that the periodic variation would be more easily smoothed. A blender with $C = 5$ now reduces the variance by a factor of eighty-one, rather than nine.

14.10 THE SURGE TANK

In the preceding sections we have studied the effectiveness of storage and blending systems in protecting a process from a variety of disturbances. Now we focus attention on the simple surge tank as a device for smoothing variations in the quality of a continuous flow stream. A surge tank merely is a vessel of

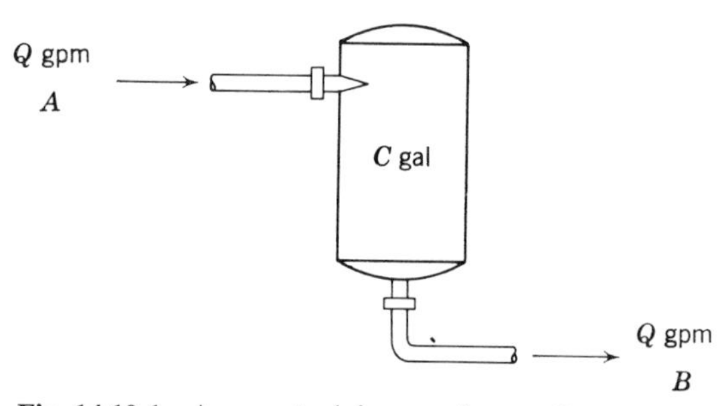

Fig. 14.10-1 A surge tank in a continuous flow stream.

some sort placed conveniently in a transportation line, to absorb both variations in the flow rate and the quality of material being transported.

In this section we analyze the surge tank illustrated in Fig. 14.10-1, assuming that the flow rate does not vary. Material enters the surge tank at a rate of Q cubic feet per hour, is well mixed, and leaves at the same rate. The quality A of the arriving material varies and we wish to estimate the variation in the quality B of the blended material. The quality of interest could be the concentration of some component.

A transient material balance over the surge tank is

$$C\frac{dB}{d\theta} = Q(B - A) \qquad (14.10\text{-}1)$$

where C is the capacity of the surge tank, and the qualities B and A both vary with time.

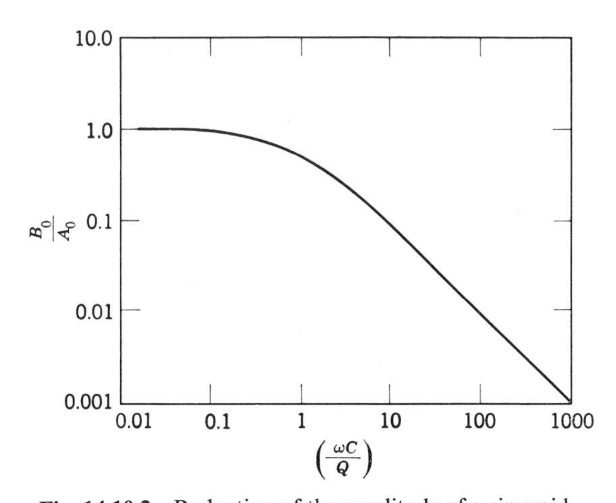

Fig. 14.10.2 Reduction of the amplitude of a sinusoid.

An input variation which leads to easy solution and which illustrates the effect of the surge tank is a simple sinusoid.

$$A = A_0 \sin(w\theta)$$

where A_0 is the amplitude and w is the frequency.

It can be shown that the steady-state solution to the material balance, Eq. 14.10-1, is also a sinusoid but with reduced amplitude and phase shifted.

$$B = B_0 \sin(w\theta + \phi)$$

Such frequency response studies are frequently employed in control studies, and here the ratio of the amplitudes is of prime interest.

$$\frac{B_0}{A_0} = \sqrt{\frac{1}{1 + (wC/Q)^2}}$$

In Fig. 14.10-2 we see the effectiveness of the surge tank in smoothing high frequency variations. A 200 gal tank would reduce the amplitude of a

1 cycle per minute sinusoidal variation in the quality of a 100 gpm stream by a factor of over two.

$$\frac{B_0}{A_0} = \sqrt{\frac{1}{1 + [(1)(200)/(100)]^2}} \approx 0.45$$

In practice a pure sinusoid would probably not appear in any flow system. Some complex combination of step changes, random variations, and periodic fluctuations would appear. Qualitatively, the surge tank would effect the same smoothing.

14.11 CONCLUDING REMARKS

We have attempted to offer a taste of the multitude of variations which might buffet a processing system and indicate an approach to engineering around them. Needless to say, our presentation was only suggestive and sketchy. The mass of problems is greater than any mass of solutions we might attempt to compile. What is important is the way of thinking, however, since, hopefully, a proper approach to any special problem will lead to a reasonable solution. We have attempted to describe a proper approach.

References

An excellent introduction to the theory of waiting lines is:

P. M. Morse, *Queues, Inventories and Maintenance*, Wiley, New York, 1958.

Applications of the theory of waiting lines include:

A. M. Stover, "Application of Queuing Theory to the Operation and Expansion of a Chemical Plant," *Chem. Eng. Progr., Symp. Ser.*, No. 42, **59** (1963).

P. D. Birkhahn, "Probability and Reality—an Ocean Going Version," *Chem. Eng. Progr., Symp. Ser.*, No. 42, **59** (1963).

N. H. Smith and D. F. Rudd, "On the Effects of Batch Time Variations on Process Performance," *Chem. Eng. Sci.*, **19** (1964).

Methods for improving the utilization of storage facilities are discussed in:

D. J. Wilde and U. Passy, "Partial Control of Linear Inventory Systems," *Chem. Eng.*, No. 2, **13** (1967).

The theory of blending is discussed in the following:

R. Shinnar, "Sizing of Storage Tanks for Off-grade Material," *Ind. Eng. Chem. Process Design Develop.*, No. 2, **6** (1967).

S. Katz, "A Statistical Analysis of Certain Mixing Problems," *Chem. Eng. Sci.*, No. 61, **9** (1958).

PROBLEMS

14.A. Feedstock is available continuously at a rate of Q pounds per hour to be used primarily in a batch process which takes a B pound batch of feedstock every A hours. If the B pounds of feedstock are not available, the batch process must remain idle until sufficient feed has accumulated.

(a) Determine the supply and process utilization factors in terms of Q, A, B, and C, where C is the capacity in pounds of a storage tank which is available.

(b) If the value of the feedstock to the batch process is C_p dollars per pound, to the overflow use is C_0 dollars per pound, and if the cost of storage is C_s dollars per pound of storage per hour, determine the economically optimal storage capacity.

14.B. This is a simple problem in probability balancing. A feedstock is to arrive in tank car quantities to a process which consists of two identical and independent reactors each capable of being charged with one tank car of feed. The tank cars arrive at random with an average of $\bar{Q}$ cars per day, and the reactors' batch times are random with each reactor averaging $\bar{Q}_r$ batches per day. There is no storage available and cars which arrive when the reactors are full are to be shunted to another process.

Assuming the system can be found in any one of the following three states: system empty, one reactor operating, two reactors operating, determine the supply and process utilization factors. What fraction of the process capacity is used?

14.C. Copper is being mined from one large pit across which the grade of ore varies. The following are representative samples taken from 100 ton truck loads when entering the coarse crusher.

Date	*Per Cent Copper*
2/1	2.86
2/9	2.49
2/12	2.32
2/14	2.94
2/15	2.73
2/23	2.90
2/23	2.61
2/24	2.05
2/28	2.30
2/29	2.55

The feed to the mill circuit requires a standard deviation of less than 0.1 per cent about the average grade. This small deviation is necessary to reduce reagent costs in the flotation unit. How large an ore hopper is needed after the primary crusher to reduce the feed grade specifications to this limit?

14.D. A two-stage batch processing system has been plagued by batch time variations, causing blocking and idleness. In the text we derived the expression shown in Fig. 14.D-1 for the productivity of the system in the case of completely random variations in batch time and no storage.

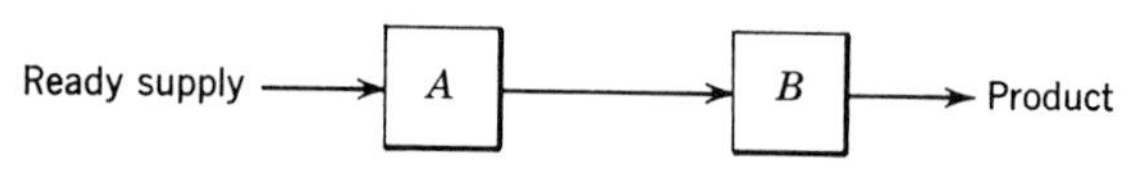

Fig. 14.D-1

$$\bar{Q}_{\text{system}} = \frac{\bar{Q}_B(1 + \bar{Q}_A/\bar{Q}_B)}{(1 + \bar{Q}_A/\bar{Q}_B + \bar{Q}_B/\bar{Q}_A)}$$

(a) One means of increasing productivity is to allow intermediate storage of the product from process A. Derive an expression for the productivity of the system when allowance is made for the storage of N batches of intermediate product. (See Fig. 14.D-2.)

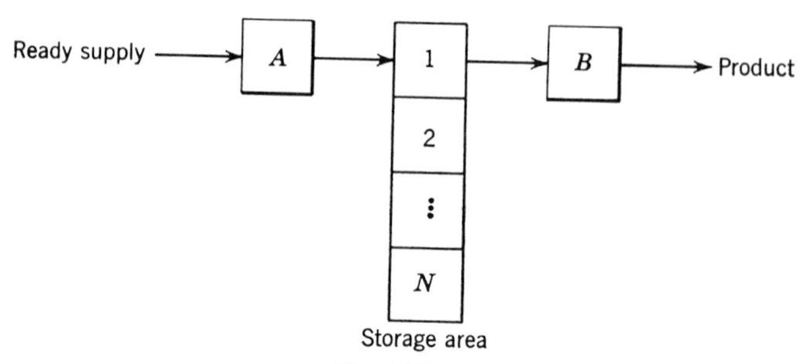

Fig. 14.D-2

(b) Another means of increasing the productivity might involve the duplication of certain of the processes. Derive an expression for the productivity of the system with process A duplicated and no storage provision. (See Fig. 14.D-3.)

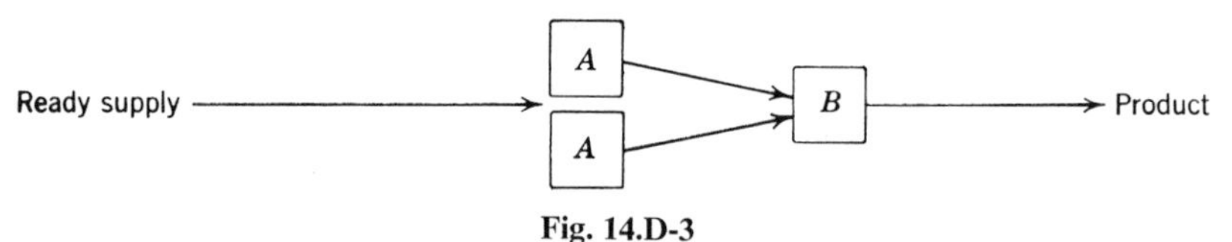

Fig. 14.D-3

(c) Derive an expression for the case of process B duplicated. (Fig. 14.D-4.)

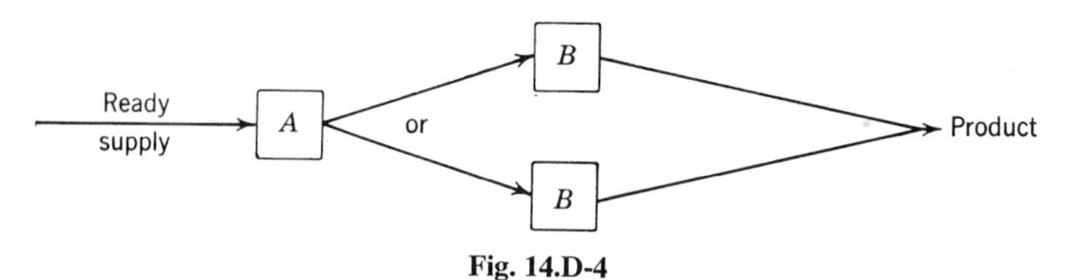

Fig. 14.D-4

(d) Show that the duplication of the original system more than doubles the productivity. (Fig. 14.D-5.)

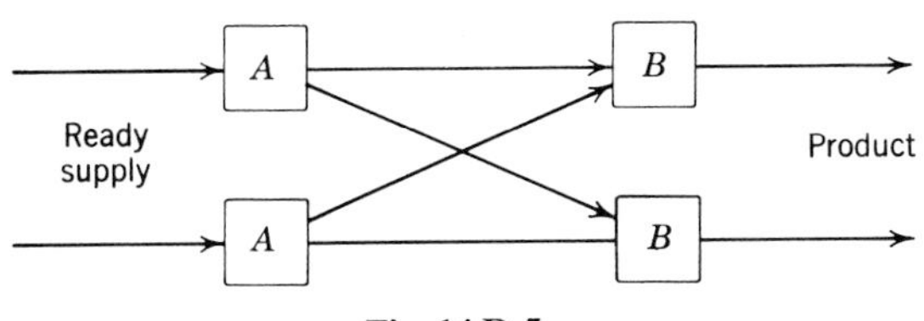

Fig. 14.D-5

14.E. A refinery receives crude oil by tankers which dock adjacent to the processing system. One 80,000 ton tanker is to arrive every three days, but weather conditions violently disrupt this schedule: the arrivals exhibit random variability and only average to one tanker every three days. It costs $C_A = \$300$ per ship hour to keep a tanker at anchor waiting for dockage.

It has also been observed that the unloading time varies from tanker to tanker once in dock, approximated here by pure randomness. Suppose that the cost of operating a given unloading facility is proportional to the average unloading rate for that facility $C_D = \$1{,}000$ per tanker per hour.

(a) Derive the following expression for the expected number of tankers at anchor, in terms of the average arrival rate, Q_A tankers per day, and the average unloading rate, Q_D tankers per day.

$$N = \frac{Q_A}{Q_D(1 - Q_A/Q_D)}$$

(b) The average wait in anchor, W, is N/Q_A days. Derive an expression for the expected operating cost for the facility, anchorage and dock, and determine the minimum cost average unloading rate.

(c) If the oil refinery which receives the variable supply of crude oil in this problem can only operate continuously at a rate of 80,000 tons per three days, how much tankage will be required to smooth the supply?

14.F. In the system of Section 14.3, consider the case when the processing time is constant rather than random. Demonstrate why the probability balance equations do not apply, and why the analysis of this problem is extremely difficult using the simple concepts of queuing theory. In the next chapter we show how this difficulty can be overcome using Monte Carlo methods.

14.G. It is frequently valuable to analyze in detail an idealized picture of the more complex problem which might present itself in practice, in order to get a feeling for the structure and form of optimal plans. Consider the problem of sizing a processing system to respond optimally to a sawtooth variation in a supply of material or energy. (See Fig. 14.G-1.)

Consider that every unit of the supply utilized yields a profit of P_p \$/unit, that each unit not utilized by this system yields a profit of P_0 \$/unit, and that the investment in the processing unit follows the usual power law

$$I = I_0 \left(\frac{Q}{Q_0} \right)^M$$

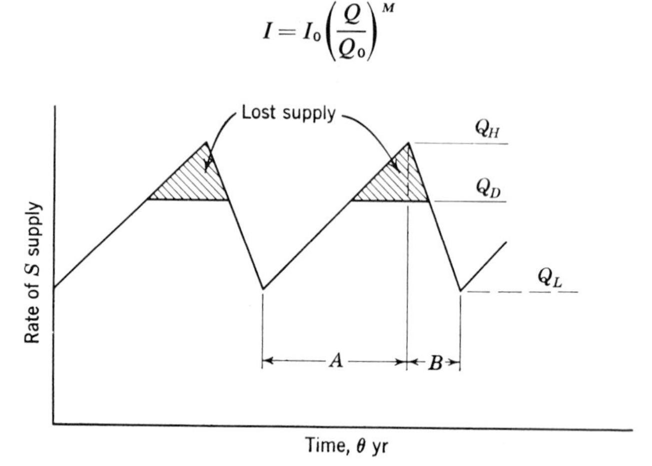

Fig. 14.G-1 A sawtooth supply variation.

(a) Show that during one complete period of operation, $A + B$ years, with the design capacity in the range $Q_H \geq Q_D \geq Q_L$, Q_a units of supply will be utilized.

$$Q_a = \left[Q_D - \frac{1}{2} \frac{(Q_D - Q_L)^2}{(Q_H - Q_L)} \right] (A + B)$$

and that the following units will be lost.

$$\left[\frac{Q_H + Q_L}{2} - Q_D + \frac{1}{2} \frac{(Q_D - Q_L)^2}{(Q_H - Q_L)} \right] (A + B)$$

(b) Show that the venture profit for this system is of the form

$$V = P_p \left[Q_D - \frac{1}{2} \frac{(Q_D - Q_L)^2}{(Q_H - Q_L)} \right] + P_0 \left[\frac{Q_H + Q_L}{2} + Q_D \frac{1}{2} \frac{(Q_D - Q_L)^2}{(Q_H - Q_L)} \right]$$
$$- iI_0 \left(\frac{Q_D}{Q_0} \right)^M$$

(c) Show that the design capacity Q_D which results in the maximum venture profit, obtained by setting the derivative of the venture profit with respect to Q_D to zero, reduces to

$$\frac{(Q_H - Q_D)}{(Q_H - Q_L)} = \frac{I(M)}{Q_D(P_p - P_0)} I_0 \left(\frac{Q_D}{Q_0} \right)^M$$

for $Q_L < Q_D \leq Q_H$

(d) Now consider the possibility that the optimal design capacity is below the sawtooth variation altogether, $Q_D < Q_L$. Show that the venture profit is then to equal

$$V = P_0 Q_D + P_0\left[\frac{(Q_H + Q_L)}{2} - Q_D\right] - iI_0\left(\frac{Q_D}{Q_0}\right)^M$$

and the resulting optimal capacity is

$$1 = \frac{1}{Q_D}\frac{i(M)}{(P_p - P_0)} I_0\left(\frac{Q_D}{Q_0}\right)^M$$

for $Q_D < Q_L$.

14.H. A hydrocracking unit is to be expanded from the present 12,000 barrels per day capacity to 18,000 barrels per day. At the same time a new hydrogen plant is to be built to augment the existing hydrogen supply to the hydrocracker. The hydrogen supplied from this new plant will be fairly steady, except when shut down. However, the existing hydrogen supply has extreme variations, fairly represented in a sawtooth fashion as shown in Fig. 14.H-1. The oil feed, primarily heating oil, varies in its quality so that the consumption of hydrogen per barrel of feed varies in a pulsed fashion. The pulses are caused by periodic inclusion with the heating oil of another feed, refuse furnace oil RFO, which is a high consumer of hydrogen.

(a) Assuming a uniform feed quality, determine a reasonable size of the new hydrogen plant. answer: 14 MMSCF/D
(b) With this reasonable hydrogen supply and assuming it constant, estimate the economically optimal way to introduce the RFO feed. Would some tankage for the RFO be justified? answer: 64 MB

There is adequate storage for heating oil but very little storage is currently available for RFO. The incremental profit for hydrocracking the RFO is \$2.00 per barrel and that for heating oil is \$0.50 per barrel. The total feed is this RFO supply plus the amount of heating oil that can be cracked with the given amount of hydrogen up to the 18,000 barrels/day limit of the expanded hydrocracker.

Costs:

Hydrogen plant $I = \$1.0\text{ MM}(Q_{H_2}/1\text{ MMSCF/day})^{0.6}$
Plant and project life $= 10$ yr
Operating costs $C_0 = C_F + (\$0.20/\text{MSCF})Q$
Tankage for RFO $I_T = \$100{,}000(S/100{,}000\text{ bbl})^{0.8}$
Marginal rate of return required \$0.20/yr-\$ invested
Hydrogen Consumptions: (a) For heating oil 2,000 SCF/bbl
(b) For RFO 3,000 SCF/bbl

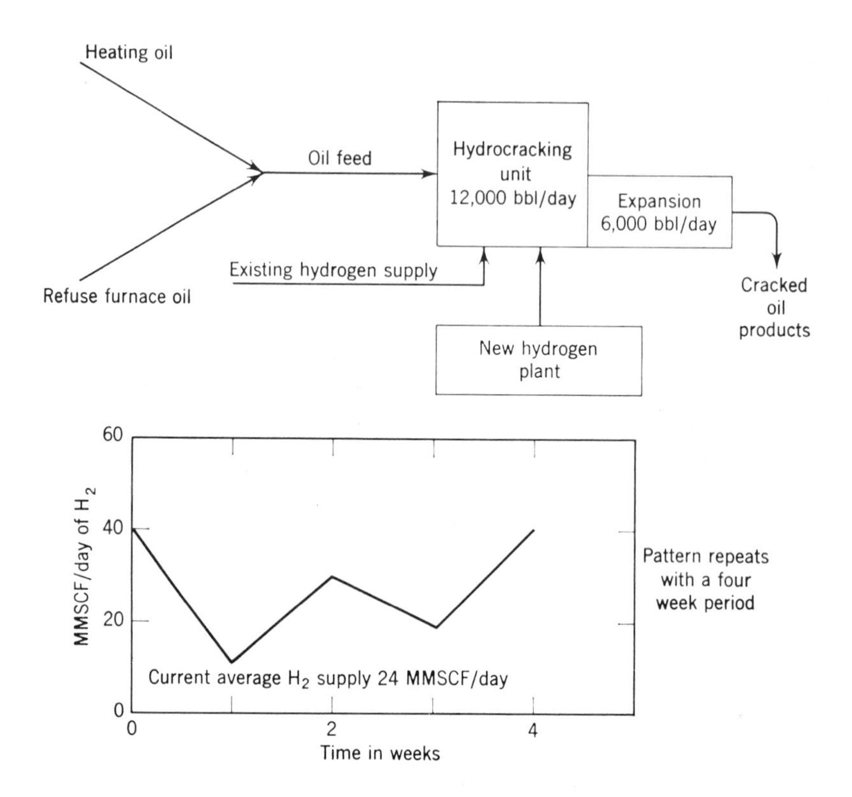
Heating oil
Oil feed
Hydrocracking unit 12,000 bbl/day
Expansion 6,000 bbl/day
Existing hydrogen supply
Refuse furnace oil
Cracked oil products
New hydrogen plant
MMSCF/day of H_2
60
40
20
0
0
2
4
Time in weeks
Pattern repeats with a four week period
Current average H_2 supply 24 MMSCF/day

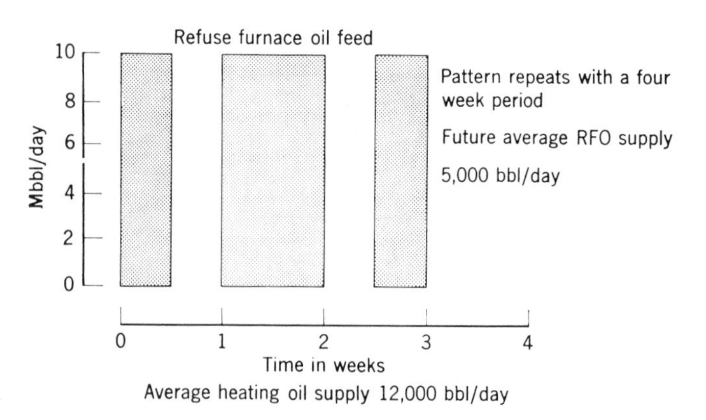
Refuse furnace oil feed
Mbbl/day
10
8
6
4
2
0
0
1
2
3
4
Time in weeks
Pattern repeats with a four week period
Future average RFO supply
5,000 bbl/day
Average heating oil supply 12,000 bbl/day

Fig. 14.H-1

14.I. Apply the methods of this chapter to the problem of equipment failure in a continuous process. For example, let pump A be the regular feed pump, and let pump B be a spare. Assume the following four random events occur at the stated average frequencies (events per year): A fails at a rate F, yr^{-1}; A is repaired at a rate R, yr^{-1}; B fails at a rate F, yr^{-1}; and B is repaired at a rate R, yr^{-1}. What are the states in which the system can be found? What is the reliability of the system with and without the spare?

15

SIMULATION

Process engineering problems can be extremely complex when they involve the need for simultaneously accounting for future developments, uncertain data, component failure, and both internal and external variability; especially when these must be accounted for during process optimization. The methods of process simulation have now been refined to the point where they are useful in interpreting these difficult processing situations.

Typical Problem

A chemicals manufacturer anticipates a need for frequent and continuing redesign of power generating capability within a large complex, as the nature and extent of processing change. The environment is extremely complex, with frequent and random outages of boilers, variable demands for power from without and within, uncertain performance characteristics, poorly forecast future, and so forth. How might a computer simulation of this environment be of use to the design engineer in planning for future design modifications?

15.1 A BALANCE BETWEEN CONVENIENCE AND REALITY

It has been suggested that the best advice to those about to undertake an extensive simulation of a processing problem is: "DON'T." While we cannot agree fully with that extreme view, it contains an element of truth. We might temper that advice to read: "Don't use a complex and involved technique to solve problems that might be solved by simple analysis." The uncertainty of the future tends to undermine the validity of over-complicated analyses.

Simulation involves the modeling of a complex situation into a simpler and more convenient form which can be observed at leisure without some of the irrelevant and troublesome side effects which might accompany the original complex situation. Now, clearly, any time we write an equation to describe a physical event or run a small scale experiment, we are engaging in modeling.

However, this does not connote simulation in the popular sense of the word. Simulation usually involves the programming of the dynamic operation of a complex system for the computer, and the study of the response of this operational simulation to a variety of environmental disturbances.

Some balance must be reached between the realism of the simulation and the ease by which the simulation can be effected. Clearly, we need not make an elaborate simulation of the operation of a proposed refinery to detect a hazardous condition, when a simple inspection of the processing plan will reveal the difficulty. On the other hand, we cannot afford the luxury of a simulation which neglects some instability which might lead to the destruction of the plant.

We illustrate the concept of simulation in process engineering by a series of examples. First, we illustrate the use of an electric analog simulator to test the design of a fire-water distribution system. Then, Monte Carlo simulation is illustrated in the design of docking facilities at the Esso Refinery, Fawley, England. A complex blending problem is analyzed next by Monte Carlo simulation. Finally, an extensive special purpose simulator on the digital computer is discussed, Dow Chemical's POWERFACTS program.

15.2 A FIRE-WATER DISTRIBUTION SIMULATOR[1]

American Oil Company of Whiting, Indiana, reports the use of an electric analog simulator to aid in planning the expansion and redesign of complex fire-fighting facilities. This is an excellent example of fitting the tool to the problem at hand.

A fire-water distribution system consists of pumping stations, a piping system, and hose stations, which might have grown throughout the years to a highly complex network as the processing complex it services grows. The addition of a crude oil topping unit here, the dismantling of an old still there, the addition of a pumping station here, and so on might so complicate the original system as to render its understanding difficult.

Will there be enough water to fight two simultaneous fires?
Should the old water lines be discarded as older refinery units are torn down?
Which pumps should be operated to fight a fire best in a given location?
Would emergency pumps be adequate should the main power source fail?
Can plugged lines be found by limited testing?
How can we train our fire-fighting crews without disrupting the operation of the refinery?

[1] L. T. Wright, *Hydrocarbon Process. Petrol. Refiner*, No. 9, **41** (1964).

Such refinery water system problems are troublesome, since they normally involve miles of piping of various sizes, connected at numerous points, supplied by pumps at various locations, and subject to demands of various amounts at various locations.

In the development of a water distribution system simulator, it was observed that there exists a strong *similarity* between the equations describing:

Turbulent flow of water through a pipe

$$\Delta p = a(F)^{1.8}$$

where Δp = pressure drop
F = flow rate
a = constant dependent on pipe length and diameter

Current flow through a light bulb below incandescence

$$\Delta V = a'(I)^{1.8}$$

where ΔV = voltage difference
I = current
a' = constant dependent on the bulb wattage

Thus, if voltage difference is interpreted as pressure drop and current is interpreted as flow rate, a network of light bulbs might simulate a network of pipes.

1 volt ≈ 1 pound per square inch
1 milliampere ≈ 100 gallons per minute

Now, by connecting the proper network of light bulbs to binding posts on a large map of the refinery, we have a convenient simulation of the fire-water distribution system. Proposed modifications of the system can be tested by simply disconnecting and connecting the proper bulbs.

The water pumps are simulated by batteries or other power sources. For example, a radio "*B*" battery plus a variable resistance offers a fairly good simulation of a pump. Or, perhaps, this simulator can use the power source of an analog computer to simulate nonlinear pump characteristics.

The flow of water through a fire hose is simulated by grounding the proper point in the model through a resistor. American Oil Company has constructed a simulator for the fire-water system at its Whiting refinery to model 11 pumps and nearly 300 segments of piping connected at some 200 points. The simulator uses 500 light bulbs.

This simulator is used for design, operation, and maintenance studies, and also for training fire-fighting crews. The fire-fighting crews can test the response of the system, at their leisure, to various fire situations, gaining experience and a "feel" for the facilities available to them, otherwise gained only during a fire.

For other applications of this electric analog simulator we quote from the report of American Oil:

Unexpected Discoveries. Some unexpected facts have been uncovered in practical refinery situations. For example, one fire-water pump, located near a remote facility was tied into the main system only by a very long line through the tank fields. It was thought useless except for a fire in the remote facility or the tank field. However, the pump has a very high shut-off head and a steep head-capacity curve. So in spite of its remote location, it can contribute to fighting fires in the main refinery. These facts were easily demonstrated on the analog, whereas they might never have been determined in the field.

In another instance, an old fire-water line had sprung a leak under a tank. Repair would have been costly. Analog studies showed this line to be of marginal value, so it was shut off and not replaced.

In still another case, a field test was made of the fire-water piping in a large tank field by opening turret fog nozzles at the far end of the field. Pressures were then read at several hydrants throughout the field. (Turret fog nozzles or fog nozzles on hoses are fairly good flow meters for such tests.)

Fire-water piping in this tank field was highly interconnected, but was hooked up to the main refinery supply by only two under-the-road crossings. Results of the field test did not agree with original analog results, so a study was made on the analog. It quickly revealed that plugging of the western supply line was the only trouble which, when simulated on the analog, would explain the test results. A field check confirmed that the line was plugged.

The good results from studies of fire-water systems indicate that extensions to such systems, for steam and process water, should be valuable.

This then describes a process engineer's "war game," pitting his creation against a reasonable environment, ironing out any difficulties before the performance really counts.

Notice that this simulation has realism and ease of handling. The realism is a direct result of the close similarity between the laws describing the operation of the piping network and the electric analog. *Such essential similarities must always be present.*

15.3 THE ESSO REFINERY BERTHING PROBLEM

B. D. Dagnall, Head of the Operations Research Section at the Esso Refinery, Fawley, England, reports on the use of simulation for the sizing of equipment to meet varying demands.[2] In particular, the use of Monte Carlo simulation

[2] B. D. Dagnall, *Brit. Chem. Eng.*, No. 11, **10** (1965).

in the redesign of berths for the ocean going tankers which bring crude oil to the Fawley refinery is discussed. This provides an excellent real-life example of Monte Carlo simulation.

At the Esso Refinery, Fawley, there were five berths for oceangoing vessels, and in 1963 only one of them, Number 5, was capable of handling vessels above 80,000 tons. It was expected that in 1964 there would be some 118 arrivals of such large tankers; on the average, one every three days. It was realized that it would be impossible to schedule the arrivals perfectly and that some large ships might have to wait at anchor for Berth 5 to clear, at an estimated waiting cost of £100 per hour per vessel. Would it be wise to modify Berth 4 by dredging and changing hose facilities, at a cost of some £70,000, to take these big ships?

The factors which prevented the analytical solution to this problem are that the arrivals are irregular, neither exactly periodic nor purely random, and that the alongside times vary from vessel to vessel. The analytical methods developed in Chapter 14 to model such variable situations are inadequate. Another approach must be taken, one which is less dependent on mathematical manipulations. As Melville has observed, there are some enterprises in which a careful disorderliness is the true method!

It had been established that the arrivals of tankers carrying crude oil might be approximated by a series of random variations superimposed on a regular arrival schedule. Thus, the irregular arrival of ships might be *simulated* by selecting time advances and delays randomly from a sample of typical advances and delays, and applying these to the regular schedule. The typical advances and delays could be established from past records for different sized vessels.

The vessels could be brought into berth only at daylight high tide (i.e., at intervals of approximately 25 hr), and the vessels would take more than 25 hr and less than 50 hr to unload. Thus, a vessel effectively would tie up the dock for two tidal days, 50 hr. Hence, a constant two day alongside time might adequately simulate the berthing.

Two additional effects were expected when Berth 4 was uprated. First, using Berth 4 for an 80,000 ton tanker prevents its normal use for smaller vessels. Second, Berth 3 has to be free when a large vessel arrives at Berth 4 to allow for maneuvering room. These were thought to be minor effects which might be neglected in the simulation; an approximate compensation of £15,000 per year was included for such interference by the use of Berth 4 for large vessels.

The simulation is simple enough to carry out by hand. Sample advances and delays can be placed in a hat and drawn at random to modify the regular schedule. Each simulated vessel can then be traced through its course of waiting and unloading; in the case of only Berth 5 available for large vessels, and in the case of both Berths 4 and 5 available. Such a simulation is shown in Table 15.3-1. For the 40 arrivals simulated, the uprating of Berth 4 saves 21 ship-days of waiting, for a yearly saving of 62 ship-days at £100 per ship-hr.

Table 15.3-1 Typical Advances and Delays in Berthing Schedules, Showing Arrival Pattern

Planned Arrival Date	Delay (+) or Advance (−)	Simulated Arrival Date	Number 5 Berth Only		Numbers 5 and 4 Berths	
			Date of Berthing	Delay (Days)	Date of Berthing	Delay (Days)
3	0	3	3		3	
6	1	7	7		7	
9	5	14	14		14	
12	0	12	12		12	
15	5	20	21	1	20	
18	1	19	19		19	
21	0	21	25	4	22	1
24	−4	20	23	3	21	1
27	1	28	28		28	
30	−2	28	30	2	28	
33	0	33	33		33	
36	3	39	39		39	
39	0	39	41	2	39	
42	1	43	45	2	43	
45	−4	41	43	2	41	
48	5	53	53		53	
51	5	56	56		56	
54	5	59	60	1	59	
57	1	58	58		58	
60	5	65	65		65	
63	−1	62	62		62	
66	2	68	68		68	
69	3	72	72		72	
72	1	73	74	1	73	
75	2	77	77		77	
78	3	81	81		81	
81	−2	79	79		79	
84	−2	82	83	1	82	
87	2	89	89		89	
90	2	92	92		92	
93	0	93	94	1	93	
96	5	101	101		101	
99	−4	95	96	1	95	
102	5	107	108	1	107	
105	1	106	106		106	
108	1	109	110	1	109	
111	3	114	114		114	
114	−2	112	112		112	
117	1	118	118		118	
120	1	121	121		121	
	Total delays			23		2

Savings by uprating Berth 4: 23 − 2 = 21 ship-days. This is in respect of 40 ships. The savings per year are therefore 21 × 118/40 = 62 ship days.

The uprating of Berth 4 seems to save some

$$\left(\frac{62 \text{ ship-days}}{\text{yr}}\right) (\pounds 100/\text{ship-hr})\ (25 \text{ hr/tidal day}) = \pounds 153{,}000 \text{ per yr.}$$

Clearly, the uprating of Berth 4 looks worthwhile, costing only £70,000 for dredging and so forth and £15,000 per year for interference with Berth 3.

This is an excellent example of a simple *Monte Carlo simulation.* Such simulations derive their name from the elements of randomness which are essential features thereof, similar to the element of chance which is an essential feature of the famous gambling house at Monte Carlo.

This particular example was performed by hand, but more extensive simulations would best be performed on a digital computer. And, even in such a manual and mechanical method of analysis as direct simulation, a theoretical background is essential. Therefore, later in this chapter we give a brief introduction to the theory of Monte Carlo methods. Notice that this example illustrates the simple solution of a problem in queuing theory which could not be solved by the probability balances of Chapter 14.

15.4 A BLENDING PROBLEM[3]

M. Tayyabkhan and T. C. Richardson, then of the Union Carbide Corporation, solved a complex blending problem by simulation. We now use their example further to illustrate a Monte Carlo method.

A chemical production complex produces a number of component products to be combined and sold as mixtures to the consumer. Each particular mix must be prepared just prior to shipment by transferring the proper amount of each component into a blender, the preparation of a mix taking one day. The average demand for the five possible mixes is shown in Table 15.4-1. On this information alone, we might conclude that 13 blenders would be adequate.

However, the tabulation of average orders per day does not contain all the information; day-to-day variations about those averages are observed in Table 15.4-2. For example, over 10 per cent of the time there are 5 orders for mix A rather than the average of 3.5 orders. Are these variations a disrupting influence, and would 13 blenders be inadequate?

The dominant features of this problem are as follows.

Variable demands for particular mixes.

The build-up of back orders should inadequate blending capacity be available.

[3] M. Tayyabkhan and T. C. Richardson, "Monte Carlo Methods," *Chem. Eng. Progr.*, January 1965.

Table 15.4-1 Average Orders per Day for Mixtures

Mixture	Average Orders per Day
A	3.5
B	2.35
C	3.5
D	1.4
E	1.75
Total for all mixtures	12.50

Table 15.4-2 Frequency Distribution of Orders per Day

Number of Orders	Percentage of Days				
	A	*B*	*C*	*D*	*E*
1	2.1	14.3	0.4	2.1	36.7
2	20.3	49.0	8.7	58.7	53.9
3	34.6	26.6	44.8	37.1	8.7
4	22.4	8.7	37.4	2.1	0.7
5	12.9	1.4	8.7		
6	4.6				
7	2.7				
8	0.4				

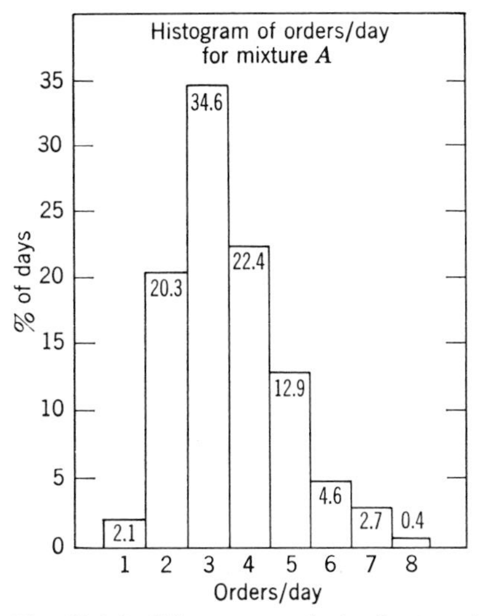

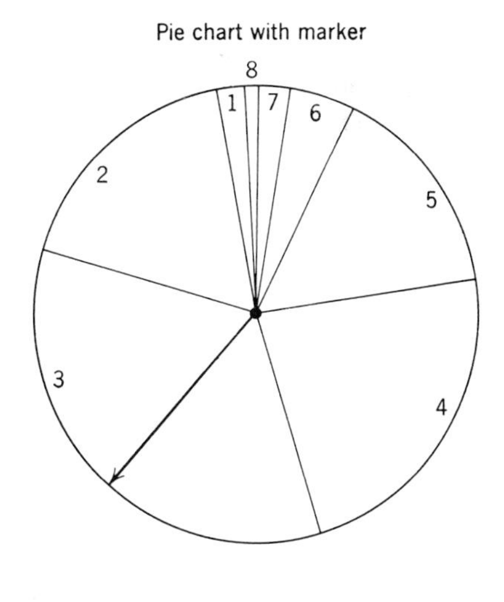

Fig. 15.4-1 Histogram and pie chart used to generate order pattern for a given mixture for a certain day.

The criterion of desirable performance might be that a certain percentage of the days should be free of back orders; this depends on the number of blenders available and on the variability of the orders.

The only unusual part of this problem which might be troublesome to simulate is the reproduction of order patterns which exhibit properties similar to the order patterns observed in Table 15.4-2. The simulation of orders is now discussed. Figure 15.4-1 contains a histogram constructed from the data in Table 15.4-2 and a pie chart based on that histogram. Should the arrow be spun randomly on the pie chart, the successive stopping points would generate a distribution of numbers of orders distributed according to the relative areas on the chart. A *uniformly distributed* random variable, the physical position of the pointer after a spin, has been mapped into a *nonuniformly distributed variable*. The number of orders can be simulated according to any distribution pattern in this way.

In practice, most large simulations are performed on the computer; programs are available for the generation of uniformly distributed random numbers. Such numbers are mapped into suitably distorted distributions by techniques not unlike the pie chart method above. Table 15.4-3 indicates the orders corresponding to a uniformly distributed random number between 000 and 999, showing how we would recover the properly biased random distribution therefrom.

Orders for, say, 2,000 days could then be simulated, and the build-up of back orders observed. The results of such simulations are shown in Table 15.4-4 for 15 mixers available. For example, observe that with 15 mixers, 91.45 per cent of the days of operation will have 15 or less orders plus back orders; such order loads can be handled by the mixers. Table 15.4-5 summarizes such limits of loading for a number of available mixers.

Notice how the use of 13 mixers, considered adequate on the basis of average order rates, results in sufficient capacity for only 38.5 per cent of the days and that 19.0 per cent of the individual orders experience delays. Table 15.4-5 would be the basis for an economic analysis to determine the optimum number of mixers. The data in this table are inaccessible by ordinary methods of analysis, and simulation is necessary to the evaluation of the effects of order variations.

Tayyabkhan and Richardson make the following observations on modifications which might increase the realism of the simulation. They suggest including:

1. Uncertainty in availability of raw material because of demand fluctuations and limited production capability.
2. Variations in batch size and mixing time for different mixtures.
3. Variations in clean-up time dependent on scheduling sequences.

Table 15.4-3 Cumulative Frequency Distribution of Orders per Day for Each Mixture, and the Assignment of the Numbers 000-999 to Different Order Levels

	Mixture *A*		Mixture *B*		Mixture *C*		Mixture *D*		Mixture *E*	
Number of Orders	Cumulative Frequency Per Cent	Assigned Numbers	Cumulative Frequency Per Cent	Assigned Numbers	Cumulative Frequency Per Cent	Assigned Numbers	Cumulative Frequency Per Cent	Assigned Numbers	Cumulative Frequency Per Cent	Assigned Numbers
1	2.1	000–020	14.3	000–142	0.4	000–003	2.1	000–020	36.7	000–366
2	22.4	021–223	63.3	143–632	9.1	004–090	60.8	021–607	90.6	367–905
3	57.0	224–569	89.9	633–898	52.8	091–527	97.9	608–978	99.3	906–992
4	79.4	570–793	98.6	899–985	90.2	528–901	100.0	979–999	100.0	993–999
5	92.3	794–924	100.0	986–999	98.9	902–988				
6	96.9	925–968			100.0	989–999				
7	99.6	969–995								
8	100.0	996–999								

Table 15.4-4 Cumulative Distribution of Orders and Back Orders per Day for 15 Mixers (a Sample of 2,000 Days). There Were No Delays 91.45 Per Cent of the Days

Number of Orders plus Back Orders	Number of Days	Per Cent	Cumulative Per Cent
6	3	0.15	0.15
7	5	0.25	0.40
8	22	1.10	1.50
9	69	3.45	4.95
10	169	8.45	13.40
11	335	16.75	30.15
12	387	19.35	49.50
13	365	18.25	67.75
14	289	14.45	82.20
15	185	9.25	91.45
16	83	4.15	95.60
17	55	2.75	98.35
18	24	1.20	99.55
19	6	0.30	99.85
20	1	0.05	99.90
21	2	0.10	100.00

Table 15.4-5 Summary of Results with Different Numbers of Mixers

Number of Mixers	Per Cent of Days with No Delay	Per Cent of Orders with No Delay
13	38.5	81.0
14	79.3	96.4
15	91.4	99.2
16	97.2	99.6
17	99.2	99.9
18	99.7	99.99
19	99.9	99.999
20	100.0	100.00

4. Availability of limited final product storage space. If these considerations were included in the model, the result might represent a realistic chemicals production unit.

With a Monte Carlo simulation of this unit, the following questions might be answered:

1. What is the proper balance between increased production capacity and increased storage facilities (raw material, intermediate and final product)? Between mixers and storage facilities?
2. How is the system affected by changes in demand characteristics?
3. What is the effect of an increase of mixing rate?

15.5 INDUSTRIAL SIMULATORS

The simulations developed in the previous sections are at the limit of simplicity; the main purpose was to introduce the field. However, many large industrial firms have developed special purpose simulators to aid in the analysis of special recurring problems. These simulators may well have taken several man-years and many thousands of dollars to develop but often pay for themselves in short order by improving decision making.

There is no need to go into great detail in describing such simulators, since they are merely larger scale models of the simple simulators and represent expansions in size rather than in concept. However, a brief explanation of the capabilities of a typical industrial simulator is in order.

The Dow Chemical Company of Midland, Michigan regularly uses a simulator called POWERFACTS to aid in the planning of its steam and power system expansions.[4] This simulator is of particular interest here, since it approximately accounts for many of the sources of uncertainty discussed in the previous chapters, such as forecasts of demand changes; daily, weekly, monthly and yearly variations in the system; the shutdown of units due to random failure or planned maintenance; and a number of familiar sources of uncertainty.

The Dow Chemical Company then had thirteen a-c turbogenerators and fourteen boilers at their Midland Division, ranging from 4,000 to 40,000 kw capacity. This steam and electric power facility accounts for a significant portion of the capital outlay in the chemical processing complex it services, and, as a result, the determination of the strategy of expansion of this facility is of extreme importance. The primary fuel is coal, although natural gas and oil are used to a limited extent. The system is tied to the public utility system

[4] R. E. Jackson, A. J. Klomparens and G. T. Westbrook "Long Range Planning," *Chem. Engr. Progr.*, No. 1, **61** (1965).

servicing the surrounding community, and power can be sold and purchased to a limited extent during periods of power excess and shortage.

The turbines and boilers both experience a significant amount of downtime, caused either by planned maintenance or random forced outages (Fig. 15.5-1).

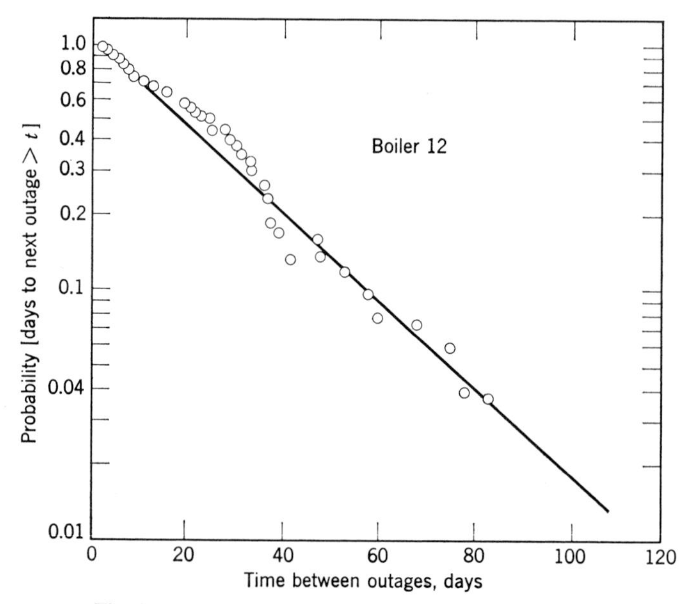

Fig. 15.5-1 Data on the random outages of boiler 12.

The random outages can be divided into two classes, forced and postponed, and the planned maintenance can be divided also into two classes, major and minor.

The real system does not operate at average conditions much of the time, since the system is frequently buffeted by equipment failures and significant patterns caused by weather variations and the seasonal demands for certain products (Fig. 15.5-2). Thus, there is frequent loading of standby boilers and turbines and demands upon the public power facilities, increasing the annual costs of power.

Also there is the problem of planning to accommodate forecast increases in the demand for the services of this facility as the chemical processing complex expands into an uncertain future.

Thus, we see that in a given industrial complex, all the factors discussed in Chapters 11 through 14 enter simultaneously and must be considered in answering the long-range planning questions.

1. What types of boilers and turbines most economically meet growing demands?

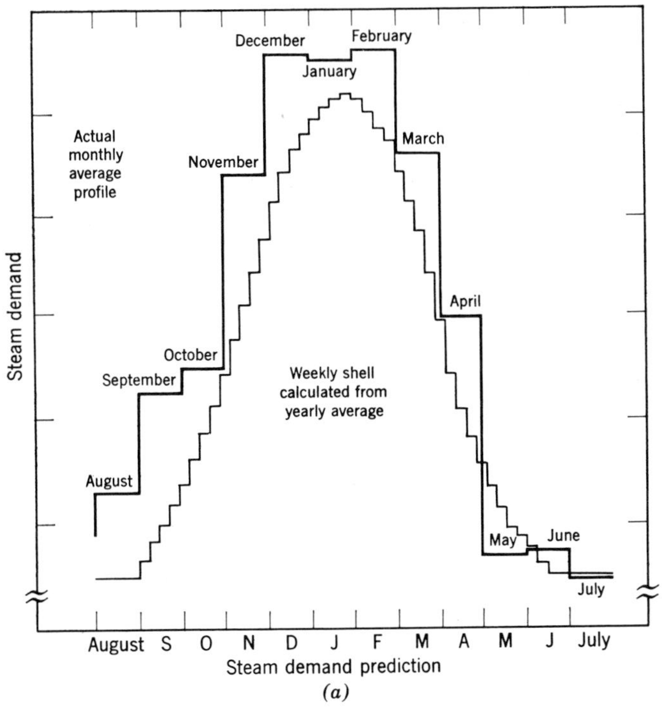

(a)

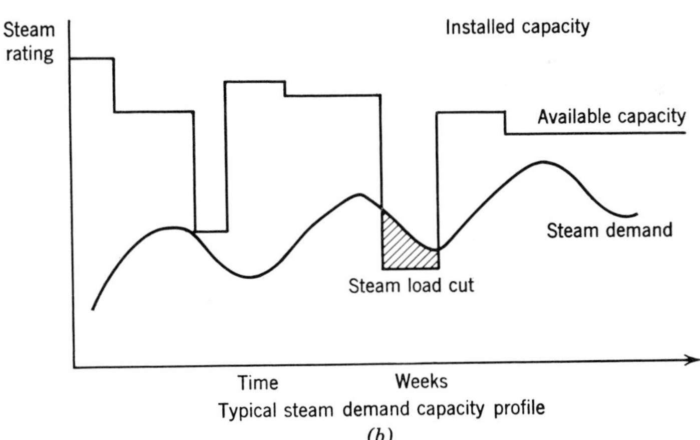

(b)

Fig. 15.5-2 Typical variation in steam demand and capacity.

2. When should these new units be installed?
3. When should the old existing facilities be retired?

The design alternatives available include:

1. A selection of boilers ranging from small, low pressure, high fuel cost units to very large, high pressure, low fuel cost units.
2. A selection from various types of turbines ranging from multiple extraction to condensing units.
3. The timing of the additions and replacements.

The POWERFACTS simulator requires data on the following variables and parameters.

A. *Boiler and turbine specifications*
   1. Efficiency, capacity, and operating pressure.
   2. Annual forecast steam and power demands.
   3. Labor, type of fuel and auxiliary steam and power requirements.
   4. Reliability constants for the Monte Carlo program.
   5. The desired annual maintenance schedule.
   6. Dispatch priority of the boilers and turbines.
   7. Future equipment installation dates.

B. *Economic factors*
   1. Capital.
   2. Fuel, labor, and maintenance unit costs.
   3. Tax rate and cost of capital.
   4. Escalation rates, if desired, on fuel, labor and materials.

The output from the model is of two types: economic results and performance results. A listing of the output data follows.

A. *Economic output data*
   1. The value of the objective function for the simulation period.
   2. Annual operating costs.
   3. Annual fuel costs.
   4. Annual purchased power costs.

B. *System performance output data*
   1. Total annual steam and power production.
   2. Frequency and severity of shortages of steam and power.
   3. Frequency and amount of reducing station use.
   4. Annual purchased and produced power.
   5. Average excess steam capacity by month.
   6. Annual production and days of operation for each boiler and turbine.

The computer flow diagram for the POWERFACTS simulator is shown in Fig. 15.5-3. The simulator has the ability to perform the following, according to the Dow report.

1. Estimate the daily steam and power demand. For example, the total steam demands are computed from weekly and daily adjustment factors applied to an annual forecast. Steam demands at various pressure levels are forecast next, based on the current and forecast steam split.
2. Estimate what equipment (both boilers and turbines) is available for operation on a given day, knowing that either random or planned shutdowns might occur for each unit in the system.
3. Study the demand and equipment status for the current day, in order to forecast what this status will be a week or two weeks hence and to arrive at some reasonable decision as to what planned or postponable outages to take.
4. Dispatch the available equipment to meet the posted demands and to calculate the resultant steam and power production performance.
5. Compute the auxiliary utilities required to operate this system. These were handled as follows.

   Steam and power. Requirements for each unit and for various miscellaneous equipment are deducted from gross production.
   Water. The basic raw material in this system is treated make-up water. This is combined with feedwater steam and preheated to a required feed temperature. Heat balances around each feedwater heater are required to obtain factors for pounds feedwater consumed per pound of steam generated.
   Fuel. Further heat balances around each unit are required to compute factors for fuel consumption in Btu per pound of steam produced.

6. Keep precise cost and production records. Utility costs are computed at the simulated year end based on total production figures on each unit. Purchased power costs are calculated each month. A rather involved procedure is required to simulate the contract governing power purchases. This cost was primarily a function of the amount purchased each month (kw-hr) plus the peak demand (kw) over the past 12-month period.
7. Finally, POWERFACTS simulates the random outages of the boilers and turbines.

POWERFACTS has been used for capacity planning, studies of the feasibility of using other types of fuels, and determining the economic potential of

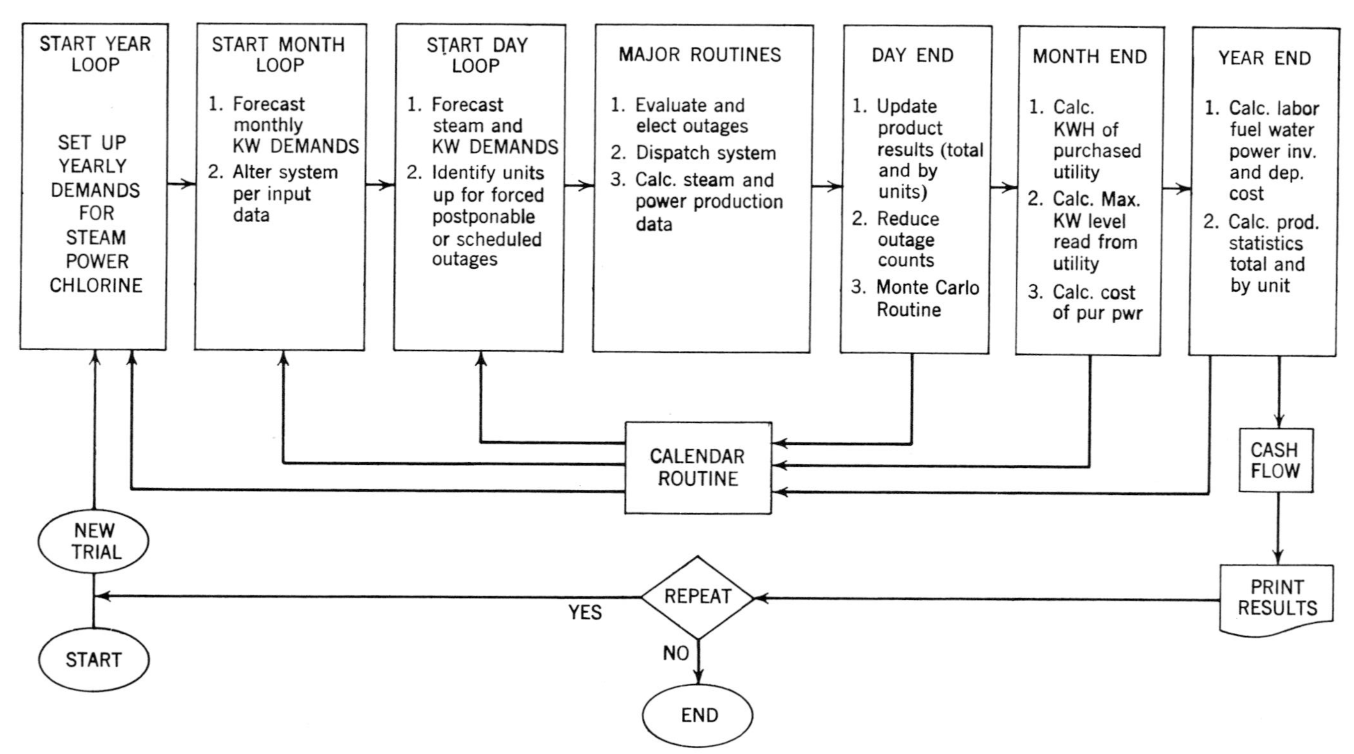

Fig. 15.5-3 The POWERFACTS simulator. (*a*) Steam demand prediction. (*b*) Typical steam demand capacity profile.

smoothing the seasonal variations in steam demand by inventorying certain seasonal products. It is reported that the Dow Chemical Company is planning similar simulations for other facets of its operations. We expect this kind of analysis to become more and more commonly used in process engineering, as processing units become steadily more interrelated.

15.6 MONTE CARLO SIMULATION THEORY

The operational simulation of a process provides access to the *probable* performance characteristics. The strategy of analysis involves the replication of a number of experiments on the simulator, the results of which are averaged to provide an estimate of the probable or average performance.

A basic question that must be answered concerns the number of runs that must be made on the simulator before the average performance can be estimated with sufficient statistical significance. If σ is the standard deviation of a single observation of a simulator run, then $\sigma/\sqrt{n}$ is an estimate of the standard deviation of the average of n observations. Thus, to improve the estimate of the average performance of a simulated process by reducing the standard deviation by one half, the number of simulation runs need to be increased by four. This can be an expensive and time consuming proposition.

However, it is possible to increase the efficiency of simulation by a number of variance reducing methods. We shall now outline the kinds of thinking involved.

Suppose, $\bar{U}_1$ and $\bar{U}_2$ are two sample averages of the objective function which describes the performance of a process. If $\bar{U}_1$ and $\bar{U}_2$ are the results of uncorrelated or random experiments on the simulator, then the standard deviations of the sum $(\bar{U}_1 + \bar{U}_2)$ and of the difference $(\bar{U}_1 - \bar{U}_2)$ are the same. However, if the two samples are from a correlated population this will not be true. If there is a positive correlation between $\bar{U}_1$ and $\bar{U}_2$ the standard deviation of the difference $\bar{U}_1 - \bar{U}_2$ will be less than before, and if there is a negative correlation between $\bar{U}_1$ and $\bar{U}_2$ the standard deviation of the sum $\bar{U}_1 + \bar{U}_2$ will be less. The variance for a sum and difference of two random variates is given in Eqs. 15.6-1 and 15.6-2 in terms of the covariance between the variates. This gives rise to methods for improving the efficiency of Monte Carlo simulations.

$$V(U_1 + U_2) = V(U_1) + V(U_2) + 2\,\mathrm{cov}(U_1, U_2) \qquad (15.6\text{-}1)$$

$$V(U_1 - U_2) = V(U_1) + V(U_2) - 2\,\mathrm{cov}(U_1, U_2) \qquad (15.6\text{-}2)$$

Notice in the berthing problem the error that would be made if only the first ten arrivals were included in the simulation. There are ten ship-days of delay for only Berth 5 and two ship-days for Berths 4 and 5. The conclusion reached from this limited sample is that a saving of

$$\frac{(118)(10-2)}{10} = 94 \text{ ship-days per year}$$

would be realized by the opening of Berth 4 to larger vessels. We saw that by sampling 40 arrivals a saving of 62 ship-days per year was predicted. Clearly, the significance of the answer depends on the method of sampling the performance of the simulator.

One method of sampling that often results in great improvements in statistical significance is the *Antithetic Variate* method of Hammersley and Morton.[5] This involves the running of two simulations where the random variations in one are negatively correlated with the random variations in the other. The results of the two variations are then averaged. The saving estimated from this new simulation is 59 ship-days per year, and according to the antithetic variate method an estimate of the true saving is

$$\frac{59+62}{2} = 60.5$$

Such programmed samplings tend to reduce the time period over which the simulation must be observed to obtain meaningful results. Moy[6] reports as much as a 70 per cent increase in the efficiency of sampling a queuing system simulator by the antithetic method.

The above has been an example of reducing the variance of the average of a sample of simulator runs by having the successive runs *negatively correlated*. However, if we are interested in the *difference* between the average performance of two processing alternatives the two runs to be compared should be *positively correlated*. For example, in the berthing problem the same arrival schedule was imposed on both situations: one dock available and two docks available. In this way the simulations are positively correlated (i.e., the result of the same series of random variations) and the difference between the two should exhibit a reduced variance. These are typical of the methods which are used to increase the statistical significance of data obtained from simulators.

[5] J. M. Hammersley and K. W. Morton, "A New Monte Carlo Technique, Antithetic Variates," *Proc. Camb. Phil. Soc.*, **52**, 449 (1956).

[6] W. A. Moy, "Practical Variance—Reducing Procedures for Monte Carlo Simulations," *Mgt. Sci.* (to appear).

15.7 SIMULATION LANGUAGES

Before the development of general process simulation programs for the computer it was necessary to build up a computer program expressly for each new processing problem. Now, much of the labor of developing a simulation can be alleviated by the use of simulation languages, which are special forms in which the processing problem can be phrased so that the computer can program the simulation.

These executive programs are designed to provide

1. A generalized structure for producing simulation models.
2. A convenient way of converting a simulation model into a computer program.
3. A rapid way of making changes in the simulation model.
4. A flexible way of obtaining useful data from the simulation.

Amongst the several simulation programs referenced at the end of the chapter are the following.

GPSS II (*G*eneral *P*urpose *S*ystem *S*imulator). Developed by IBM to simulate complex queuing problems.

GASP (*G*eneral *A*ctivity *S*imulation *P*rogram). Developed by the United States Steel Corporation, and used to simulate such situations as an open-hearth steel making shop, the transportation system in a steel mill, a steel plate heat-treating and annealing shop.

GSP (*G*eneral *S*imulation *P*rogram). Developed by the United States Steel Company Ltd. and oriented towards manufacturing plants.

CSL (*C*ontrol and *S*imulation *L*anguage). Developed by IBM United Kingdom Ltd. and Esso Petroleum Company Ltd.

CHEVRON. A heat and material balancing simulator developed by Chevron Research Corporation.

CHIPS (*Ch*emical *E*ngineering *I*nformation *P*rocessing *S*ystem). Developed by Service Bureau Corporation, a subsidiary of IBM.

CHEOPS (*Ch*emical *E*ngineering *Op*timization *S*ystem). Developed to improve the efficiency of chemical process designing by the Shell Oil Company, and used on the design of processes for the manufacture of isoprene, ethylene oxide, polypropylene, ethyl alcohol, and others.

PACER (*P*rocess *A*ssembly *C*ase *E*valuator *R*outine). Developed by P. T. Shannon and recently applied to the simulation of a sulfuric acid process and a styrene plant.

FLEXIBLE FLOWSHEET. One of the earliest programs developed by the M. W. Kellogg Company.

SPEEDUP (*S*imulation *P*rogramme for the *E*conomic *E*valuation and *D*esign of *U*nsteady-state *P*rocesses). Developed expressly to organize complex process design problems by R. W. H. Sargent and A. W. Westerberg.

PEDLAN (*P*rocess *E*ngineering *D*esign *Lan*guage). Developed by Mobil Oil Company.

The development of simulation executive languages is an area of study in which most major processing companies are engaged, and a study of the literature cited at the end of the chapter is recommended. We strongly recommend the text, Crowe, Hamielec, Hoffman, Johnson, Shannon, and Woods, *Chemical Plant Simulation*, as an introduction to the simulation of chemical processes.

References

The following are suggested as general references to the principles and practices of simulation.

K. D. Tocher, *The Art of Simulation*, D. Van Nostrand, Princeton, N.J., 1963.
Jay Forrester, *Industrial Dynamics*, The MIT Press, Cambridge, 1961.
J. M. Hammersley and D. C. Handscomb, *Monte Carlo Methods*, Wiley, New York, 1964.
T. H. Naylor, J. F. Balinty, D. S. Burdick, and K. Chu, *Computer Simulation Techniques*, Wiley, New York, 1966.

For the principles which lead to the development of process simulations, the following are recommended.

D. Himmelblau and K. Bishoff, *Process Analysis and Simulation*, Wiley, New York, 1967.
R. G. E. Franks, *Mathematical Modeling in Chemical Engineering*, Wiley, New York, 1967.
C. M. Crowe et al., *Chemical Plant Simulation*, Mc Master University, 1969.
A. Carlson, *Analog Simulation in Chemical Engineering*, Wiley, New York, 1967.

These refer to specific articles of interest either on applications or on explanations of special simulation programs.

D. F. Boyd, H. S. Krasnow, and A. C. R. Petit, "Simulation of an Integrated Steel Mill," *IBM Systems Journal*, No. 1, III (1964).
J. N. Buxton and J. G. Laski, "Control and Simulation Language," *The Computer Journal*, V (1962).
E. Efron and G. Gordon, "A General Purpose Digital Simulator and Examples of Its Application," *IBM Systems Journal*, No. 1, III (1964).
G. Gordon "A General Purpose Systems Simulator," *IBM Systems Journal*, I (1962).

D. H. Kelly and J. W. Buxton, "Monte Code—An Interpretative Program for Monte-Carlo Simulation," *The Computer Journal*, V (1962).
K. D. Tocher, "A Review of Simulation Languages," *Operational Research Quarterly*, XVI (1965).
D. E. Freeman, "Programming Languages Ease Digital Simulation," *Control Engineering*, November 1964.
E. B. Dahlin and R. N. Linebarger, "Digital Simulation Applied to Paper Machine Dryer Studies," 6th ISA Pulp and Paper Symposium, May 1965.
W. L. Godfrey and R. D. Benham, "Analog Simulation of Countercurrent Crystallization Process," *Simulation*, No. 1, **4** (1964).
H. A. Lindahl, "Improve Refinery Operations with Process Simulation," *Chem. Eng. Progr.*, No. 4, **61** (1965).
N. R. Amundson and A. J. Pontinen, "Multi-Component Distillation Calculations on a Large Digital Computer," *Ind. Eng. Chem.*, No. 5, **50** (1962).
R. F. Baddour, P. L. T. Brian, B. A. Longeais, and J. P. Emery, "Steady State Simulation of an Ammonia Converter," *Chem. Engr. Sci.*, **20** (1965).
N. H. Smith and D. F. Rudd, "On the Effects of Batch Time Variations on Process Performance," *Chem. Eng. Sci.*, **19** (1964).
P. T. Shannon, "The Integrated Use of the Digital Computer in Chemical Engineering Education," *Chem. Eng. Educ.* (March 1963).
P. T. Shannon *et al.*, "Computer Simulation of a Sulfuric Acid Plant," *Chem. Eng. Progr.*, **62**, 6 (1966).
R. W. H. Sargent and A. W. Westerberg, "Speed-up in Chemical Engineering Design," *Trans. Inst. Chem. Engrs. (London)*, **42** (1964).
E. Singer, "Simulation and Optimization of Oil Refinery Design," *Chem. Eng. Progr., Symp. Ser.*, No. 37, **58** (1962).
Dion G. Steward, "A Survey of Computer-Aided Chemical Process Design," *Chem. Eng. Progr.*, Vol. 64, No. 4, 1968.

PROBLEMS

15.A. Crude oil is to arrive at a refinery complex in 40,000-ton tankers at the average rate of one tanker every two days. However, advances and delays in the schedule are to be expected; it is equally probable that a given tanker will arrive on schedule, one day advanced or one day delayed. When a tanker arrives it can be unloaded in just under one day by the continued pumping of its contents into storage tanks. It is the responsibility of the refinery engineers to provide sufficient storage to prevent the additional delay of the tankers.

The oil is processed continuously by the refinery at a maximum rate of 22,000 tons per day, except when unexpected outages occur at which time the processing rate drops to 10,000 tons per day. The outages occur at random on the average of one every 15 operating days and are of one day duration.

What storage facilities should be provided for the crude oil?

(a) Establish an operational simulation of this processing problem with one day as the time interval.

(b) Simulate the operation of the process for 30 days, and determine the maximum amount of oil that must be stored during that time period, and the amount of surplus oil that will be available for piping to the other refinery.

(c) Perform an antithetic variate simulation to the simulation performed in (b) by interchanging the advances and delays in the tanker arrival schedule. The average of the maximum amount of storage required in (b) and (c) is a more nearly accurate estimate of the probable maximum storage needed.

(d) Perform a sufficient number of simulations to assess the accuracy of the estimates made in (c).

15.B. The process in Fig. 15.B-1 produces a batch of product each day; if the product is high in the content of a certain component, it can be sold at a profit of $100 per batch; if the product has a composition within a medium range, the batch is worth $1,000; and if the composition is on the low side, the batch is worth $300.

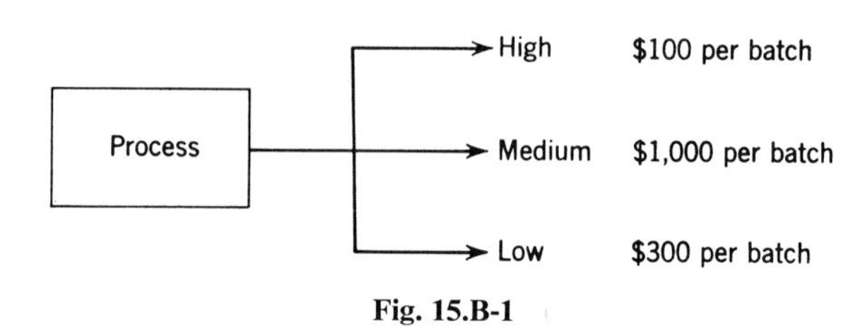

Fig. 15.B-1

Out of ten batches on the average there will be six high, one medium, and three low, and there is no correlation between the quality of successive batches. The expected profit from the process is then

$$365 \frac{\text{days}}{\text{year}} [0.6(100) + 0.1(1{,}000) + 0.3(300)] = \$91{,}000 \text{ per year}$$

However, it is possible to blend a batch of high quality and a batch of low quality to produce a batch of medium quality. This suggests the design modification shown in Fig. 15.B-2 which allows for the storage of two batches and the in-line blending of material.

For example, if a batch of high quality is in storage and the process generates a batch of low quality, the two batches can be blended in-line to prepare two batches of medium quality. How much would you be willing to invest in such a storage and blending system?

This problem is distinguished by the fact that an operating policy for the blender must be established before simulation can begin and the optimal policy is not known initially. Simulate the operation of the system using the following policies. Which seems to be the best policy?

(a) Never sell a low, unless there are two lows in storage, and a third low is produced and must be sold.
(b) Strive to maintain a high and low in storage.
(c) Strive to keep two highs in storage.
(d) Strive to maximize the empty storage by blending whenever possible.

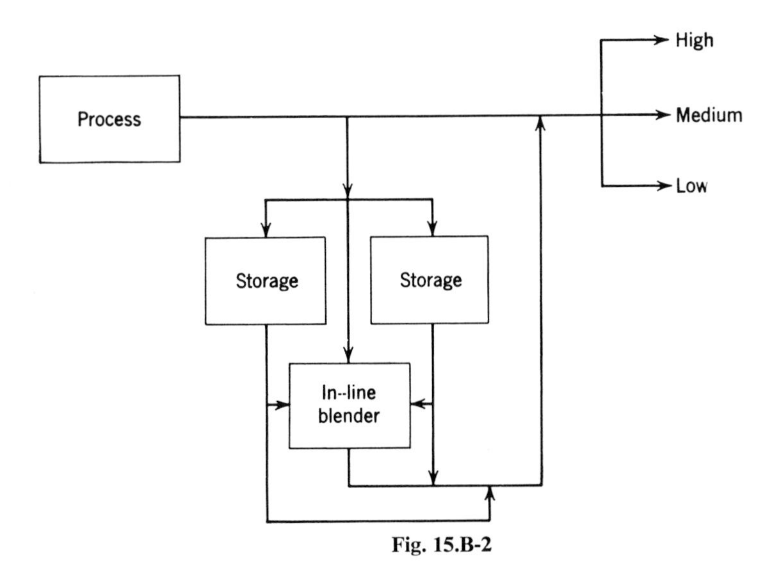

Fig. 15.B-2

15.C. The inplant hydrogen rate to a hydrocracker varies randomly and often cannot sustain the design crude oil feed to the hydrocracker. Therefore, plans are being made to build on-site storage capacity so that oil feed may be stored during periods of short hydrogen supply, and the stored oil may be added above the design capacity during periods of excess hydrogen supply. What storage do you recommend?

The design feed rate of oil to the hydrocracker is 13,500 barrels per day, each barrel consuming 2,000 SCF of hydrogen. The average arrival rate of the hydrogen is 25×10^6 SCF/day. The profit derived from the upgraded crude oil is $1.00 per barrel. Any crude oil not sent to the hydrocracker or stored is sold at cost to a neighboring oil refinery. The investment in tankage is $1.00 per barrel capacity, and the firm has an economic structure similar to that of an ordinary chemical manufacturer.

Hint. You might approximate the hydrogen rate variations by daily batch arrivals of 5×10^6 SCF, the number of batches arriving varying randomly between 10 and 40×10^6 SCF/day.

15.D. A polyethylene plant is producing a batch of polymer every eight hours. The batch is stored for 18 hours in rundown bins until the laboratory results on the quality control test are known. About 80 per cent of the production is acceptable and is transferred then to load out or to additional process steps.

When a batch misses on the low side of the specifications, the batch following the one currently being made can be made on the high side to permit cross blending the low and the high. After the lab tests confirm this, the two batches are in-line blended by simultaneous transfer in the air conveying system. A double-size on-spec batch is thus produced. For the cases considered below, how many bins are required?

CASE A. *The case of unconditional post-bin acceptance* (*UPBA*). This is the case in which there is 100 per cent certainty that when a batch is ready to be transferred from the bin it occupies, it can be accepted by the downstream process, whether it be load out or the next processing step.

CASE A–1. *The nonrandom appearance case.* It is assumed in the nonrandom appearance case that every fifth batch coming from the polyethylene process is below specification.

CASE A–2. *The random appearance case.* it is assumed in the random appearance case that the below specification batches appear in a uniformly distributed random fashion. There is still, however, a 20 per cent chance that any particular batch is bad.

CASE B. *The case of possible post-bin nonacceptance* (*PPBN*). In the PPBN case the possibility of breakdown of the downstream facilities is accounted for. It is assumed that because of system failure and switching difficulties there is a certain probability, also randomly distributed, that even though a batch is satisfactory and is ready to be transferred from its bin, it cannot be accepted either by shipping or by the next process. When this occurs it is further assumed that it takes seven hours before the batch can be subsequently transferred from its bin.

Further assumptions include:

1. If all of the bins are full at the time a batch is coming out of the process, either one of the batches already residing in the bins or the new batch must be dumped. The proper choice as to which batch to throw out will depend upon the existing situation and the case (either A-1, A-2 or B) considered.
2. When a batch is dumped it is considered to be nonrecoverable and a total loss.
3. The emptying of a bin is done instantaneously or at least in such a way that handling problems do not arise when a batch already in residence must be removed from a bin to make room for an oncoming batch.
4. The probability of accepting a double-size batch by the downstream process is the same as that for a single-size batch.

AUTHOR INDEX

SUBJECT INDEX